Lecture Notes in Artificial Intelligence   954

Subseries of Lecture Notes in Computer Science
Edited by J. G. Carbonell and J. Siekmann

Lecture Notes in Computer Science

Edited by G. Goos, J. Hartmanis and J. van Leeuwen

T0230415

Springer

Berlin
Heidelberg
New York
Barcelona
Budapest
Hong Kong
London
Milan
Paris
Tokyo

Gerard Ellis Robert Levinson
William Rich John F. Sowa (Eds.)

Conceptual Structures: Applications, Implementation and Theory

Third International Conference
on Conceptual Structures, ICCS '95
Santa Cruz, CA, USA, August 14-18, 1995
Proceedings

Springer

Series Editors

Jaime G. Carbonell, Carnegie Mellon University, USA

Jörg Siekmann, University of Saarland, DFKI, Germany

Volume Editors

Gerard Ellis
Computer Science Department, Royal Melbourne University of Technology
GPO Box 2476V, Melbourne, Victoria 3001, Australia

Robert Levinson
Department of Computer and Information Sciences, University of California
225 Applied Sciences Building, Santa Cruz, CA 95064, USA

William Rich
IBM Santa Teresa Laboratory
555 Bailey Av. L16/H120, San Jose, CA 95141, USA

John F. Sowa
SUNY Binghamton
21 Palmer Avenue, Croton-on-Hudson, NY 10520, USA

Cataloging-in-Publication Data applied for

Die Deutsche Bibliothek - CIP-Einheitsaufnahme

Conceptual structures : applications, implementation and
theory ; proceedings / Third International Conference on
Conceptual Structures, ICCS '95, Santa Cruz, CA, USA, August
14 - 18, 1995. Gerard Ellis ... (ed.). - Berlin ; Heidelberg ; New
York : Springer, 1995
  (Lecture notes in computer science ; Vol. 954 : Lecture notes in
  artificial intelligence)
  ISBN 3-540-60161-9
NE: Ellis, Gerard [Hrsg.]; International Conference on Conceptual
  Structures <3, 1995, Santa Cruz, Calif.>; GT

CR Subject Classification (1991): I.2, G.2.2, H.2.1

ISBN 3-540-60161-9 Springer-Verlag Berlin Heidelberg New York

© Springer-Verlag Berlin Heidelberg 1995
Printed in Germany

Typesetting: Camera ready by author
SPIN 10486494   06/3142 – 5 4 3 2 1 0   Printed on acid-free paper

Preface

Conceptual structures are a modern treatment of Charles Sanders Peirce's Existential Graphs, developed in 1896 as a graphic notation for classical logic with higher order extensions. Peirce viewed existential graphs as "his luckiest discovery" and "a logic of the future".

John Sowa showed that conceptual graphs can be mapped to classical predicate calculus or order sorted logic, and are thus seen as a (graphic) notation for logic. However, it is the topological nature of formulas (topology was a field Peirce helped develop) which conceptual graphs make clear, and which can be exploited in reasoning and processing. Conceptual graphs are intuitive because they allow humans to exploit their powerful pattern matching abilities to a larger extent than does the classical notation. Conceptual graphs can be viewed as an attempt to build a unified modeling language and reasoning tool. Conceptual graphs can model data, functional, and dynamic aspects of systems. They form a unified diagrammatic tool which can integrate entity-relationship diagrams, finite state machines, petri nets, and dataflow diagrams.

These proceedings include the best papers presented at the Third International Conference on Conceptual Structures (ICCS'95), held at the University of California at Santa Cruz, California, August 14-18, 1995. The annual ICCS conference is the primary forum for reporting progress in conceptual graph research. The papers in the proceedings can be broadly classified into the following categories: applications, natural language, programming in conceptual graphs, machine learning / knowledge acquisition, hardware and implementation, graph operations, and ontologies and conceptual graph theory.

This conference was sponsored by:
- IBM, Santa Teresa Laboratory, San Jose
- University of California at Santa Cruz
- Royal Melbourne University of Technology, Australia.
- AAAI (American Association of Artificial Intelligence).

We thank all the individuals and institutions who contributed to making this conference a success. We thank all the authors and the editorial staff of Springer-Verlag for making this book a valuable contribution in the knowledge representation research field.

May 1995   G. Ellis, R.A. Levinson, W. Rich, J.F. Sowa

Organizing Committee

Program Chair	Local Arrangements Chair	Finance Chair
Gerard Ellis	Robert Levinson	Bill Rich
Royal Melbourne Inst. of Tech	Univ. of California	IBM San Jose.
Australia	Santa Cruz	California

Honorary Chair : John Sowa, State Univ. of New York

Program Committee

Hassan Aït-Kaci	(Simon Fraser Univ., Canada)
Harmen van den Berg	(Univ. of Twente, Netherlands)
Duane Boning	(MIT, USA)
Boris Carbonneill	(LIRMM, France)
Michel Chein	(LIRMM, France)
Key Sun Choi	(KAIST, Korea)
Peter Creasy	(Univ. of Queensland, Australia)
Walling Cyre	(Virginia Tech, USA)
Harry Delugach	(Univ. of Alabama in Huntsville, USA)
Judy Dick	(Univ. of Maryland, USA)
Peter Eklund	(Univ. of Adelaide, Australia)
Bruno Emond	(Univ. du Quebec a Hull, Canada)
Norman Foo	(Syndey Univ., Australia)
Brian Gaines	(Univ. of Calgary, Canada)
Adil Kabbaj	(Univ. de Montreal, Canada)
Fritz Lehmann	(GRANDAI Software, USA)
Dickson Lukose	(Univ. of New England, Australia)
Craig McDonald	(Charles Sturt Univ., Australia)
Guy Mineau	(Laval Univ., Canada)
Jens-Uwe Moeller	(Univ. of Hamburg, Germany)
Bernard Moulin	(Laval Univ., Canada)
Marie Laure Mugnier	(LIRMM, France)
Jonathan Oh	(Univ. of Missouri at Kansas City, USA)
Heike Petermann	(Univ. Hamburg, Germany)
Heather Pfeiffer	(New Mexico State Univ., USA)
James Slagle	(Univ. of Minnesota, USA)
Bill Tepfenhart	(AT&T Bell Laboratories, USA)
Eileen Way	(SUNY Binghamton, USA)
Michel Wermelinger	(Univ. Nova de Lisboa, Portugal)
Mark Willems	(Vrije Univ. Amsterdam, Netherlands)
Walter Wilson	(IBM T. J. Watson Research Center, USA)
Vilas Wuwongse	(Asian Institute of Technology, Thailand)

Auxiliary Reviewers

Alex Bejan (IBM, USA), Phil Kime (Edinburgh Univ., UK).
Maurice Pagnucco (Sydney Univ., Australia), Tao Lin (CSIRO, Australia).

Table of Contents

Invited Papers

Applications

Natural Language

Natural Language - continued

Programming in Conceptual Graphs

Machine Learning / Knowledge Acquisition

Hardware and Implementation

Hardware and Implementation - continued

Graph Operations

Ontologies and Theory

Author Index

Syntax, Semantics, and Pragmatics of Contexts

John F. Sowa
Philosophy and Computers and Cognitive Science
State University of New York at Binghamton

Abstract. The notion of context is indispensable in discussions of meaning, but the word *context* has often been used in conflicting senses. In logic, the first representation of context as a formal object was by the philosopher C. S. Peirce; but for nearly eighty years, his treatment was unknown outside a small group of Peirce afficionados. In the early 1980s, three new theories included related notions of context: Kamp's *discourse representation theory*; Barwise and Perry's *situation semantics*; and Sowa's *conceptual graphs*, which explicitly introduced Peirce's approach to the AI community. More recently, John McCarthy and his students have begun to use a closely related notion of context as a basis for organizing and partitioning knowledge bases. Each of the theories has distinctive, but complementary ideas that can enrich the others, but the relationships between them are far from clear. This paper analyzes the semantic foundations of these theories and shows how McCarthy's ist(*c,p*) predicate can be interpreted in terms of the semantic notions underlying the others.

1. Theories of Contexts

In the AI literature, the term *context* has been applied to a profusion of ideas that have not been clearly distinguished. Some of them concern the syntactic representation of contexts; others refer to the semantic relationship of a linguistic context to a physical situation; and still others introduce pragmatic notions concerning the purpose or use of a context in various applications. Each of these major areas can be subdivided further. Syntactically, there are three distinct aspects of context:

1. A mechanism for grouping, associating, or packaging information that can be named and referenced as a single unit.

2. The contents of that package, which have been called anything from quoted formula to microtheory.

3. The permissible operations on the information in the package and the constraints on importing and exporting information into and out of a package.

All three of these notions represent syntactic mechanisms for representing and manipulating logical formulas without any consideration of their relationship to the real world, a possible world, or some model of the world. Much of the controversy about contexts results from the lack of a formal semantics that relates these operations to a Tarski-style model. Even an informal semantics that displays the intuitive meaning of contexts in terms of real-world objects and situations would be helpful as a guide to further analysis and formalization.

Some of the confusion about contexts results from an ambiguity in the English word. Dictionaries list two major senses of the word *context*:

- The basic meaning is a section of the linguistic text or discourse that surrounds some word or phrase of interest.

- The derived meaning is a nonlinguistic situation, environment, domain, setting, background, or milieu that includes some entity, subject, or topic of interest.

These two informal senses suggest intuitive criteria for distinguishing the various functions of contexts:

- *Syntax.* The syntactic function of context is to group, delimit, or package "a section of linguistic text." Formally, a context behaves like the QUOTE operator in Lisp together with the parentheses that delimit the portion of text that is quoted.

- *Semantics.* The quoted text of a context refers to something, which may be a physical entity or situation, a mathematical construction, or some other expression in a natural or artificial language.

- *Pragmatics.* The word *interest*, which occurs in both senses of the English definition, suggests some reason or purpose for distinguishing "a section of linguistic text" or "a nonlinguistic situation." That purpose constitutes the pragmatics or the reason why the text is being quoted. In Lisp, the QUOTE operator blocks the execution of the standard Lisp interpreter to allow nonstandard operations to be performed for some other purpose. In logic, a quote blocks the standard rules of inference and allows the definition of new rules for some special purpose.

As this analysis indicates, the notion of context is intimately connected with a complex of related ideas. Much of the confusion results from which of them happens to be called a *context*: some people apply the word to the package; and others to the information contained in the package, to the thing that the information is about, or to the possible uses of either the information or the thing. The ideas themselves may be compatible, but they must be carefully distinguished and sorted out.

These intuitive criteria provide a basis for analyzing John McCarthy's (1993) "Notes on Formalizing Context" and relating the ideas to the other theories. McCarthy's basic notation is the predicate ist(c,p), which may be read "the proposition p is true in context c." In his dissertation written under McCarthy's direction, R. V. Guha (1991) applied McCarthy's approach to the problem of partitioning a large, monolithic knowledge base into a collection of smaller, more modular *microtheories*. Guha implemented the microtheories in the Cyc system (Lenat & Guha 1990), in which they have become a fundamental mechanism for organizing and structuring a knowledge base. McCarthy and Buvač (1994) have also applied contexts and the *ist* predicate to the analysis and representation of natural language discourse.

Although McCarthy, Guha, and Buvač have shown that the *ist* predicate can be a powerful tool for building knowledge bases and analyzing discourse, they have not clearly distinguished the syntax of contexts and propositions from their semantic relationship to some domain of discourse. In fact, the *ist* predicate itself mixes the syntactic notion of containment (*is-in*) with the semantic notion of truth (*is-true-of*). To clarify these relationships, it may be helpful to analyze the *ist* predicate as a conjunction of three more primitive predicates, *is-in, refers-to,* and *describes*:

$$\text{ist}(c,p) \equiv (\exists x{:}\text{Entity})(\text{is-in}(c,p) \wedge \text{refers-to}(c,x) \wedge \text{describes}(x,p)).$$

According to this analysis, the proposition *p* is true in context *c* if and only if there exists some entity *x* such that *p* is in *c, c* refers to *x,* and *p* describes *x*. The formula distinguishes the abstract context *c* from some nonlinguistic entity *x,* which represents the "situation, environment, domain, setting, background, or milieu" associated with *c*. The predicate *is-in* represents the syntactic relationship of *c* to *p*; and the predicates *refers-to* and *describes* represent the semantic relationships of *c* and *p* to the external entity *x*. McCarthy, Guha, and Buvač have primarily considered the syntactic operations associated with the *is-in* component of the *ist* predicate. To justify those operations, the semantics of the *refers-to* and *describes* components must also be addressed.

Much of the controversy about contexts results from the abundance of notation and terminology in different theories, their application to diverse phenomena, and the lack of communication between the different schools of thought. The purpose of this paper is to emphasize the underlying similarities and to promote cross-fertilization of ideas. The following five theories will be considered:

1. Charles Sanders Peirce (1885) invented the modern algebraic notation for predicate calculus; but a dozen years later, he developed an alternate notation, which he called *existential graphs* (Roberts 1973). Although Peirce's algebra and graphs had equivalent expressive power, the graphic structure served as a heuristic aid that led him to explore operations and applications that were overlooked by logicians who used only the algebraic notation. In particular, Peirce's graphic notation for contexts was isomorphic to the *discourse representation structures* (DRSs) invented by Hans Kamp eighty years later. His rules of inference were based on operations of *iterating* and *deiterating* information to and from contexts in a way that resembles John McCarthy's *lifting rules*.

2. Hans Kamp (1981) developed *discourse representation theory* (DRT) to express the logical constraints on anaphoric references in natural language. Because of the difficulty of expressing those constraints in the algebraic notation for logic, Kamp introduced the graphic DRS notation, which allowed a simpler formulation of his rules. Significantly, the nested contexts in Kamp's DRSs are isomorphic to the nest of contexts in Peirce's EGs, even though Kamp had no previous knowledge of them. Kamp deserves credit for discovering the constraints on anaphora in DRT, but DRSs and EGs are equally suitable for expressing those constraints.

3. Jon Barwise and John Perry (1983) developed *situation semantics* as a theory of meaning in natural language. Unlike Montague's approach (1975), which related the semantics of language to potentially infinite models of the real world or possible worlds, Barwise and Perry adopted finite *situations* as their basis. Each situation is a bounded region of space-time containing physical objects and processes, as well as other situations. A great deal of research has been done within the paradigm of situation semantics (Barwise et al. 1991),

including efforts to merge it or at least reconcile it with DRT (Cooper & Kamp 1991). An important question is how and whether it can be merged or reconciled with McCarthy's contexts as well.

4. John Sowa (1984) developed *conceptual graphs* as a system of logic and reasoning based on the semantic networks of AI and the existential graphs of C. S. Peirce. The nodes called *concepts* correspond to typed, quantified variables in a sorted predicate calculus. A *context* is a defined as a concept of type Proposition, whose referent field contains one or more conceptual graphs that state the proposition. Later papers (Sowa & Way 1986; Sowa 1991) used the CG contexts to represent Kamp's DRSs and Barwise and Perry's situations. In a paper on "Crystallizing Theories out of Knowledge Soup," Sowa (1990) proposed the use of contexts for partitioning a knowledge base into a collection of smaller "chunks" that could be assembled into theories appropriate to any particular application. In his dissertation, Guha (1991) cited the knowledge soup paper, which he said was "in the same spirit as the work described in this document."

5. John McCarthy is one of the founding fathers of AI, whose collected work (McCarthy 1990) has frequently inspired and sometimes revolutionized the application of logic to knowledge representation. His work on context, although published later than the previous four approaches, has grown out of ideas based on his earlier work. McCarthy's *ist* predicate is the key to relating that work to the ongoing research in the other paradigms. If the *ist* predicate can be defined in terms of the other theories, then any results obtained in one approach can be translated to any of the others. Besides defining contexts, McCarthy has been emphasizing his *lifting rules* for importing and exporting information into and out of the quoted text or package. Such rules, which resemble Peirce's rules of iteration and deiteration, are essential for allowing quoted information to be unquoted and used.

Besides these five theories, there is a long history of related ideas in logic, philosophy, linguistics, and AI. The most important ones for a theory of context include indexicals, possible worlds, metalanguage, belief revision, and ontology revision. Yet the various ideas and theories were developed by people with different intuitions, which they applied to different problems of knowledge representation. Trying to unify and clarify those intuitions by defining the terms of one theory in those of another runs the risk of distorting the insights of both. This paper will explore the implications of these definitions to determine whether the benefits of unification and clarification outweigh any possible distortions of the original insights.

2. Peirce's Contexts

First-order predicate calculus was independently invented by Gottlob Frege (1879) and Charles Sanders Peirce (1885). Frege used a tree notation, which no one else ever adopted. But Peirce developed an algebraic notation, which through the textbook by Ernst Schröder (1890) and with a change of symbols by Giuseppe Peano became the modern system of predicate calculus. Long before Bertrand

Russell learned logic from Frege and Peano, it had become a flourishing subject based on the Peirce-Schröder foundations.

The early history of modern logic is a fascinating tale that has been recounted by Roberts (1973) and Houser et al. (1995). The main point for this paper is that the man who invented the common algebraic notation for logic later abandoned it for a graph representation, which he called his "chef d'oeuvre" and "the luckiest find of my career." With the *existential graphs* that he invented in 1897, Peirce developed an aspect of logic that was largely ignored by the mathematical logicians of the twentieth century. In relating Peirce's later logic and philosophy to situation semantics, Burke (1991) said "Peirce anticipated in his own way some of the concerns of situation theory (or rather, he happened to be working before it went out of fashion to wrestle with such concerns)." A century later, those concerns are back in fashion, and Peirce is once again in the avant garde of modern logic.

The three primitives of existential graphs (EGs) include the *oval enclosure*, which delimits a context, the *line of identity*, which corresponds to an existentially quantified variable, and *juxtaposition*, which represents conjunction. The default interpretation of an oval with no other qualifiers is negation of the graphs nested inside. Existence, conjunction, and negation provide a complete representation for all of first-order logic. As an example, the middle diagram in Figure 1 is an existential graph for the sentence *If a farmer owns a donkey, then he beats it.*

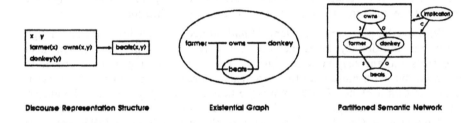

Discourse Representation Structure · Existential Graph · Partitioned Semantic Network

Figure 1. Three representations for "If a farmer owns a donkey, then he beats it."

The EG in Figure 1 has two ovals, which represent negations. It also has two lines of identity, represented as linked bars: one line, which connects *farmer* to the left side of *owns* and *beats*, represents an existentially quantified variable ($\exists x$); the other line, which connects *donkey* to the right side of *owns* and *beats* represents another variable ($\exists y$). When Figure 1 is translated to the algebraic notation, *farmer* and *donkey* map to monadic predicates; *owns* and *beats* map to dyadic predicates. The implicit conjunctions can be represented with Peano's symbol $\wedge$:

$$\sim(\exists x)(\exists y)(\text{farmer}(x) \wedge \text{donkey}(y) \wedge \text{owns}(x,y) \wedge \sim\text{beats}(x,y)).$$

Peirce called a nest of two ovals, as in Figure 1, a *scroll*, which he used to represent material implication, since $\sim(p \wedge \sim q)$ is equivalent to $p \supset q$. Using the $\supset$ symbol, the above formula may be rewritten as

$$(\forall x)(\forall y)((\text{farmer}(x) \wedge \text{donkey}(y) \wedge \text{owns}(x,y)) \supset \text{beats}(x,y)).$$

The algebraic formula with the ⊃ symbol illustrates a peculiar feature of logic in comparison with natural languages: in order to preserve scope, the implicit existential quantifiers in the phrases *a farmer* and *a donkey* must be moved to the front of the formula and be translated to universal quantifiers. This puzzling feature of logic has posed a problem for linguists. In his discourse representation structures, Hans Kamp (1981) resolved it by introducing a new symbol for implication with different scoping rules. The diagram on the left of Figure 1 shows a DRS for the donkey sentence. The two boxes connected by an arrow represent the English pair *if-then*. The variables x and y in the antecedent box have implicit existential quantifiers; Kamp defined the scoping rules for the DRS to include consequent box within the scope of the antecedent. As in existential graphs, conjunction is implicitly shown by juxtaposition. Altogether, the DRS may be read *If there exists a farmer x and a donkey y and x owns y, then x beats y.*

Although the DRS and EG notations look quite different, they are exactly isomorphic: they have the same three primitives and exactly the same scoping rules for variables or lines of identity. What makes this coincidence remarkable is that in the dozens of notations for semantic networks in the 1960s and 1970s, no one else rediscovered Peirce's conventions. The notation that comes closest is the *partitioned semantic network* by Gary Hendrix (1975), which is illustrated in the rightmost diagram of Figure 1. Like Peirce and Kamp, Hendrix took the existential quantifier as the default, represented conjunction by juxtaposition, and used a graphic enclosure for partitioning contexts. But unlike Peirce and Kamp, Hendrix allowed overlapping contexts: the two overlapping boxes in Figure 1 represent the antecedent and the consequent of the implication.

With overlapping contexts, Hendrix had no need for scoping rules. Although the farmer and donkey nodes each occurred only once, the overlap allowed them to occur simultaneously in both contexts. Yet the overlapping contexts proved to be unwieldy: with more than three contexts, it became impossible to draw partitioned nets on a plane. Furthermore, the nesting of clauses in natural languages has a more direct mapping to the Peirce-Kamp nested contexts than to Hendrix's overlapping contexts. Kamp's rules for resolving anaphora in DRSs could be stated equally well in terms of EGs, but not in terms of overlapping contexts.

Besides notation, Peirce defined rules of inference for EGs, which in many respects are the simplest and most elegant inference rules ever devised for any version of logic. A typical theorem that requires 43 steps to prove with Russell and Whitehead's rules of 1910 takes only 8 steps with Peirce's rules of 1897. Peirce's rules are a generalization of *natural deduction*, which Gerhard Gentzen discovered 40 years later. Like Gentzen, Peirce took the empty set as his only axiom, but Peirce's proofs are simpler than Gentzen's because the nesting of contexts eliminates the bookkeeping needed for making and discharging assumptions — the most error-prone aspect of Gentzen's system. For further discussion of these points, see Roberts (1973), Sowa (1984, 1993), and Houser et al. (1995).

In discussing modality, Peirce imagined the graphs drawn on "a book of separate sheets, tacked together at points." The upper sheet represents "a universe of existing individuals," while the other sheets "represent altogether different universes with which our discourse has to do." Graphs on those sheets may represent "conceived propositions which are not realized." Peirce said that a necessarily true proposition could be considered as replicated on all the sheets in the book, while a possible proposition might occur on only one. In the algebraic notation, Peirce used the symbol ω as an index for "a state of things," to which he applied a universal quantifier for necessity and an existential quantifier for possibility. With his interpretation of necessity as truth "under all circumstances," Peirce was following Leibniz and anticipating Kripke.

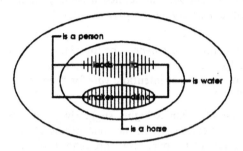

Figure 2.   EG for "You can lead a horse to water, but you can't make him drink."

In 1906, Peirce introduced colors or *tinctures* to represent modalities. Figure 2 shows one of Peirce's examples, but with shading instead of the original red for possibility. The graph contains four ovals: the outer two are associated to form a scroll for *if-then*; the inner two represent possibility (shading) and impossibility (shading inside a negation). The outer oval may be read *If there exists a person, a horse, and water*; the next oval may be read *then it is possible for the person to lead the horse to the water and not possible for the person to make the horse drink the water.*

The notation _leads_to_ represents the triadic predicate leads-to(x,y,z), and _makes_drink_ represents makes-drink(x,y,z). In the algebraic notation with $\Diamond$ for possibility, Figure 2 maps to the following formula:

$\sim(\exists x)(\exists y)(\exists z)(\text{person}(x) \land \text{horse}(y) \land \text{water}(z)$
$\land \sim(\Diamond\text{leads-to}(x,y,z) \land \sim\Diamond\text{makes-drink}(x,y,z))\ ).$

With the symbol $\supset$ for implication, this formula becomes

$(\forall x)(\forall y)(\forall z)((\text{person}(x) \land \text{horse}(y) \land \text{water}(z))$
$\supset (\Diamond\text{leads-to}(x,y,z) \land \sim\Diamond\text{makes-drink}(x,y,z))\ ).$

This version may be read *For all x, y, and z, if x is a person, y is a horse, and z is water, then it is possible for x to lead y to z, and not possible for x to make y drink z.*

As a systematic way of representing the kinds of contexts, Peirce adopted the traditional heraldic tinctures, which were classified as metal, color, or fur. He applied that three-way distinction to actual, modal, and intentional contexts:

1. Metal: *argent, or, fer,* and *plomb.* Peirce used *argent* (white background) for "the actual or true in a general or ordinary sense," and the other metallic tinctures for "the actual or true in some special sense." A statement about the

physical world, for example, would be actual in an ordinary sense. Peirce also considered mathematical idealizations, such as Cantor's hierarchy of infinite sets, to be "actual," but not in the same sense as ordinary physical entities.

2. Color: *azure, gules, vert,* and *purpure.* Peirce distinguished four basic modalities: *azure* for logical possibility (dark blue) and subjective possibility (light blue); *gules* for objective possibility; *vert* for "what is in the interrogative mood"; and *purpure* for "freedom or ability." Each of these modalities could be combined with negation: he defined *necessary* in the usual way as *not-possibly-not* and *obligatory* as *not-freedom-not* (an anticipation of deontic logic).

3. Fur: *sable, ermine, vair,* and *potent.* The four furs correspond to propositional attitudes: *sable* for "the metaphysically, or rationally, or secondarily necessitated"; *ermine* for purpose or intention; *vair* for "the commanded"; and *potent* for "the compelled."

Peirce's three-way classification is highly suggestive, but incomplete. He wrote that the complete classification of "all the conceptions of logic" was "a labor for generations of analysts, not for one." But throughout his analyses, he clearly distinguished the logical operators represented by the graphs, from the tinctures, which, he said, do not represent

> differences of the *predicates,* or *significations* of the graphs, but of the predetermined objects to which the graphs are intended to refer. Consequently, the Iconic idea of the System requires that they should be represented, not by differentiations of the Graphs themselves but by appropriate visible characters of the surfaces upon which the Graphs are marked.

It seems that Peirce did not consider the tinctures to be part of logic itself, but of the metalanguage for describing how the logic applies to the universe of discourse:

> The nature of the universe or universes of discourse (for several may be referred to in a single assertion) in the rather unusual cases in which such precision is required, is denoted either by using modifications of the heraldic tinctures, marked in something like the usual manner in pale ink upon the surface, or by scribing the graphs in colored inks.

By 1906, mathematical logic based on Peirce's algebraic notation had become a flourishing field of research, and his graphs were ignored. There were several reasons for the neglect: the notation and terminology were unfamiliar; most logicians, who had a strong background in mathematics, had already found the algebraic notation congenial to their tastes; and significantly, Peirce's novel applications of his graph logic to modality, intentionality, and metalanguage were outside the main interests of the logicians of his time. Today, however, Peirce's contributions are central to research on contexts:

1. Representation of contexts by nests of enclosures, which separate or partition groups of propositions of different modal status.

2. First-order logic based on three operators: existence (line of identity), conjunction (juxtaposition), and negation (oval enclosure on a white background).

3. Sound and complete rules of inference for first-order logic based on operations of drawing or erasing graphs and importing or exporting graphs into and out of contexts.

4. Tinctures for distinguishing the purpose or "nature" of a context from its logical operators (for which he used only the basic three — existence, conjunction, and negation).

5. A three-way classification of the use of contexts for representing actuality (metal), modality (color), or intentionality (fur).

6. The use of graphs as a metalanguage for talking about graphs.

7. Complete statement of the rules of inference for existential graphs in existential graphs themselves.

Peirce's later writings, although fragmentary, incomplete, and mostly unpublished, are no more fragmentary and incomplete than many modern publications about contexts. Although (or perhaps because) he did not use the word *context*, Peirce was more consistent in distinguishing the syntax (oval enclosures), the semantics ("the universe or universes of discourse"), and the pragmatics (the tinctures that "denote" the "nature" of those universes).

3. Contexts in Conceptual Graphs

Conceptual graphs are extensions of existential graphs with new features based on the semantic networks of AI and the linguistic research on *thematic roles* and *generalized quantifiers*. The primary difference is in the treatment of lines of identity. In existential graphs, the lines serve two different purposes: they represent existential quantifiers, and they show how the arguments are connected to the relations. In conceptual graphs, those two functions are split: boxes called *concepts* contain the quantifiers, and arcs marked with arrows show the connections of arguments to circles called *conceptual relations*. This separation of functions has several important consequences:

- Concepts have a place to represent a type label for each quantifier. Conceptual graphs therefore correspond to a typed or sorted logic, unlike the untyped existential graphs.

- Concepts may also contain a name or other specification of the *referent* of the concept, as in [Cat: Yojo], where the type is Cat and the particular individual is named Yojo. The area on the left of the colon is called the *type field*, and the area on the right is called the *referent field*.

- When an existential graph or an untyped formula in predicate calculus is mapped to a conceptual graph, the type label τ may be used to mark the *universal type*, which is a supertype of all others. The type τ imposes no restrictions on the quantifier or referent. The concept [τ: Yojo], for example, would represent an entity named Yojo whose type was unknown or unspecified.

- The arrows or numbers on the arcs of conceptual relations distinguish the arguments more clearly than Peirce's unlabeled lines. For dyadic relations, the arrow pointing towards the circle is the first argument, and the arrow pointing away is the second argument. For relations with more than two arguments, the arcs are numbered 1, 2, ..., *n*; the arrow on the *n*-th arc points away from the circle, and the other arcs point towards the circle.

- In an existential graph, any point on a line of identity could be considered as a separate quantified variable. Conceptual graphs concentrate the point of quantification in the concepts rather than the lines. A blank referent field represents the quantifier ∃, but the universal quantifier ∀ or other *generalized quantifiers* may also occur in the referent field of a concept.

- Peirce's lines of identity could cross context boundaries, but a concept may only occur in a single context. When two concepts in different contexts refer to the same individual, they must be associated by *coreference labels* or by a dotted line called a *coreference link*.

To illustrate these features, Figure 3 shows two equivalent conceptual graphs for the donkey sentence. The CG on the left uses the basic notation with the ¬ symbol to mark negation and with dotted lines for coreference links. The CG on the right uses an extended notation with the types If and Then defined as negated propositions and with coreference shown by the labels x and y. The concepts marked *x and *y are the *defining nodes* with implicit existential quantifiers, and the concepts marked ?x and ?y are *bound nodes* within the scope of *x and *y.

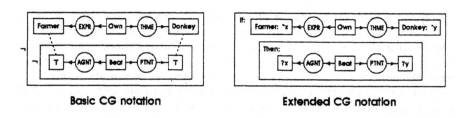

Basic CG notation   Extended CG notation

Figure 3. Two conceptual graphs for "If a farmer owns a donkey, then he beats it."

The scoping rules for the CGs in Figure 3 are the same as the rules for the DRS and EG in Figure 1. The CGs may be read *If a farmer x owns a donkey y, then x beats y*. The circles represent dyadic conceptual relations, where the arrow pointing towards the circle marks the first argument and the arrow pointing away marks the second argument. The two CGs in Figure 3 would correspond to the following formula in algebraic notation:

$$\sim(\exists x\!:\!\mathrm{Farmer})(\exists y\!:\!\mathrm{Donkey})(\exists z\!:\!\mathrm{Own})(\mathrm{expr}(z,x) \wedge \mathrm{thme}(z,y)$$
$$\wedge \sim(\exists w\!:\!\mathrm{Beat})(\mathrm{agnt}(w,x) \wedge \mathrm{ptnt}(w,y))\ ).$$

This representation follows C. S. Peirce and Donald Davidson in reifying verbs with the *event variables* z and w. The dyadic relations represent the thematic roles or case relations used in linguistics: experiencer (EXPR); theme (THME); agent (AGNT); and patient (PTNT). For convenience, there is also a linear notation that makes CGs easier to type:

¬[[Farmer: *x]←(EXPR)←[Own]→(THME)→[Donkey: *y]
   ¬[[?x]←(AGNT)←[Beat]→(PTNT)→[?y]]], .

[If: [Farmer: *x]←(EXPR)←[Own]→(THME)→[Donkey: *y]
   [Then: [?x]←(AGNT)←[Beat]→(PTNT)→[?y]]].

In the graphic form of Figure 1, coreference may be shown by dotted lines, but coreference labels must be used in the linear notation.

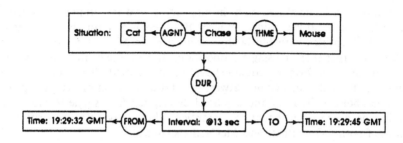

Figure 4. CG for "A cat chased a mouse for an interval of 13 seconds from 19:29:32 GMT to 19:29:45 GMT."

In CGs, a context is defined as a concept whose referent field contains nested conceptual graphs. Since every context is also a concept, it can have a type label, coreference links, and attached conceptual relations. In Figure 4, the graph for a cat chasing a mouse is nested inside a concept of type Situation. The conceptual graph in the inner context describes the situation. Attached to that context is the relation DUR for *duration*, which is linked to a concept for an interval of 13 seconds. The relations FROM and TO show that the interval lasted from 19:29:32 GMT to 19:29:45 GMT.

When a conceptual graph occurs in a context of type Graph, it is used as a literal; in a context of type Proposition, it states a proposition; in a context of type Situation, it describes a situation. The following concept, by itself, may be read *A situation of a cat chasing a mouse.*

[Situation: [Cat]←(AGNT)←[Chase]→(THME)→[Mouse]].

This graph may be considered an abbreviation for the following graph, which says that there exists a situation whose description (DSCR) is the proposition that a cat is chasing a mouse:

[Situation]→(DSCR)→[Proposition: [Cat]←(AGNT)←[Chase]→(THME)→[Mouse]].

In most applications, the abbreviated form shown in Figure 4 is used. When necessary, the type label of the context determines how the nested graph is interpreted. This use of the type label corresponds to *type coercion* in programming languages.

To relate Figure 4 to McCarthy's *ist* predicate, first expand the abbreviations; then translate the expanded graph to predicate calculus:

$(\exists s\text{:}\mathtt{Situation})(\exists i\text{:}\mathtt{Interval})$
$\quad(\mathtt{duration}(s,i)\ \wedge\ \mathtt{measure}(i,\mathtt{13sec})\ \wedge\ \mathtt{from}(i,\mathtt{19:29:32GMT})\ \wedge\ \mathtt{to}(i,\mathtt{19:29:45GMT})$
$\quad\wedge\ \mathtt{dscr}(s,$
$\quad\quad(\exists x\text{:}\mathtt{Cat})(\exists y\text{:}\mathtt{Mouse})(\exists z\text{:}\mathtt{Chase})(\mathtt{agnt}(z,x)\ \wedge\ \mathtt{thme}(z,y))\ )).$

Now it is possible to relate parts of this formula to the earlier definition of the *ist* predicate:

$$\mathtt{ist}(c,p)\ \equiv\ (\exists x\text{:}\mathtt{Entity})(\mathtt{is\text{-}in}(c,p)\ \wedge\ \mathtt{refers\text{-}to}(c,x)\ \wedge\ \mathtt{describes}(x,p)).$$

The entity x can be identified with the situation s, and the proposition p with the proposition that a cat is chasing a mouse. But there is no entity in this translation that can be identified with McCarthy's context c. In McCarthy's sense, c is supposed to be larger than the single proposition. In fact, the context may be so rich that no finite set of formulas could exhaust its full content. This discussion raises a question about the distinction between a total description of everything that is knowable about a situation and a partial description in a single formula. That issue has been central to situation semantics, which may help to clarify the semantics of both CG contexts and McCarthy's contexts.

4. Situations, Contexts, and Intentions

Barwise and Perry (1983) developed *situation semantics* as a reaction against the potentially infinite models of Kripke's and Montague's modal and intensional logics. Each situation is a finite configuration of some aspect of the world in a limited region of space and time. It may be a static configuration that remains unchanged for a period of time, or it may include processes and events that are causing changes. It may include people and things with their actions and thoughts; it may be real or imaginary; and its time may be present, past, or future. A situation may be as large as the solar system or as small as an atom, and it may contain nested situations for smaller or more detailed aspects of the world.

As an illustration of the way situations are related to partial descriptions, Figure 5 shows a concept of type Situation, which is linked to two images and a description. The image relation (IMAG) links the situation to two different kinds of images of that situation: a picture and the associated sound. The description relation (DSCR) links it to a proposition that describes some aspect of the situation, which is linked by the statement relation (STMT) to three different statements of the proposition in three different languages: an English sentence, a conceptual graph, and a formula in the Knowledge Representation Language (KIF) by Genesereth and Fikes (1992).

As Figure 5 illustrates, the proposition stated by the English sentence "A plumber is carrying a pipe" or by its translation to some version of logic represents a tiny fraction of the total information available. Both the sound image and the picture image capture information that is not in the sentence, but even they are only partial representations. A picture may be worth a thousand words, but a situation can be worth a thousand pictures.

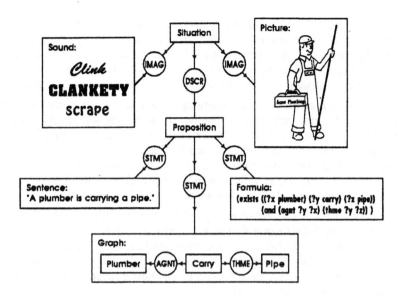

Figure 5. A situation of a plumber carrying a pipe

To relate a situation to the information it contains, Keith Devlin (1991) introduced the term *infon* for an abstract information object that is closely related to what McCarthy, Sowa, and many others call propositions. Devlin's basic formula is $s \models \sigma$, where s is a situation and σ is an infon that is semantically entailed by s. Devlin's construction that comes closest to what McCarthy calls a context is the set of all infons entailed by a situation: $\{\sigma \mid s \models \sigma\}$. If the DSCR relation is interpreted as entailment of a partial description, McCarthy's context c may be considered a complete description of a situation:

$$\text{completeDscr}(s,c) \equiv c = \{\sigma \mid s \models \sigma\}.$$

For Figure 5, the context c would include propositions or infons saying that the plumber is carrying a toolbox, he works for Acme Plumbing Co., he is dragging the pipe with a clankety noise, etc. The predicate is-in(c,p) could then be interpreted as provability.

With this analysis, the following formula may be adopted as a definition of the *ist* predicate in terms of situations and infons:

$$\text{ist}(c,p) \equiv (\exists s{:}\text{Situation})(\text{provable}(c,p) \wedge \text{completeDscr}(s,c) \wedge \text{dscr}(s,p)).$$

The syntactic predicate is-in(c,p) is interpreted as provable(c,p). The two semantic predicates are defined in terms of entailment: refers-to(c,s) is completeDscr(s,c); and dscr(s,p) is $s \models p$. In a pure first-order framework, $s \models p$ would be redundant, since it would be implied by $c \vdash p$ and the definition of c. Including it, however, could accommodate nonstandard logics in which provability is not equivalent to entailment.

This analysis does not solve all the problems concerning McCarthy's contexts, since there are still many open issues concerning situations, including their content, boundaries, and interpretation. However, those issues are remarkably similar for both McCarthy's contexts and for Barwise and Perry's situations. Relating the two theories does not automatically solve the problems for either one, but it helps researchers collaborate in addressing them. Peirce would have defined a context as a kind of sign, to which he would apply his basic principle of semiotics:

A sign, or *representamen*, is something which stands to somebody for something in some respect or capacity. It addresses somebody, that is, creates in the mind of that person an equivalent sign, or perhaps a more developed sign. That sign which it creates I call the *interpretant* of the first sign. The sign stands for something, its *object*. It stands for that object, not in all respects, but in reference to a sort of idea, which I have sometimes called the *ground* of the representamen. (CP 2.228)

Peirce's notion of sign was broad enough to include situations, contexts, propositions or infons, and their expression in any language, including English and logic. His notion of *ground* is crucial: it acknowledges that some agent's purpose, intention, or "conception" is essential for determining the scope of a situation. But as Peirce said, the complete classification of all those conceptions is "a labor for generations of analysts, not for one."

References

Barwise, Jon, Jean Mark Gawron, Gordon Plotkin, & Syun Tutiya, eds. (1991) *Situation Theory and its Applications*, CSLI, Stanford, CA.

Barwise, Jon, & John Perry (1983) *Situations and Attitudes*, MIT Press, Cambridge, MA.

Burke, Tom (1991) "Peirce on truth and partiality," in Barwise et al. (1991) pp. 115-146.

Cooper, Robin, & Hans Kamp (1991) "Negation in situation semantics and discourse representation theory," in Barwise et al. (1991) pp. 311-333.

Devlin, Keith (1991) "Situations as mathematical abstractions," in Barwise et al. (1991) pp. 25-39.

Frege, Gottlob (1879) *Begriffsschrift*, translated in Jean van Heijenoort, ed. (1967) *From Frege to Gödel*, Harvard University Press, Cambridge, MA, pp. 1-82.

Genesereth, Michael R., & Richard E. Fikes (1992) *Knowledge Interchange Format*, Reference Manual, Version 3.0, Report Logic-92-1, Computer Science Department, Stanford University.

Guha, R. V. (1991) *Contexts: A Formalization and Some Applications*, technical report ACT-CYC-423-91, MCC, Austin, TX.

Houser, N., D. D. Roberts, & J. Van Evra, eds. (1995) *Studies in the Logic of Charles Sanders Peirce*, Indiana University Press, Bloomington.

Kamp, Hans (1981) "Events, discourse representations, and temporal references," *Langages* **64**, 39-64.

Lenat, Douglas B., & R. V. Guha (1990) *Building Large Knowledge-Based Systems*, Addison-Wesley, Reading, MA.

McCarthy, John (1990) *Formalizing Common Sense*, Ablex, Norwood, NJ.

McCarthy, John (1993) "Notes on formalizing context," *Proc. IJCAI-93*, Chambéry, France, pp. 555-560.

McCarthy, John, & Saša Buvač (1994) *Formalizing Context*, Technical Note STAN-CS-TN-94-13, Stanford University. Available from http://sail.stanford.edu.

Peirce, Charles Sanders (1885) "On the algebra of logic," *American Journal of Mathematics*, vol. 7, pp. 180-202. Reprinted in Peirce (W) vol. 5.

Peirce, Charles Sanders (CP) *Collected Papers of C. S. Peirce*, ed. by C. Hartshorne, P. Weiss, & A. Burks, 8 vols., Harvard University Press, Cambridge, MA, 1931-1958.

Peirce, Charles Sanders (W) *Writings of Charles S. Peirce*, vols. 1-5, Indiana University Press, Bloomington, 1982-1993.

Roberts, Don D. (1973) *The Existential Graphs of Charles S. Peirce*, Mouton, The Hague.

Schröder, Ernst (1890-1895) *Vorlesungen über die Algebra der Logik*, 3 vols., Teubner, Leipzig. Reprinted by Chelsea Publishing Co., Bronx, NY, 1966.

Sowa, John F. (1984) *Conceptual Structures: Information Processing in Mind and Machine*, Addison-Wesley, Reading, MA.

Sowa, John F. (1990) "Crystallizing theories out of knowledge soup," in Z. W. Ras & M. Zemankova, eds., *Intelligent Systems: State of the Art and Future Directions*, Ellis Horwood, New York, p. 456-487.

Sowa, John F. (1991) "Towards the expressive power of natural language," in J. F. Sowa, ed. (1991) *Principles of Semantic Networks: Explorations in the Representation of Knowledge*, Morgan Kaufmann Publishers, San Mateo, CA, pp. 157-189.

Sowa, John F. (1993) "Logical foundations for representing object-oriented systems," *Journal of Experimental and Theoretical AI*, vol. 5, nos. 2&3, pp. 237-261.

Sowa, John F., & Eileen C. Way (1986) "Implementing a semantic interpreter for conceptual graphs," *IBM Journal of Research and Development* **30:1**, pp. 57-69.

Ontology Revision

Norman Foo
Knowledge Systems Group,
Department of Computer Science
University of Sydney, Sydney, NSW 2006,
Australia
email: norman@cs.su.oz.au

ABSTRACT

Knowledge systems as currently configured are static in their concept sets. As knowledge maintenance becomes more sophisticated, the need to address issues concerning dynamic concept sets will naturally arise. Such dynamics is properly called ontology revision, or in the simpler case, expansion. A number of sub-disciplines in artificial intelligence, philosophy and recursion theory have results that are relevant to ontology expansion even though their motivations were quite different. More recently in artificial intelligence ontologies have been explicitly considered. This paper is partly a summary of early results, and partly an account of ongoing work in this area.

Keywords: ontology, concept formation, theoretical term, predicate invention, theory change, induction, type hierarchy, action.

1. INTRODUCTION

Recent developments in the axiomatics of belief revision has shed light on what it means to rationally change one's beliefs. The AGM postulates [GAR] are representative of several in the genre. The crux of the AGM system is the notion of *minimal change* as the manifestation of rationality when an agent is confronted by new facts. The framework, however, is essentially one in which all potential facts and theories reside within a given ontology, expressed in a fixed first-order (indeed in effect a propositional) language. Here I want to outline the background, circumstances and opportunities that would attend a theory of *ontology revision* in which linguistic change is allowed. While it is premature to axiomatize such a theory, some of its landscape is now visible, and its potential impact on the practice of knowledge representation is already being felt.

As this is a preliminary exploration of new territory, no formal definitions and theorems will be given. Instead, I will endeavor to highlight results both old and new that will play a role in any investigation of ontologies.

Ontology *expansion* occurs when new information can be accommodated by the refinement, abstraction or addition of concepts, possibly relations or predicates. *Revision* is necessary when old ontologies have to be discarded. An understanding of the former is essential for dealing with the latter, and this paper will be mainly devoted to expansion. I believe that the AGM system provides many pointers to the direction in which work on ontologies should proceed, and if so the role of expansion has to be clarified first.

Section 2 reviews material from the philosophy of science where theoretical terms are linguistic expansions that have more than pragmatic import. Section 3 presents well-known results from model theory and some related but less familiar ones from computer science concerning expressibility of concepts. Section 4 examines the related notions of abstraction and refinement as ontology change. Section 5 recalls results from recursion theory that impinge upon this topic. Section 6 applies some of these ideas to a case study of type hierarchy change. Section 7 explains how the logic of actions can be used to detect ontological inadequacy. Section 8 rounds off the technical discussion with a description of the role of induction in this area.

2. PHILOSOPHY

The need to expand ontology via the introduction of new linguistic terms — *concept formation* as it is called in machine learning and psychology — is closely related to the classical problem of *theoretical terms* in the philosophy of science [NAG]. In the early development of this philosophy, terms of a scientific theory were partitioned into the *observational* and the *theoretical* variety. The former were supposed to refer to entities that were directly or indirectly measurable, and the latter to those that were not. Logical empiricists like Russell felt it necessary to eliminate theoretical terms by reducing them to observational ones via *definitions*, e.g.,

$$(2.1) \qquad T(X) \leftrightarrow \phi(O_1(X), \cdots, O_n(X)).$$

where $T(X)$ is a theoretical term, ϕ is a formula and $O_i(X)$ are observational terms. For example, $T(X)$ might be "X is soluble", and the rhs is the formal equivalent of "If X is placed in water then X dissolves". In this example we see that "observational" includes an "experiment", viz., X is placed in water. If one regards the introduction of theoretical terms to be a manifestation of ontology expansion, by this means it was hoped that this expansion is in principle *eliminable* since every occurrence of such theoretical terms can be replaced by its definition which consists only of observational terms. This proved to be a vain hope for two reasons. The first is that there is no way in which all conceivable experiments and observations can be delimited once and for all, so definitions as above have to be re-formulated as time progresses. The second is illustrated by the example of solubility. Since a material implication is used to define it, any substance that will never be placed in water will by definition be soluble!

Carnap's attempted solution [CAR] was ingenious, but as Hempel [HEM] showed, it essentially gave up on the elimination of theoretical terms. As I will suggest later how it can be used in a modern context to explore new concepts, a brief description of it is convenient. A theoretical term is related in Carnap's proposal to observational or empirical terms by two formulas of the form:

(2.2) $$P_1(X) \rightarrow (P_2(X) \rightarrow Q(X))$$
(2.3) $$P_3(X) \rightarrow (Q(X) \rightarrow P_4(X))$$

where the P's are observational and Q is theoretical. In many cases the P_1 is identical with P_3, and represents an experimental proposal. Also, P_2 and P_4 may coincide, and can be interpreted as an experimental result. The problem with the Rusellian proposal is thus avoided, but notice now that while the elimination of Q yields the *empirical prediction*

(2.4) $$(P_1(X) \wedge P_3(X)) \rightarrow (\neg P_2(X) \vee P_4(X))$$

this in no sense replaces Q. Further, Q may feature in many other similar formulas without damage to the two above, so its function as a theoretical term is *open*.

3. MODEL THEORY

Developments in model theory not only revealed the inexpressibility of a number of familiar concepts, but also introduced proof techniques that can be extended to our approaches to ontology expansion. There were a number of parallel developments, the most striking of which was the compactness theorem [END]. Finiteness and closed (topological) space were among the many that were shown using compactness to be not expressible in first-order logic. More recently, Gaifman et.al. [GOW] extended model-theoretic arguments to show that global properties may not be expressible, in contrast to local ones that are. This contrast appears to be pervasive. One consequence of compactness, for instance, is the inability to confine models of Peano arithmetic to the standard numbers (i.e., those "reachable" from 0 by a finite number of applications of the successor function). Therefore, the notion of *transitive closure* is not first-order expressible since the standard numbers can be regarded as the transitive closure of 0 under the successor. The impact of this on knowledge representation is the correct semantics that one can give to, say, Horn clauses of the form:

(3.1) $$descendant(X,Y) \leftarrow child(X,Z) \wedge descendant(Z,Y)$$

where child(X,Z) means X is a child of Y, etc., and the predicate descendant(_,_) is regarded as an ontological expansion. Re-interpreting child(X,Z) as successor(X,Z) in Peano arithmetic (where successor is written relationally) shows that {Y | descendant(0,Y)} is the set of standard natural numbers, and therefore first-order inexpressible. The *computational* interpretation of {Y | descendant(0,Y)} is however not problematic as it is identified in logic programming with the least fixed point [LLO] of the operator that uses ground child(_,_) facts as the basis for the recursive computation encoded by the formula above regarded as a rule.

The admission of *recursive definitions* of new terms (like descendant) only complicates the Rusellian notion of eliminability to the extent that we must assume standard domains. In such domains, every occurrence of the new term can in principle be replaced by a finite number of occurrences (in some logical combination) of old terms. However, this assumption also has the side-effect of *completing* the one way implication above into a bi-implication as suggested by Clark [CLA]. Clark completion will be discussed again below for type hierarchies. Recent work on recursive theoretical terms can be found in [SHS].

Expressibility is also called definability in the literature [END]. If a language was chosen to represent knowledge, and subsequently an informal concept arises, how might one know whether it is expressible or definable in the given language? Padoa introduced the idea of implicit definability of concept C in terms of those in an underlying language L by saying that C is so definable if any interpretation of L in some domain determines uniquely the interpretation of C. This is given an explicit characterization in Beth's theorem [B&J] where implicit definability is identified with the expansion of L and its logic by the defining formula for C, viz.,

(3.2) $C(X) \leftrightarrow \phi(X)$

The close relationship between this and Russell-like definitions is unmistakable. The formula ϕ is in L, and cannot contain C unlike the recursive definitions common in modern knowledge representation and computer science. It is not clear how recursive relations can be implicitly defined.

4. REFINEMENT AND ABSTRACTION

In the design of knowledge systems and software specifications, and in the evolution of object-oriented models, refinement and abstraction are pre-eminent techniques. Both of these are manifestations of ontology expansion. I will briefly re-examine these techniques from this perspective as a step toward the eventual axiomatization of the process of ontology revision. This will be by example, one from knowledge representation and the other from data structures.

An abstraction L′ of L is technically a mapping h from the models of L to those of L′ that preserve the properties that are *not* being abstracted. In practice, h is often just a homomorphism [END], and this will be so (for example) when a clause in L is replaced by a new predicate and connected to it by a Beth-like defining formula. A theorem of Lyndon's [LYN] shows that homomorphisms preserve positive sentences (in a logic with only the $\wedge$, $\vee$ and $\neg$ connectives), so the rhs of defining formulas can be quite general.

An example of the use of such an abstraction might be the absorption into nondeterministic conditions of multiple cases for action preconditions:

(4.1)   if light(X) then can_pickup(X)
(4.2)   if $\neg$glued(X) then can_pickup(X)

may be replaced by the abstraction:

(4.3)   liftable(X) $\leftrightarrow$ light(X) $\vee$ $\neg$glued(X)
(4.4)   if liftable(X) then can_pickup(X).

This is almost trite, but it is easy to imagine the use of this to abstract fairly complex predicates in cnf or dnf form. Indeed, a lot of machine learning abstractions are confined to the search for cnf and dnf abstractions of this type.

Are such abstractions ontology expansions? In a strict sense, no; for they are really Russell-like definitions and therefore eliminable. It can be shown that any expansion of this kind is *conservative*, i.e., there is nothing in the old vocabulary that can be proved using them that cannot be proved without them. But the convenience and succinctness they provide may be pragmatic gains. For instance, in specifying a state that enables something to be picked up, it is sufficient (if the original conditions are complete) to say that it be made liftable without prescribing exactly how (there are three ways to achieve it). In other words, abstractions are tolerant of partial specifications.

Refinements introduce ontologies in the converse sense. That is, the mapping now goes from L' to L where L' is "a more detailed" description of L. With this perspective, in the above example liftable would be in L and the new predicates light and glued would be refinements of it in L'. For a data structures example suppose we have an algebraic description of the "pop" and "push" operations on a stack in the form of an equation:

(4.5) $pop(push(S,A)) = S$.

A refinement of this might be a linked list implementation with the description:

(4.6)   pop(List, A) = Tail(List)
(4.7)   push(List, A) = Cons(A, List)
(4.8)   Tail(Cons(A, List)) = List

and other equations that specify Cons, Tail and Head on lists. If we "solve out" the these list operations from the refined specifications (4.6) - (4.8) what is usually obtained is the original specification (i.e., it is a theorem of the expanded system), suggesting the conservative nature of refinement. (The "usually" qualification refers to the assumption that all terms in the refined equations only have standard interpretations, and technically this is equivalent to the use of an ω-rule.) Refinements, even when they are conservative, do introduce novel ontologies. This is easily seen in the two examples above, since there is no limit to the kinds or number of new predicates and axioms that can be introduced.

5. RECURSION THEORY

A well-known theorem of Craig's [CRA] has sometimes been used by people sympathetic to the strict logical empiricist program to argue for the eliminability of theoretical terms, and hence for the non-necessity of ontological expansion beyond observational predicates. Craig himself provided disclaimers similar to the one I will explain. A scientific theory T is *effective* to the extent that there is a way to systematically enumerate its predictions or theorems. In the parlance of recursion theory [END] [B&J] this is to say that the theory T is *recursively enumerable*. A theory such as T is *recursively axiomatizable* if there is a subset Δ of T such that (i) T = Cn(Δ) and (ii) Δ is decidable. Cn is the logical consequence relation, and decidability of S means that there is an effective method to tell if a formula is in S or not. Craig's theorem showed that T is recursively enumerable if and only if it is recursively axiomatizable. To understand the philosophical import of this result it is necessary to see an outline of its proof. Let the recursive enumeration of T be A_1, A_2, A_3, etc. Consider the set $\Delta = \{A_1, A_2 \wedge A_2, A_3 \wedge A_3 \wedge A_3, \ldots \}$. It is not hard to see that T = Cn(Δ) and that Δ is recursive. Hence T is recursively axiomatizable, and moreover its axiom set Δ does *not* involve any extraneous terms beyond those already in T. It is this latter remark that apparently justifies the assertion that no theoretical terms are ever necessary. As Nagel [NAG], Craig himself, and others have pointed out, this is wrong because the set Δ so constructed has to be the completed infinity of *all* theorems (predictions, experiments, observations) which is impossible to achieve in practice. So, while it is an interesting result, Craig's theorem is more of a curiosity than it is relevant to discussions on ontology expansion.

Such fundamental objections cannot be raised against a couple of other important results that all workers in ontologies should know. The first is by Kleene [KLE]. What he showed was that if a theory T is recursively axiomatizable, then it is finitely axiomatizable by the addition of a finite number of new predicates. Finite axiomatizability of T means that it has a set Δ as above that is finite (hence decidable). Now, since all interesting scientific theories have a finite number of axioms (or axiom schemes), this suggests that interesting scientific theories may arise from those that are merely recursively axiomatizable by finite ontology expansion. Later, Craig and Vaught [C&V] strengthened this result to show that for many theories finite axiomatizability is obtainable by the addition of only *one* new predicate symbol.

Because these proofs are long and elaborate, I will outline an alternative derivation for a version of Kleene's result that is more accessible to computer scientists and use it to indicate pragmatic objections to the Kleene, and Craig and Vaught results. Suppose we are given a recursively axiomatizable theory T. Buchi [BUC] [B&J] showed how to use first-order logic to simulate computations by Turing Machines. Since recursive axiomatization amounts to the existence of an effective decision procedure for axioms, there is a Turing Machine M that realizes it. Buchi's translation of M into a finite set Γ of first-order formulas can be regarded as a compression of the axioms of T by introducing a finite number of new predicate symbols (encoding the states, tape symbols, tape cells, head position, time). Then Cn(Γ) is a conservative extension of T in the same sense as in section 4, i.e., any theorems of Γ that is entirely in the vocabulary of T is already a theorem of T. Now, while this is of theoretical importance, in practice the additional new predicates, as indicated above, do not have what we may call predictive or explanatory significance. Introduced predicates should stand on their own as in Carnap's Q in (2.2) and (2.3) above, and in (7.1) below. Turing Machine cell numbers are too far removed from T. How notions like "relevance", "explanatory significance", and "too far removed" can be formalized is an interesting open question.

6. TYPE HIERARCHIES

In this section I consider the application of some of the ideas on ontology expansion to the problem of type hierarchy change as the result of new knowledge. Although this is done by example the principles adduced in it point clearly to generalizations for other kinds of hierarchy change.

A simplified biological way of distinguishing mammals from other vertebrates, before the arrival at Australia of western zoologists, was by saying that they are the vertebrates which are hairy, give milk, and are placental (give birth to their young live and developed). A sentence of logic that expresses this is:

(6.1)   mammal ← hairy ∧ gives_milk ∧ placental.

In practice, the way this was applied to abstractions (as in section 4) using *property lists*, is to regard this as a definition or characterization. Logically this is equivalent to closing the above by including the reversal of the implication, changing the "if" to "if and only if", viz.

(6.2)   mammal $\leftrightarrow$ hairy $\wedge$ gives_milk $\wedge$ placental.

This is an example of the well-known *closed world assumption* (CWA), and in the particular case where the sentence is Horn as above, the CWA realized by completing the "if" to "if and only if" is known as the Clark completion [CLA]. The CWA is a nonmonotonic [REI] rule, and as with all nonmonotonic rules it is defeasible in the light of new knowledge. Historically, as is well-known, this new knowledge came to light when monotremes (e.g., duck-billed platypus) and marsupials (e.g., kangaroos) were examined by zoologists. While they had hair and gave milk, they were not placental in different ways. Monotremes laid eggs while marsupials gave birth to young that had to undergo further development in pouches outside the womb. Moreover they sometimes satisfied the other two properties of mammals in somewhat novel ways (e.g. milk is secreted through the entire body by the platypus).

Figure 1 shows the taxonomy within which mammals sit. The existing ontology had to be modified, but how? There are two putative ways to do it. If we wish to preserve (6.2) which is the result of the CWA applied to (6.1), then the modification in figure 2 is an obvious one. Call this ontology refinement by splitting. What his amounts to is a decision that (6.1) should be replaced by three new classifier rules:

(6.3)   placental $\leftarrow$ hairy $\wedge$ gives_milk $\wedge$ developed_baby.
(6.4)   monotreme $\leftarrow$ hairy $\wedge$ gives_milk $\wedge$ lays_egg.
(6.5)   marsupial $\leftarrow$ hairy $\wedge$ gives_milk $\wedge$ foetus$-$in$-$pouch.

In this splitting, the explicit concept of a "mammal" is foregone. The second way is to jettison (6.2), i.e., admit that the CWA is wrong. Figure 3 shows this alternative, which is tantamount to replacing (6.2) with the weakened rule

(6.6)   mammal $\leftarrow$ hairy $\wedge$ gives_milk

and introducing new rules (closable by the CWA)

(6.7)   placental $\leftarrow$ mammal $\wedge$ developed_baby
(6.8)   monotreme $\leftarrow$ mammal $\wedge$ lays_egg
(6.9)   marsupial $\leftarrow$ mammal $\wedge$ embryo_in_pouch.

In this version the concept of a "mammal" is of course different from the original one. Modern biology rejects the first alternative and accepts the second on the basis of functional and genetic grounds. Many of these are coarsely reflected in properties such as "hairy" above, and there is a simple way to then justify the choice made by biology. A metric can be defined between two objects using the symmetric difference in their property lists [FO1]. Taxonomic trees should minimize the semantic difference [FGRT] in such trees between objects that have small metrical distance. If this is done, figure 3 will be the choice. But for this we need to have *global* knowledge, in this case the categories (here, reptiles, fishes, etc.) at the same level as the category being refined. The seemingly innocent remark on the relationship between semantic distance and symmetric difference is the informal expression of a theorem awaiting proof.

Logically the distinction between (6.3) to (6.5) and (6.6) to (6.9) *disappears* if they are all replaced by their CWA completions. Then "mammal" in (6.6) is really a *definition*, indeed what is called an eliminable definition [HEM] [NAG], that has the character of a weak theoretical term. It *abstracts* the property "hairy $\wedge$ gives_milk" as significant enough to merit a term "mammal". (If history were different, it is conceivable that "mammal" might truly have been such an abstraction.) This shows that a logical theory alone is not sufficient to indicate taxonomic structure once CWA closure is applied.

These alternative ways of refining ontologies are common enough to deserve the respective names of *Splitting* and *Sub-class Creation*. As a rule, splitting results when a CWA applied to classifying properties is retained in the light of evidence that conflicts with its categorization.

Belief revision can be used in accounting for deciding which trees to adopt when there are alternatives. In belief revision we minimize change in the light of new beliefs. If taxonomic trees should have near-uniform differences in properties between objects at a given level, then in the above two alternatives, minimal change in such uniformity from the tree in figure 1 would favor that in figure 3.

This example also illustrates another way to think about ontology change from the viewpoint of *second-order logic*. The predicate "mammal" was weakened in the passage to figure 3. Its *denotation* was expanded in the process. Ways to reason about alternative denotations for predicate symbols are inherently second-order, and indeed the circumscription (see next section) schema can be regarded as a second-order constraint on denotations.

7. ACTIONS

Here I will outline our recent work on detecting the need for ontology expansion. The setting is a dynamic system that has a partial theory of constraints or laws, and of how actions can effect state change. Shapiro's MIS [SHA] or Sandewall's game-theoretic model of experimentation [SAN] are reasonable formalizations of such settings. Perhaps the easiest way to motivate this work is by means of an example [FO2] that featured very early in this work. Consider a homogeneous uniform blocks world in which one block can be on two blocks as in figures 4.1 and 4.2, assume that the only predicate used to capture this world is on(_,_). It is clear that the situation in either figure 4.1 or 4.2 is captured by the ground facts S1 = {on(a,b),on(a,c),on(b,table),on(c,table)}. S1 cannot separate the two physical situations as displayed, yet an action such as remove(b) will reveal (if ordinary physics holds) their difference, viz., from figure 4.1 block a will topple, while from figure 4.2 it will not.

For simplicity we will assume actions are deterministic. The key point in this example is that the state description S1 is inadequate to the extent that it conflates models that can be distinguished by experiments admissible in the model. If blocks cannot be removed this inadequacy will never be detected. Formally, the notion of state description adequacy with respect to an action (experiment) is the dynamic analog of the isomorphism criterion for definable predicates [END]. Suppose S is a state description and A an action. Actions map *models* to *models*, i.e., they are semantic, even though attempts may be made to provide the logic with action calculi that are sound and complete. S is an A-adequate description if for any two models M1 and M2 of S, A(M1) = A(M2). In particular therefore, every state description that renders an action A nondeterministic is A-inadequate. This idea may be called the *dynamorphism* version of the cited isomorphism condition for definability. It says that the vocabulary of S has to be expanded to fix the looseness in action A.

While the technicalities are actually more elaborate than this, the basic insights are the same. Now, once inadequacy is detected in this sense, how might the ontology be expanded to correct it? It is possible to exploit ontological expansion versions of Shapiro's MIS to do this. However it is instructive to see how some approaches explained above can also be tailored to do this. Consider (2.2) and (2.3) when P_1 is P_3 and P_2 is P_4. Then these can be combined and re-written as $P_1 \rightarrow (P_2 \leftrightarrow Q)$. Now, in the blocks world situations of figure 4 let P_1 mean remove block b, and let P_2 mean that block a topples. Then the "theoretical" term Q of Carnap's is exactly what we are looking for since it is a "quality" or "property" that will distinguish between the toppling of block a and its non-toppling. Again, the formalization of this is a bit more elaborate since the Carnap formula does not incorporate directly a time sequence. However, using a language similar to the situation calculus this is easy to rectify. The Carnap formula sought is:

(7.1) $remove(b,S1) = S2 \rightarrow (Q \leftrightarrow topple(a,S2))$.

The problem now is to determine "solutions" for this Q. Many approaches are possible, but all can be described as *learning* or *ontology expansion*. A habitual way to do this without extensive induction is to use *circumscription* [MCC] to determine the constants (variables if S1, S2, b, a, and c are regarded as generic names) that have to feature in Q. This is tantamount to believing that a finite set of outcomes for remove(b,S1) are sufficient for all learning to be complete and that information content in Q should be minimized. Of course, in this example, Q would be Q(S1,b,a) meaning that b supports a in state S1. Kwok [K&F] and Gibbon [GF1] [GF2] are among my colleagues who have worked with me to pursue these ideas in diverse areas including Mendelian genetics; in particular Kwok showed that the number of Q's needed is a logarithm of the degree of nondeterminism of action A, and Gibbon showed that extending Shapiro's MIS framework to account for ontological expansion is quite natural.

It is appropriate to note that this kind of ontological expansion is related to the so-called "hidden variable" problem in theories of physics.

8. INDUCTION

Concept formation by induction is an age-old dream. Computational approaches to induction relevant to ontology expansion are of two kinds. The first comes from theorem proving, and the second from machine learning. I will briefly describe both.

Resolution [C&L] is a standard technique in mechanical theorem proving. The basic idea is simple. Given two clauses C1 (= L $\vee$ D1) and C2 (= ¬L $\vee$ D2), to resolve C1 and C2 using L as the selected literal is to "cancel" L and merge D1 and D2 to give the resolvent clause D1 $\vee$ D2. Resolution preserves satisfiability, so that if ever the null (and therefore unsatisfiable) clause is generated as a resolvent, the original set is unsatisfiable.

Muggleton and Buntine [MGB] conceived the process of inverse resolution to induce clauses from which resolvents may be derived. In so inverting resolution he proposed two schemes, called the V and the W for obvious reasons once the diagrams that illustrate them are revealed. We show these as figure 5. The V operator takes an input clause C and a resolvent clause D and produces a candidate input clause denoted by ? in the diagram. C and ? should resolve to give D. In most applications, D is a singleton literal, and for these cases ? is easy to determine. The V operator does not involve predicate addition. In the W operator case predicate invention is in fact the rule rather than the exception. Here the known clauses are D1 and D2, and the W operator has to find C1, C2 and C3. In this situation there could be a literal that was in the C's but not in the D's. Such a literal will be "invented" by the process of inverse resolution. In practice, restrictions are placed on the syntactic forms of the clauses, e.g., that they be Horn, or singleton literals, etc., to control the search space.

Such invented literals may be viewed as ontology expansions.

In machine learning, Quinlan's FOIL [QUI] attempts to construct recursive Horn definitions (like (3.1) above) of observed positive and negative ground literals. This technique is closer to classical induction than inverse resolution in that it requires a lot of ground literals to ensure its reliability. Because it expands ontology by providing a new term that is recursive, it presumes a fixed point semantics for it as explained in section 3. We have recently [F&T] showed that FOIL can be generalized to guarantee the least fixed point if it converges.

Stahl [STA] contains a good exposition of how the recursion-theoretic and inductive techniques are related as well as a number of complexity results on predicate invention.

9. CONCLUSION

Ontology change is at the frontier of knowledge systems research. Whenever the incremental re-design of object classes is contemplated the issue of what is a rational ontology change has to be addressed. There is no set of accepted principles for it at the moment, but it is clear from our preliminary investigations that inputs from areas as diverse as the philosophy of science, nonmonotonic reasoning, and belief revision are necessary. I have tried in this paper to outline the main ideas, to suggest fruitful avenues to explore, and to indicate how we ourselves have been applying these ideas to type hierarchy revision and the diagnosis of action logics. A lot more experience is required before postulates and principles can be established.

10. ACKNOWLEDGEMENTS

I am indebted to my colleagues in the Knowledge Systems Group for many discussions pertinent to this work. In particular Rex Kwok, Greg Gibbon, Abhaya Nayak, Maurice Pagnucco and Pavlos Peppas have all contributed to the understanding of many slippery issues in this research.

11. REFERENCES

[B&J]   G. Boolos and R. Jeffery, "Computability and Logic", 2ed, Cambridge University Press, 1980.

[BUC]   J.R. Buchi, "Turing Machines and the Entscheidungsproblem", Math. Annalen, 148, 1962, pp 201-213.

[CAR]   R. Carnap, "Testability and Meaning", Philosophy of Science, III, 1936, pp 419-471; IV, 1937, pp 1-40.

[KLE]   S.C. Kleene, "Finite Axiomatizability of Theories in the Predicate Calculus Using Additional Predicate Symbols", Memoirs of the American Mathematical Society, no. 10, 1952, pp 27-68.

[K&F]   R. Kwok and N. Foo, "Detecting, Diagnosing and Correcting Faults in a Theory of Actions", Proceedings of the Second Automated Reasoning Day, Bribie Island, 1994, 5p.

[LLO]   J. Lloyd, "Mathematical Foundations of Logic Programming", Springer Verlag, 1984.

[LYN]   R. Lyndon, "Properties Preserved Under Homomorphisms", Pacific Journal of Mathematics, 9, 1959, pp 143-154.

[MCC]   J. McCarthy, "Applications of Circumscription to Formalising Commonsense Knowledge", Artificial Intelligence, 28, 1, 1986, pp 89-116.

[MGB]   S. Muggleton and R. Buntine, "Machine Invention of First-Order Predicates by Inverting Resolution", Fifth International Conference on Machine Learning, 1988, Morgan Kaufmann, pp 339-352.

[NAG]   E. Nagel, "The Structure of Science", Routledge and Kegan Paul, 1961.

[QUI]   J.R. Quinlan, "Learning Logical Definitions from Relations", Machine Learning, vol.5, no. 3, 1990, pp 239-266.

[REI]   R. Reiter, "Nonmonotonic Reasoning", Annual Review of Computer Science, 2, 1987, pp 147-186.

[SAN]   E. Sandewall, "Features and Fluents", Oxford University Press, 1994.

[SHA]   E. Shapiro, "Inductive Inference of Theories from Facts", TR 192, Dept of Computer Science, Yale University, 1981.

[SHS]   W-M. Shen and H.A. Simon, "Fitness Requirements for Scientific Theories Containing Recursive Theoretical Terms", British Journal of Philosophy, 44, 1993, pp 641-652.

[STA]   I. Stahl, "On the Untility of Predicate Invention in Inductive Logic Programming", Proceedings of the European Conference on Machine Learning, ECML-94, 1994, pp 272-286.

[C&L]   C.L. Chang and R.C.T. Lee, "Symbolic Logic and Mechanical Theorem Proving", Academic Press, 1973.

[CLA]   K. Clark, "Negation as Failure", in Logic and Databases, ed. H. Gallaire and J. Minker, Plenum Press, 1978, pp 293-322.

[CRA]   W Craig, "On Axiomatizability Within A System", The Journal of Symbolic Logic, vol. 18, no. 1, pp 30-32.

[C&V]   W. Craig and R.L Vaught, "Finite Axiomatizability Using Additional Predicates", The Journal of Symbolic Logic, vol. 23, no. 3, pp 289-308.

[END]   H. Enderton, "A Mathematical Introduction to Logic", Academic Press, 1972.

[FO1]   N. Foo, "Comments on Defining Software by Continuous Smooth Functions", IEEE Transactions on Software Engineering, vol. 19, no. 3, 1993, pp 307-309.

[FO2]   N. Foo, "How Theories Fail - A Preliminary Report", Proceedings National Conference on Information Technology, Penang, Malaysia, 1991, pp 244-251.

[F&T]   N. Foo and T. Tang, "An Inductive Principle for Learning Logical Definitions from Relations", Proceedings of the Seventh Australian Joint Conference on Artificial Intelligence, World Scientific Press, 1994, pp 45-52.

[FGRT] N. Foo, B. Garner, A. Rao, and E. Tsui, "Semantic Distance in Conceptual Graphs", in "Current Directions in Conceptual Structure Research", (eds) P Eklund, T Nagle, J Nagle and L Gerhotz, Ellis Horwood, 1992, pp 149-154.

[GOW]   H. Gaifman, D. Osherson and S. Weinstein, "A Reason for Theoretical Terms", Erkenntnis, 32, 1990, pp 149-159.

[GAR]   P. Gardenfors, "Knowledge in Flux", Bradford Books, MIT Press, 1988.

[GF1]   G. Gibbon and N. Foo, "Predicate Discovery in a Model Identification Framework", Proceedings of the Sixth Australian Joint Conference on Artificial Intelligence, World Scientific Press, 1993, pp 65-70.

[GF2]   G. Gibbon and N. Foo, "A General Framework for Concept Formation", Proceedings of the Seventh Australian Joint Conference on Artificial Intelligence, World Scientific Press, 1994, pp 53-59.

[HEM]   C. Hempel, "Fundamentals of Concept Formation in Empirical Science", International Encyclopedia of Unified Science, vol. 2, no. 7, University of Chicago Press, 1952.

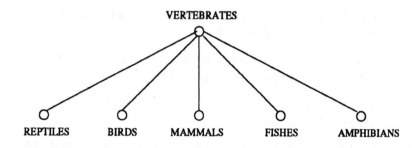

FIGURE 1   ORIGINAL TAXONOMY

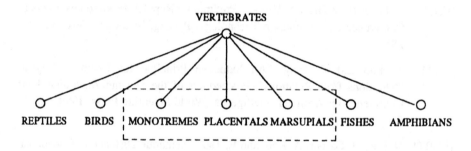

FIGURE 2   REFINEMNENT BY SPLITTING

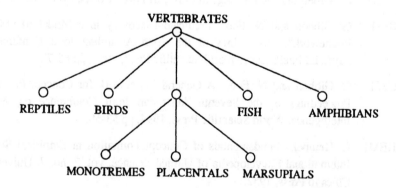

FIGURE 3   REFINEMENT BY SUBCLASSES

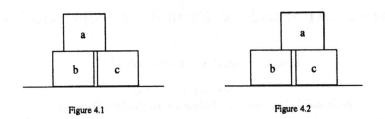

Figure 4.1   Figure 4.2

FIGURE 4   ONTOLOGICAL INADEQUACY OF THE ON(_,_) PREDICATE

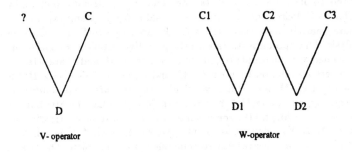

V- operator   W-operator

FIGURE 5   INVERSE RESOLUTION OPERATORS

A Triadic Approach to Formal Concept Analysis

Fritz Lehmann[1] and Rudolf Wille[2]

[1] *GRANDAI* Software, Irvine
[2] Fachbereich Mathematik, Technische Hochschule Darmstadt

Abstract. *Formal Concept Analysis*, developed during the last fifteen years, has been based on the dyadic understanding of a concept constituted by its extension and its intension. The pragmatic philosophy of Charles S. Peirce with his three universal categories, and experiences in data analysis, have suggested a triadic approach to Formal Concept Analysis. This approach starts with the primitive notion of a *triadic context* defined as a quadruple (G, M, B, Y) where G, M, and B are sets and Y is a ternary relation between G, M, and B, i.e. $Y \subseteq G \times M \times B$; the elements of G, M, and B are called *objects*, *attributes*, and *conditions*, respectively, and $(g, m, b) \in Y$ is read: the object g has the attribute m *under* (or *according to*) the condition b. A *triadic concept* of a triadic context (G, M, B, Y) is defined as a triple (A_1, A_2, A_3) with $A_1 \times A_2 \times A_3 \subseteq Y$ which is maximal with respect to component-wise inclusion. The triadic concepts are structured by three quasiorders given by the inclusion order within each of the three components. In analogy to the dyadic case, we discuss how the ordinal structure of the triadic concepts of a triadic context can be analysed and graphically represented. A basic result is that those structures can be understood order-theoretically as *"complete trilattices"* up to isomorphism.

1   Triadic Contexts

Formal Concept Analysis views concepts as means of intersubjective understanding in situations of purpose-oriented action. The formalization of concepts and concept systems shall especially support the interpretation and communication of conceptual relationships in different situations. This requires that the formalization of concepts has to be transparent and simple, but also comprehensive so that all main aspects of a concept may have explicit references in the formal model.

The extensional and the intensional aspect, which were made explicit first by the "Port Royal Logic" [AN62] in the 17th century, are basic for the developed dyadic approach to Formal Concept Analysis. This approach starts with the primitive notion of a *formal context* as a triple (G, M, I) where G is a set of (*formal*) *objects* and M is a set of (*formal*) *attributes*, while I is a relation between G and M indicating when an object g of G has a certain attribute m of M. Formal contexts are often described by a two-dimensional cross table with a cross in position (g, m) if the relationship gIm holds, i.e., if the object g has the attribute m. A *formal concept* of a formal context (G, M, I) is defined as a pair (A, B) with $A \subseteq G$, $B \subseteq M$, $A = \{g \in G \mid gIm$ for all $m \in B\}$ and $B = \{m \in M \mid gIm$ for all $g \in G\}$, so the *extent* A contains exactly each object of G which has all the attributes of B, and the *intent* B contains exactly each attribute of M which is valid for all the objects of A. It turns out that the formal concepts of (G, M, I) are exactly the pairs (A, B) of subsets of G and M which are maximal with respect to the component-wise set inclusion in satisfying $A \times B \subseteq I$. If (G, M, I) is described by a cross table, this means that, under suitable permutations of the rows and the columns of the cross table, the concept (A, B) is represented by a maximal rectangle full of crosses. With respect to the *subconcept-superconcept-relation*, the concepts of a formal context (G, M, I) form a complete lattice called the *concept lattice* of (G, M, I).

Although the dyadic approach has been successful in many applications (see e.g. [Wi87],[Wi89a],[Wi89b],[Wi92a]), there have been situations suggesting an extension of formal concepts by a third component (cf. [Gi81],[Mar92], [KOG94]). Such a triadic approach to Formal Concept Analysis is also demanded by the pragmatic philosophy of Charles Sanders Peirce with his three universal categories (cf. [Wi94a], [Wi94b]) outlined further in the following.

Peirce developed a system of *"categories"* (as alternatives to Aristotle's and Kant's categories) and founded certain fundamental philosophical distinctions on this. In his lectures on pragmatism (1903) he gave the following description of his categories:

- Category the First is the Idea of that which is such as it is regardless of anything else. That is to say, it is a *quality* of feeling.
- Category the Second is the Idea of that which is such as it is as being Second to some First, regardless of anything else, and in particular regardless of any Law, although it may conform to a law. That is to say, it is *Reaction* as an element of the phenomenon.
- Category the Third is the Idea of that which is such as it is as being a Third, or Medium, between a Second and a First. That is to say, it is *Representation* as an element of the phenomenon. [Pe35;5.66]

Instead of the dyadic relation between objects and attributes used in Formal Concept Analysis up to now, the triadic approach is based on the triadic relation saying that *the object g has the attribute m under the condition b*. How should such a relationship be viewed through Peirce's three categories? The answer is not obvious because Peirce wrote about his categories over more than thirty years in a great variety of different explanations. The quoted description of the three categories allows us to interpret the triadic relationship as follows: *the object g is a First as some suchness to which the attribute m is a Second as some accident while the condition b is a Third as some medium between g and m.* In different situations, the Third as medium may be understood more specifically as relation, mediation, representation, interpretation, evidence, evaluation, modality, meaning, reason, purpose, condition etc. concerning a present connection between an object and an attribute.

Although the given interpretation of the triadic relationship seems to be convincing, the description of the First as *quality* and the Second as *reaction* conflicts with this interpretation. Indeed, in Peirce's writings, the Firstness is connected and explained more often with the term 'quality' than with any other term (cf. [Kr60], p. 11) and the examples of First are mainly taken from the area of feelings as, for instance, the feeling of Red. The Secondness is explained by the reaction caused by facts which struggle with human feelings and thoughts. These explanations reflect Peirce's dominating interest in epistemology and phenomenology. Under this view the human subject is First and the objective world is Second. But the speech acts of intersubjective communication should be viewed differently. This becomes clear in the different understandings of a *predication* which combines a subject and a predicate: epistemologically, the subject is understood as an instance of the concept decribed by the predicate (or the subject concept is subsumed under the predicate concept); semantically, the predicate is understood to be assigned to the object denoted by the subject term (cf. [We89]). These two understandings of predication were extensively discussed in the Middle Ages as identity and inherence theory of predication with the copula 'is' (cf. [Ma79]). Already the etymology of the word 'attribute' indicates that the dyadic predication *'the object g has the attribute m'* should be understood semantically, i.e., m is accidental to g. This semantic interpretation is compatible with Peirce's general understanding of the First as *suchness* [Pe35;1.303] and the Second as *otherness* [Pe35;1.295].

The given interpretation of the dyadic predication also underlies the description of *concept* in [AN62] which is based on the distinction of extension and intension. The objects of the extension are considered as things which exist by themselves and the attributes of the intension are viewed as parts which determine things but cannot be without them. This object-dependent role of attributes is basic for concept theories until today which is witnessed, for instance, by the German standards DIN 2330 [DIN79]. The dependency suggests to interpret also the dyadic relationship between an object and an assigned attribute as instance of Peirce's second category. Especially in Peirce's investigations of logic, Secondness is the general character of dyadic relations and Thirdness is the general character of triadic relations. Thus, the triadic relationship between an object, an attribute and a condition may be interpreted as an instance of Peirce's third category too.

The triadic approach to Formal Concept Analysis is based on a formalization of the triadic relation connecting formal objects, attributes and conditions. Since real situations can only be analysed within restricted contexts, *Triadic Concept Analysis* is founded on a formal notion of triadic contexts which allows set-theoretical formalizations. A *triadic context* is defined as a quadruple (G, M, B, Y) where G, M, and B are sets and Y is a ternary relation between G, M, and B, i.e. $Y \subseteq G \times M \times B$; the elements of G, M, and B are called (*formal) objects, attributes,* and *conditions,* respectively, and $(g, m, b) \in Y$ is read: the object g has the attribute m *under* the condition b (the relational notation $b(g, m)$ might also be used for $(g, m, b) \in Y$). Just as dyadic contexts are often described by two-dimensional cross tables, triadic contexts may be represented by three-dimensional cross tables. Figures 1, 3, and 5 below show examples of such cross tables for triadic contexts.

Formal objects, attributes, and conditions may formalize entities in a wide range, but in the triadic context they are understood in the role of the corresponding Peircean category. In particular, the formal conditions may formalize - as mentioned above - relations, mediations, representations, interpretations, evidences, evaluations, modalities, meanings, reasons, purposes, conditions etc. If real data are described by a triadic context, the names of the formal objects, attributes, and conditions yield the elementary bridges to reality which are basic for interpretations (cf. [Wi92b]). For theoretical developments it is often convenient to use K_1, K_2, and K_3 instead of G, M, and B; the alternative symbols indicate that the elements of the component K_i are seen in the role of an instance of Peirce's i-th category.

2 Triadic Concepts

Concepts are understood as units of thought. This means that a concept tends to be *homogeneous* and *closed*. If a concept is viewed through the triadic paradigm then, for the purpose of formalization, a concept should be seen as a combination of objects, attributes, and conditions which is homogeneous and closed. Homogeneity is attained if each object has each attribute under each condition within the concept. Closure is attained if the concept is maximal with respect to this property. Therefore, we define a *triadic concept* of a triadic context (G, M, B, Y) as a triple (A_1, A_2, A_3) with $A_1 \subseteq G$, $A_2 \subseteq M$, and $A_3 \subseteq B$ such that the triple (A_1, A_2, A_3) is maximal with repect to component-wise set inclusion in satisfying $A_1 \times A_2 \times A_3 \subseteq Y$, i.e., for $X_1 \subseteq G$, $X_2 \subseteq M$, and $X_3 \subseteq B$ with $X_1 \times X_2 \times X_3 \subseteq Y$, the containments $A_1 \subseteq X_1$, $A_2 \subseteq X_2$, and $A_3 \subseteq X_3$ always imply $(A_1, A_2, A_3) = (X_1, X_2, X_3)$. If (G, M, B, Y) is described by a three-dimensional cross table, this means that, under suitable permutations of rows, columns, and layers of the cross table, the triadic concept (A_1, A_2, A_3) is represented by a maximal rectangular box full of crosses. For a particular triadic concept (A_1, A_2, A_3), the components A_1, A_2, and A_3 are called the *extent*, the *intent*, and the *modus* of (A_1, A_2, A_3), respectively.

As in the dyadic case, derivation operators are useful for the construction of triadic concepts within a triadic context. For the description of derivation operators, it is convenient to denote the underlying triadic context alternatively by $\mathbb{K} := (K_1, K_2, K_3, Y)$. For $\{i, j, k\} = \{1, 2, 3\}$ with $j < k$ and for $X \subseteq K_i$ and $Z \subseteq K_j \times K_k$, the (i)-*derivation operators* are defined by

$$X \mapsto X^{(i)} := \{(a_j, a_k) \in K_j \times K_k \mid a_i, a_j, a_k \text{ are related by } Y \text{ for all } a_i \in X\},$$

$$Z \mapsto Z^{(i)} := \{a_i \in K_i \mid a_i, a_j, a_k \text{ are related by } Y \text{ for all } (a_j, a_k) \in Z\}.$$

This definition yields the derivation operators of the dyadic contexts defined by

$$\mathbb{K}^{(1)} := (K_1, K_2 \times K_3, Y^{(1)}),$$
$$\mathbb{K}^{(2)} := (K_2, K_1 \times K_3, Y^{(2)}),$$
$$\mathbb{K}^{(3)} := (K_3, K_1 \times K_2, Y^{(3)})$$

where $gY^{(1)}(m,b) :\iff mY^{(2)}(g,b) :\iff bY^{(3)}(g,m) :\iff (g,m,b) \in Y$. For the construction of triadic concepts, further derivation operators are needed. For $\{i,j,k\} = \{1,2,3\}$ and for $X_i \subseteq K_i$, $X_j \subseteq K_j$, and $A_k \subseteq K_k$, the (i,j,A_k)-*derivation operators* are defined by

$$X_i \mapsto X_i^{(i,j,A_k)}$$
$$:= \{a_j \in K_j \mid a_i, a_j, a_k \text{ are related by } Y \text{ for all } (a_i, a_k) \in X_i \times A_k\},$$
$$X_j \mapsto X_j^{(i,j,A_k)}$$
$$:= \{a_i \in K_j \mid a_i, a_j, a_k \text{ are related by } Y \text{ for all } (a_j, a_k) \in X_j \times A_k\}.$$

This definition yields the derivation operators of the dyadic contexts defined by

$$\mathbb{K}_{A_k}^{ij} := (K_i, K_j, Y_{A_k}^{ij})$$

where $(a_i, a_j) \in Y_{A_k}^{ij}$ if and only if a_i, a_j, a_k are related by Y for all $a_k \in A_k$. In case of $(1,2,3) = (i,j,k)$, the relationship $(a_1, a_2) \in Y_{A_3}^{12}$ means that the object a_1 has the attribute a_2 under all conditions a_3 with $a_3 \in A_3$.

If one wants to get a triadic concept having a given object set X_1 in its extent, one may first generate a dyadic concept in $\mathbb{K}_{A_3}^{12}$ with X_1 in its extent and then extend it to a triadic concept using the corresponding (3)-derivation operator in $\mathbb{K}^{(3)}$. This is formally performed by forming the triadic concept

$$(X_1^{(1,2,A_3)(1,2,A_3)}, X_1^{(1,2,A_3)}, (X_1^{(1,2,A_3)(1,2,A_3)} \times X_1^{(1,2,A_3)})^{(3)}).$$

In words, one first determines the set of all attributes which all objects of X_1 have under all conditions of A_3; secondly, X_1 is extended to the set of all objects having all those attributes under all conditions of A_3; thirdly, A_3 is extended to the set of all conditons under which each of the derived objects has each of the derived attributes. Of course, this construction of a triadic concept may analogously be applied to other choices of $X_i \subseteq K_i$ and $A_k \subseteq K_k$ for $i \neq k$ in $\{1,2,3\}$. It should also become clear that a triple (A_1, A_2, A_3) with $A_i \subseteq K_i$ for $i = 1,2,3$ is a triadic concept of $\mathbb{K}$ if and only if $A_i = (A_j \times A_k)^{(i)}$ for all $\{i,j,k\} = \{1,2,3\}$ with $j < k$. Note that there are always the extremal triadic concepts $\mathfrak{o}_1 := ((K_2 \times K_3)^{(1)}, K_2, K_3)$, $\mathfrak{o}_2 := (K_1, (K_1 \times K_3)^{(2)}, K_3)$, and $\mathfrak{o}_3 := (K_1, K_2, (K_1 \times K_2)^{(3)})$.

Next we discuss how the set $\mathfrak{T}(\mathbb{K})$ of all triadic concepts of the triadic context $\mathbb{K} := (K_1, K_2, K_3)$ is structured. The most natural structure is given by the set inclusion in each of the three components of the triadic concepts. For each $i \in \{1,2,3\}$, one obtains a quasiorder $\lesssim_i$ and its corresponding equivalence relations $\sim_i$ defined by

$$(A_1, A_2, A_3) \lesssim_i (B_1, B_2, B_3) :\iff A_i \subseteq B_i \quad and$$
$$(A_1, A_2, A_3) \sim_i (B_1, B_2, B_3) :\iff A_i = B_i \quad (i = 1,2,3).$$

The three quasiorders satisfy the following *antiordinal dependencies* (cf. [Wi95]): For $\{i,j,k\} = \{1,2,3\}$, $(A_1, A_2, A_3) \lesssim_i (B_1, B_2, B_3)$ and $(A_1, A_2, A_3) \lesssim_j (B_1, B_2, B_3)$ imply $(A_1, A_2, A_3) \gtrsim_k (B_1, B_2, B_3)$ for all triadic concepts (A_1, A_2, A_3) and

(B_1, B_2, B_3) of $\mathbb{K}$. This becomes clear if one visualizes the two triadic concepts by maximal boxes within the three-dimensional cross table. It follows that, for $i \neq j$, the relation $\sim_i \cap \sim_j$ is the identity on $\mathfrak{T}(\mathbb{K})$, i.e., a triadic concept is uniquely determined by two of its components. Furthermore, $\lesssim_{ij} := \lesssim_i \cap \lesssim_j$ is an order on $\mathfrak{T}(\mathbb{K})$. Let us denote the equivalence class of $\sim_i$ which contains the triadic concept (A_1, A_2, A_3) by $[(A_1, A_2, A_3)]_i$. The quasiorder $\lesssim_i$ induces an order $\leq_i$ on the factor set $\mathfrak{T}(\mathbb{K}) / \sim_i$ of all equivalence classes of $\sim_i$ which is characterized by $[(A_1, A_2, A_3)]_i \leq_i [(B_1, B_2, B_3)]_i \iff A_i \subseteq B_i$. Clearly, $(\mathfrak{T}(\mathbb{K}) / \sim_1, \leq_1)$ can be identified with the ordered set of all extents of $\mathbb{K}$, $(\mathfrak{T}(\mathbb{K}) / \sim_2, \leq_2)$ with to the ordered set of all intents of $\mathbb{K}$, and $(\mathfrak{T}(\mathbb{K}) / \sim_3, \leq_3)$ with the ordered set of all modi of $\mathbb{K}$. For such an ordered set one cannot expect special properties in general because every ordered set with smallest and greatest element is isomorphic to the ordered set of all extents (resp. intents, modi) of some triadic context as shown in [Wi95]. Therefore the extents (resp. intents, modi) need not to form a *closure system* as in the dyadic case.

For analysing conceptual relationships within triadic contexts $\mathbb{K} := (K_1, K_2, K_3)$, it is basic to investigate the relational structures $\underline{\mathfrak{T}}(\mathbb{K}) := (\mathfrak{T}(\mathbb{K}), \lesssim_1, \lesssim_2, \lesssim_3)$. There is already a computer program which determines the triadic concepts and the relational structure $\underline{\mathfrak{T}}(\mathbb{K})$ for a given triadic context (see [GK94]). The program is based on a nested use of *Ganter's Algorithm* for dyadic contexts (see [Ga87]). The extented algorithm for determining the triadic concepts is explicitly discussed in [KOG94].

3 Triadic Diagrams

In this section we discuss how the 'triadic' relational structure $\underline{\mathfrak{T}}(\mathbb{K})$ of a triadic context $\mathbb{K} := (K_1, K_2, K_3)$ may be represented graphically. For this, the relational structure can be understood as a combination of two types of structures: the *geometric structure* $(\mathfrak{T}(\mathbb{K}), \sim_1, \sim_2, \sim_3)$ and the *ordered structures* $(\mathfrak{T}(\mathbb{K}) / \sim_i, \leq_i)$ for $i = 1, 2, 3$. The geometric structure gives rise to a 'partial 3-net' $(\mathfrak{T}(\mathbb{K}), \bigcup_{i=1}^{3} \{[(A_1, A_2, A_3)]_i \mid (A_1, A_2, A_3) \in \mathfrak{T}(\mathbb{K})\})$ in which the classes of different equivalence relations meet in at most one element. This suggests to represent the three equivalence relations $\sim_1$, $\sim_2$, and $\sim_3$ by three systems of parallel lines in the plane and to locate the elements of one equivalence class on one line of the corresponding parallel system. Unfortunately, such a representation cannot be performed by straight lines in general because that would never admit a violation of the so-called 'Thomsen Condition' (see [KLST71], pp.250 ff.) which might occur in triadic contexts (see [WZ95]). But it seems that the Thomsen Condition and its ordinal generalizations (cf. [WiU95]) is only seldom violated by real data wherefore we concentrate in this section on graphical representations of the geometric structure by straight lines. The ordered structures are represented as usual by line diagrams (Hasse diagrams). Some examples will show how the graphical representations of the geometric and the ordered structures may be combined into a *triadic diagram* of the whole relational structure.

Let us first consider an elementary type of 'triadic' relational structures whose

	1					2					3					4					5				
	1	2	3	4	5	1	2	3	4	5	1	2	3	4	5	1	2	3	4	5	1	2	3	4	5
1	X	X	X	X	X	X	X	X	X	X	X	X	X	X	X	X	X	X	X	X	X	X	X	X	
2	X	X	X	X	X	X	X	X	X	X	X	X	X	X	X	X	X	X	X		X	X	X		
3	X	X	X	X	X	X	X	X	X	X	X	X	X	X		X	X	X			X	X			
4	X	X	X	X	X	X	X	X	X		X	X	X			X	X				X				
5	X	X	X	X		X	X	X			X	X				X									

Fig. 1. The triadic 5-chain context $\mathbb{K}_5^c$

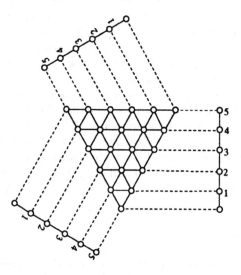

Fig. 2. A triadic diagram of the triadic 5-chain $\underline{\mathfrak{T}}(\mathbb{K}_5^c)$

ordered structures are only chains. The most regular subtype of those structures are given by the triadic *n-chain contexts* $\mathbb{K}_n^c := (\{1, \ldots, n\}, \{1, \ldots, n\}, \{1, \ldots, n\}, Y_n^c)$ and $(x_1, x_2, x_3) \in Y_n^c :\Longleftrightarrow x_1 + x_2 + x_3 \le 2n$. The triadic concepts of those contexts form as geometric structure a regular triangle pattern as it is indicated by the following example. In Figure 1, the triadic 5-chain context $\mathbb{K}_5^c$ is described by a cross table in which the rows represent objects, the columns attributes, and the subtables conditions. A *triadic diagram* of $\underline{\mathfrak{T}}(\mathbb{K}_5^c)$ is shown in Figure 2. The geometric structure of the triadic concepts is represented by the triangular pattern in the center of the diagram. The circles represent the triadic concepts and the lines the equivalence classes, i.e., the horizontal lines those of $\sim_1$, the lines ascending to the right those of $\sim_2$, and the lines ascending to the left those of $\sim_3$. The perforated lines indicate the connection to the extent diagram on the right, to the intent diagram on the lower left, and to the modus diagram above. A circle of the line diagram on the right represents the extent consisting of those

	1			2			3		
	1	2	3	1	2	3	1	2	3
1		×	×	×	×	×	×	×	×
2	×	×	×	×		×	×	×	×
3	×	×	×	×	×	×	×	×	

Fig. 3. The triadic power set context $\mathbb{K}^b_{\{1,2,3\}}$

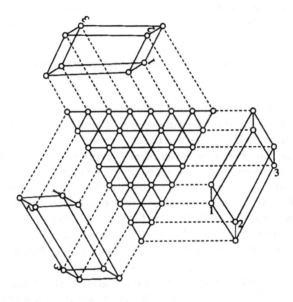

Fig. 4. The power set trilattice $\underline{\mathfrak{T}}(\mathbb{K}^b_{\{1,2,3\}})$

objects whose signs are attached to this circle or a circle below. The intents and modi can analogously be read from the diagram where the intents get larger from the upper left to the lower right and the modi get larger from the upper right to to the lower left. Using the given information, all triadic concepts can be completely determined by the triadic diagram. For instance, the next circle vertically above the lowest circle connects horizontally with the extent $\{1, 2\}$, to the lower left with the intent $\{1, 2, 3, 4\}$, and to the upper left with the modus $\{1, 2, 3, 4\}$; hence it represents the triadic concept $(\{1, 2\}, \{1, 2, 3, 4\}, \{1, 2, 3, 4\})$.

Another elementary type of triadic contexts are the triadic *power set contexts* defined for arbitrary sets S by $\mathbb{K}^b_S := (S, S, S, Y^b_S)$ with $Y^b_S := S^3 \setminus \{(x, x, x) \mid x \in S\}$. The triadic concepts of $\mathbb{K}^b_S$ are exactly the triples $(X_1, X_2, X_3) \in \mathfrak{P}(S)^3$ with $X_1 \cap X_2 \cap X_3 = \emptyset$ and $X_i \cup X_j = S$ for $i \neq j$ in $\{1, 2, 3\}$. In Figure 3, the triadic power set context $\mathbb{K}^b_{\{1,2,3\}}$ is described by a cross table. A triadic diagram

	a						b						c						d					
	1	**2**	**3**	**4**	**5**	**6**	**1**	**2**	**3**	**4**	**5**	**6**	**1**	**2**	**3**	**4**	**5**	**6**	**1**	**2**	**3**	**4**	**5**	**6**
A	×	×					×	×					×	×					×	×	×		×	
B	×												×	×		×			×				×	
C	×	×	×	×	×	×	×	×	×	×	×	×	×	×					×	×	×			×

A: Superordinate stereotypes
B: Unspecified
C: Subtypes
a: Recall
b: Impression Formation
c: Behavior Prediction
d: Evaluation

1. Physical appearance
2. Political beliefs
3. Attitudes
4. Behavior
5. Traits
6. Situations

Fig. 5. A triadic context of experimental data

of $\mathfrak{T}(\mathbb{K}^b_{\{1,2,3\}})$ is shown in Figure 4. The triadic diagram is analogously read as the diagram of Figure 2; the only difference is that not every intersection point in the triangular pattern represents a triadic concept.

The last example, based on experimental data, is a triadic context taken from [KOG94] and desribed by the cross table in Figure 5. The objects of the triadic context are person categories named by 'A: Superordinate stereotypes', 'B: Unspecified', and 'C: Subtypes', the attributes are attribute types named by '1: Physical appearance', '2: Political beliefs', '3: Attitudes', '4: Behavior', '5: Traits', and '6: Situations', and the conditions are goals named by 'a: Recall', 'b: Impression Formation', 'c: Behavior Prediction', and 'd: Evaluation'. A cross in the table indicates that the object/attribute/condition-triple was rated above the median in the considered study. A triadic diagram of the triadic data context is shown in Figure 6. It seems to be quite typical for real data that the eleven triadic concepts spread rather loosely in the relatively large triangular pattern. In the modus diagram above the name 'a: Recall' occurs twice since there is not a smallest modus containing the condition 'a: Recall'. This might happen because modi (resp. intents, extents) need not to form a closure system which was already mentioned in Section 2. As the diagram shows, there are two triadic concepts with minimal modus containing the condition 'a: Recall', namely $(\{C\}, \{1, 2, 3, 4, 5, 6\}, \{a, b\})$ and $(\{A, B, C\}, \{1\}, \{a, c, d\})$.

4 Concept Trilattices

The relational structures $\mathfrak{T}(\mathbb{K}) := (\mathfrak{T}(\mathbb{K}), \lesssim_1, \lesssim_2, \lesssim_3)$ play an analogous role in triadic concept analysis as the *concept lattices* in the dyadic case. Therefore the investigation of $\mathfrak{T}(\mathbb{K})$ should identify the algebraic operations which are the triadic analogues to infima and suprema. Since those operations have the aim of producing triadic concepts from others, it seems natural to reach the operations via the construction of triadic concepts described in Section 2. This

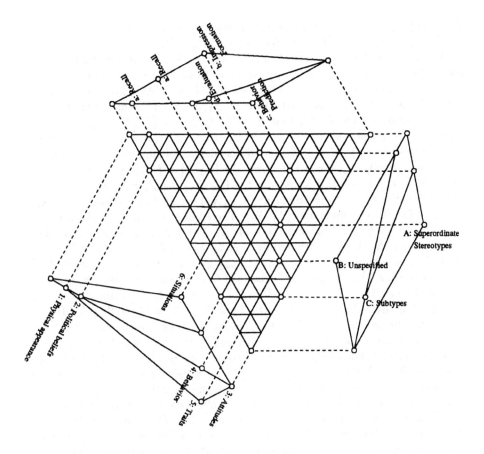

Fig. 6. A triadic diagram of the 'triadic' relational structure of the context in Figure 5

construction indicates that one needs two subsets of triadic concepts to produce another triadic concept. Let $\{i, j, k\} = \{1, 2, 3\}$ and let $\mathfrak{X}_i$ and $\mathfrak{X}_k$ be sets of triadic concepts of $\mathbb{K}$; furthermore, let $X_i := \bigcup\{A_i \mid (A_1, A_2, A_3) \in \mathfrak{X}_i\}$ and $X_k := \bigcup\{A_k \mid (A_1, A_2, A_3) \in \mathfrak{X}_k\}$. Then the *ik-join* of the pair $(\mathfrak{X}_i, \mathfrak{X}_k)$ is defined to be the triadic concept

$$\nabla_{ik}(\mathfrak{X}_i, \mathfrak{X}_k) := (B_1, B_2, B_3) \text{ with}$$
$$B_i := X_i^{(i,j,X_k)(i,j,X_k)},$$
$$B_j := X_i^{(i,j,X_k)},$$
$$B_k := (X_i^{(i,j,X_k)(i,j,X_k)} \times X_i^{(i,j,X_k)})^{(k)}.$$

If the k-th-components of the triadic concepts in $\mathfrak{X}_i$ are all equal to X_k, then $\nabla_{ik}(\mathfrak{X}_i, \mathfrak{X}_k)$ is just the supremum of $\mathfrak{X}_i$ in the ordered set $(\{(B_1, B_2, B_3) \in \mathfrak{T}(\mathbb{K}) \mid$

$B_k = X_k\}, \lesssim_i)$ and, because of the antiordinal dependency, the infimum of $\mathfrak{X}_i$ in the dually ordered set $(\{(B_1, B_2, B_3) \in \mathfrak{T}(\mathbb{K}) \mid B_k = X_k\}, \lesssim_j)$.

For the further development of Triadic Concept Analysis it is useful to understand the operations ∇_{ik} on the purely order-theoretic level. Here we can only sketch the order-theoretic treatment elaborated in [Wi95]. First a *triordered set* is defined as a relational structure $(S, \lesssim_1, \lesssim_2, \lesssim_3)$ for which the relations $\lesssim_i$ are quasiorders on S such that $\lesssim_i \cap \lesssim_j \subseteq \gtrsim_k$ for $\{i, j, k\} = \{1, 2, 3\}$ and $\sim_1 \cap \sim_2 \cap \sim_3 = id_S$ where $\sim_i := \lesssim_i \cap \gtrsim_i$ $(i = 1, 2, 3)$. It immediately follows that $\sim_i \cap \sim_j = id_S$ for $i \neq j$. The triadic analogues of suprema and infima can order-theoretically be defined as follows: For $\{i, j, k\} = \{1, 2, 3\}$ and $X_i, X_k \subseteq S$, an element u of S is called an *ik-bound* of (X_i, X_k) if $u \gtrsim_i x$ for all $x \in X_i$ and $u \gtrsim_k x$ for all $x \in X_k$; an *ik*-bound u of (X_i, X_k) is called an *ik-limit* of (X_i, X_k) if $u \gtrsim_j v$ for all *ik*-bounds v of (X_i, X_k). There exists at most one *ik*-limit u of (X_i, X_k) with $u \lesssim_k v$ for all *ik*-limits v of (X_i, X_k) in $\underline{S}$; this element u is called the *ik-join* of (X_i, X_k) and denoted by $\nabla_{ik}(X_i, X_k)$. Now, a *complete trilattice* is defined to be a triordered set $\underline{L} := (L, \lesssim_1, \lesssim_2, \lesssim_3)$ in which the *ik*-joins exist for all $i \neq k$ in $\{1, 2, 3\}$ and all pairs of subsets of L. In a complete trilattice $\underline{L}$, the extremal element $0_i := \nabla_{jk}(L, L)$ $(= \nabla_{kj}(L, L))$ is uniquely determined by $0_i \lesssim_i x$ for all $x \in L$ (note that also $0_i = \nabla_{ij}(\emptyset, L) = \nabla_{ik}(\emptyset, L) = \nabla_{ji}(\emptyset, \emptyset) = \nabla_{ki}(\emptyset, \emptyset)$).

By the *Basic Theorem* of Triadic Concept Analysis (proved in [Wi95]), the relational structure $\mathfrak{T}(\mathbb{K})$ of a triadic context $\mathbb{K}$ is a complete trilattice whose order-theoretical *ik*-joins are exactly the *ik*-joins defined by using the derivation operators. Since $\mathfrak{T}(\mathbb{K})$ is a complete trilattice, it is called the *concept trilattice* of the triadic context $\mathbb{K}$. Conversely, every complete trilattice $\underline{L} := (L, \lesssim_1, \lesssim_2, \lesssim_3)$ is isomorphic to a concept trilattice of a suitable triadic context for which one can choose $(L, L, L, Y_{\underline{L}})$ with $Y_{\underline{L}} := \{(x_1, x_2, x_3) \in L^3 \mid$ there exists an $u \in L$ with $u \gtrsim_i x_i$ for $i = 1, 2, 3\}$. Thus, the concept trilattices are up to isomorphism the complete trilattices (as the concept lattices are up to isomorphism the complete lattices in the dyadic case). Hence, on the purely order-theoretic level, the concept trilattices are generally understood, but the detailed study of those structures is just at the beginning.

References

[AN62]   A. Arnaud, P. Nicole: La logique ou l'art de penser. Paris 1662.

[DIN79]   Deutsches Institut für Normung: DIN 2330; Begriffe und Benennungen: Allgemeine Grundsätze. Beuth, Berlin, Köln 1979.

[Ga87]   B. Ganter: Algorithmen zur Formalen Begriffsanalyse. In: B. Ganter, R. Wille, K.E. Wolff: Beiträge zur Begriffsanalyse. B.I.- Wissenschaftsverlag, Mannheim 1987, 241–254.

[GK94]   B. Ganter, R. Krauße: TRICEPT - Programm zur triadischen Begriffsanalyse. Institut für Algebra, TU Dresden 1994.

[GW95]   B. Ganter, R. Wille: Formale Begriffsanalyse: Mathematische Grundlagen. Springer, Heidelberg (in preparation)

[Gi81]   G. Gigerenzer: Messung als Modellbildung in der Psychologie. Reinhardt Verlag, München, Basel 1981.

43

[KLST71]   D.H. Krantz, R.D. Luce, P. Suppes, A. Tversky: Foundations of measurement. Vol 1. Academic Press, San Diego 1971.

[Kr60]   P. Krausser: Die drei fundamentalen Strukturkategorien bei Charles S. Peirce. Philosophia Naturalis 6 (1960), 3–31.

[KOG94]   S. Krolak-Schwerdt, P. Orlik, B. Ganter: TRIPAT: a model for analyzing three-mode binary data. In: H.-H. Bock, W. Lenski, M.M. Richter (eds.): Information systems and data analysis. Springer, Berlin-Heidelberg 1994, 298–307.

[Ma79]   J. Malcolm: A reconsideration of the identity and inherence theories of the copula. J. Hist. Philos. 17 (1979), 383–400.

[Mar92]   R. Marty: Foliated Semantic networks: concepts, facts, qualities. Computers & Mathematics with Applications 23 (1992), 697–696; reprinted in: F. Lehmann (ed.): Semantic networks in artificial intelligence. Pergamon Press, Oxford 1992.

[Pe35]   Ch. S. Peirce: Collected Papers. Harvard Univ. Press, Cambridge 1931–35.

[We89]   H. Weidemann: Prädikation. In: Historisches Wörterbuch der Philosophie. Band 7. Schwabe, Basel 1989, 1194–1208.

[Wi82]   R. Wille: Restructuring lattice theory: an approach based on hierarchies of concepts. In: I. Rival (ed.): Ordered sets. Reidel, Dordrecht-Boston 1982, 445–470.

[Wi87]   R. Wille: Bedeutungen von Begriffsverbänden. In: B. Ganter, R. Wille, K.E. Wolff (Hrsg.): Beiträge zur Begriffsanalyse. B.I.-Wissenschaftsverlag, Mannheim 1987, 161–211.

[Wi89a]   R. Wille: Lattices in data analysis: how to draw them with a computer. In: I. Rival (ed.): Algorithms and order. Kluwer: Dordrecht, Boston 1989, 33–58.

[Wi89b]   R. Wille: Knowledge acquisition by methods of formal concept analysis. In: E. Diday (ed.): Data analysis, learning symbolic and numeric knowledge. Nova Science Publ., New York, Budapest 1989, 365–380.

[Wi92a]   R. Wille: Concept lattices and conceptual knowledge systems. Computers & Mathematics with Applications. 23 (1992), 493–515.

[Wi92b]   R. Wille: Begriffliche Datensysteme als Werkzeug der Wissenskommunikation. In: H.H. Zimmermann, H.-D. Luckhardt, A. Schulz (Hrsg.): Mensch und Maschine - Informationelle Schnittstellen der Kommunikation. Universitätsverlag Konstanz, Konstanz 1992, 63–73.

[Wi94a]   R. Wille: Plädoyer für eine philosophische Grundlegung der Begrifflichen Wissensverarbeitung. In: R. Wille, M. Zickwolff (Hrsg.): Begriffliche Wissensverarbeitung: Grundfragen und Aufgaben. B.I.-Wissenschaftsverlag, Mannheim 1994, 11–25.

[Wi94b]   R. Wille: Restructuring mathematical logic: an approach based on Peirce's pragmatism. In: P. Agliano, A. Ursini (eds.): International Conference on Logic and Algebra, Siena 1994 (to appear)

[Wi95]   R. Wille: The basic theorem of triadic concept analysis. Order (to appear)

[WZ95]   R. Wille, M. Zickwolff: Grundlegung einer Triadischen Begriffsanalyse (in preparation)

[WiU95]   U. Wille: Geometric representation of ordinal contexts. Doctoral thesis, Univ. Gießen 1995.

Automatic Integration of
Digital System Requirements using Schemata

R. Y. Kamath and W. R. Cyre

Virginia Tech
The Bradley Department of Electrical Engineering
Blacksburg VA 24061-0111

Abstract. Natural language and diagrams of various types are commonly used for specifications on digital systems containing descriptions and requirements. The objective of the research reported here is to develop an algorithm for the automatic integration of requirements using schemata to aid in the automatic detection and joining of common references (coreferences) to objects in natural language and other specifications. This paper describes a rule-based algorithm for the integration of requirements which are expressed as conceptual graphs. The algorithm uses design knowledge in the form of schemas to detect coreferences and perform joins. The algorithm is demonstrated by a small example.

Index Keywords: natural language processing, digital system design, schemata, coreferences, requirements integration.

1. Introduction

Once knowledge has been acquired from various source notations, such knowledge has to be integrated as a prelude to further processing. Knowledge integration requires identifying when two references have the same external referent. This process is known as coreference detection. This paper deals with the problem of coreference detection in the domain of digital system requirements and presents an algorithmic solution. The solution uses a set of rules and special conceptual graphs known as schemas to detect coreferent concepts in a pair of conceptual graphs.

The design of digital systems begins with the development of a set of specifications outlining the requirements of the desired system. These specifications are usually composed of block diagrams, timing diagrams, flow-charts and natural language. The effort reported here is part of a project called the Automated Specifications Interpreter (ASPIN) being developed to increase efficiency and accuracy in creating and interpreting requirements of digital systems. The ASPIN system takes various forms of informal specifications as input and creates formal engineering models . Simulation, analysis and further synthesis can then be performed with these formal models. The formal models of the specifications are expressed in the VHSIC hardware description language (VHDL) [1].

The objective of the work described in the current paper is the integration of conceptual graphs representing individual requirements in a system specification into a single aggregate "specification" graph representing the entire system. A special kind of conceptual graphs known as schemas are used to aid in the integration process. The schema-driven integrator algorithm takes the conceptual graphs of individual requirements as input, selects schemas from its schema database as necessary, uses an integrating algorithm to detect identical concepts across requirement graphs and integrates the requirement graphs into a single aggregate specification conceptual graph.

2. Related Research

Integration of requirements involves joining coreferent concepts across requirements graphs. This section considers methods of joining graphs. It is shown that join definitions such as maximal join are underconstrained for knowledge integration. This is because mechanical join algorithms join comparable concepts without checking whether their attributes and their relations with other concepts are consistent.

Mugnier and Chein [5] present algorithms for projection [2] operation on conceptual graphs. A join on two maximally extended compatible projections [2] is known as a maximal join. Although the task of requirements integration needs an operation similar to a maximal join, a maximal join does not produce correct results in this application. The following example illustrates this point.

Consider the following conceptual graphs. Let us assume that one of them is a schema and the other is a requirement graph.

Schema :

[1 : UART : *]-
     (part) -> [2 : register : *1] -
               (size) -> [3 : length : @ 7bit],
     (part) -> [4 : register : *2] -
               (size) -> [5 : length : @ 8bit],,.

Requirement Graph :

[1 : UART : #1]-
     (part) -> [2 : register : #2] -
               (size) -> [3 : length : @ 8bit],,.

Now, suppose the requirement graph causes the schema for the UART to be invoked. A possible maximal join of the two graphs obtained by extending the join of Concept 1 of the schema with Concept 1 of the requirement graph is given below. The join is incorrect. The correct join would involve the join of Concept 4 of the schema to Concept 2 of the requirement graph. A consistency check would prevent the incorrect join. However, a consistency check is not part of a maximal join operation. Therefore a maximal join operation does not necessarily give a valid integrated graph.

```
[ 1 : UART : #1 ]-
        ( part ) -> [ 2 : register : #2 ] -
                            ( size ) -> [ 3 : length : @ 7bit ]
                            ( size ) -> [ 4 : length : @ 8bit ],
        ( part ) -> [ 5 : register : *2 ] -
                            ( size ) -> [ 6 : length : @ 8bit ],,.
```

Shankaranarayanan and Cyre [7] consider the problem of detection of coreferent object concepts but not action or state concepts. This paper extends the problem to detection of coreferent concepts of all types found in the digital design domain. Schemata are used to aid this process.

3. Notation and Theory

This section defines some of the terms used to describe the research effort presented in this paper. In particular, the reader needs to understand the meaning of the terms "restrictive relations", "attributes", "incompatible attributes", "coherences" and "schemas" in order to understand the integration rules and algorithm presented in section 4. It would be helpful to refer to Table 3.1 when perusing the integration rules of section 4.

The restrictive relations of a concept are used as the primary determiners of the context of a concept. The context of concepts are used in deciding whether the references of two concepts are coreferent. If it is determined that the restrictive relations do not provide sufficient information to determine coreference, the coherences of the concepts are used as additional contextual information.

3.1 References : Sentences in natural language are composed of words and phrases which are called *expressions*. Expressions which identify real world entities, actions or abstractions are called *references*. The entities, actions or abstractions referred to are known as referents of those references. References can introduce referents or can point to referents that have already been introduced (*anaphoric*) or to referents that are introduced later (*cataphoric*) [6].

3.2 Definitions and Coreferences. Definition [7] of a referent (entity or action) is said to occur when a reference occurs for the first time in the system specification under consideration. The reference in this case is said to be a *definitional reference* or simply a definition. When a reference recurs, it is said to be a non-definitional or *recurring reference* or *coreference*. Two references pointing to the same referent are called coreferences [7]. For example :

The *status register* is eight bits long.   (3.1)

Execution of a program loads the *status register*.   (3.2)

In the two example sentences presented above, the reference "status register" occurs for the first time in sentence 3.1, it is being introduced and is therefore an definition. The occurrence of "status register" in the sentence 3.2 is a reference to "status register" in sentence 3.1 and is therefore an anaphoric reference.

A conceptual reference consists of a concept with all its attached relations. This concept is called the *head of the reference*. All the relations attached to the head act as modifiers of the head of the reference. If two references are coreferent, the heads of the references will also be called coreferent.

When the conceptual graph of a sentence is represented as a directed tree, the root concept of this tree is known as the *root concept of the sentence*.

3.3 Restrictive Relations [7] : Relations which restrict the generality of a concept by relating it to other concepts are known as restrictive relations. Restrictive relations give more information about the referent alluded to by the reference and are, therefore, invaluable in determining coreferences. Some relation types identified as restrictive relations are shown in Table 3.1.

3.4 Attributes and Incompatible Attributes : Concepts which describe the characteristics of an object concept are called its attributes. A concept is said to have incompatible attributes if all the following conditions are true:
>   i) It has two restrictive relations with identical type labels
>   ii) These type labels are subtypes of the type "attr"
>   iii) The targets of the two relations are not coreferent.

3.6 Coherences [7] : A coherence of a concept in a sentence is a chain from the root concept of the sentence to the head of the reference. The coherence of the head of a reference is used by the integrating algorithm to detect coreferences if the information available from restrictive relations is insufficient.

3.7 Schemas : A schema [8] is a conceptual graph representing the expected context of a concept. Schemas provide design knowledge that may be used in joining concepts and to support reasoning. For example, the nodes of the schema for a register concept may indicate an expected size, logic family, power consumption, cost, value contained etc. A schema for a ROM is shown below.

schema for *rom(x)* is
[rom : *x]-
        (struct) -> [array]
        (part) -> [word]
        (size) -> [length : @ words]
        (agnt-1) -> [apply]
        (contain) -> [value]
        (--) -> [ports],.

When deciding whether two concepts of type *rom* are coreferent, the schema for *rom* helps to supply a list of the possible restrictive which would comprehensively define the context of the concept. Using this knowledge, we can decide whether the two concepts are or are not coreferent or whether additional information would be useful in making this decision.

3.8 Join Ambiguity : The join ambiguity of a join of a requirement graph or schema concept is the number of possible coreferences of the concept in the aggregate specification conceptual graph which can be determined with the available information. The type label of a candidate for coreference of a requirement graph concept must match the concept type or be a supertype of it. If a schema concept is under consideration, the type label of a candidate for coreference must match the concept type or be its subtype.

3.9 Integration : Integration is the process of detecting and joining coreferent concepts across requirements graphs and schemas.

4. Integration Rules and Algorithm

A digital system can be described in terms of the devices which constitute the system and the interaction between these devices. The integration algorithm considers requirement graphs one-at-a-time and joins coreferent concepts across requirement graphs to build up a specification graph. Since each device is typically assigned an identifier or a name, the integration algorithm begins by joining coreferent device concepts between a partial specification graph and a requirement graph. Once a join has been obtained, the algorithm extends the join by joining all coreferent concepts between the requirement graph and the specification graph and, thus, obtains a new specification graph.

To determine if the current reference is coreferent with any concept of the aggregate specification graph, the following four rules are applied. Rule 1 uses the names and identifiers associated with two concepts to decide whether they are coreferent. Rule 2 deals with coreference of object concepts and their related attribute concepts. Rule

Table 3.1. Some Restrictive Relation Types

Type	Abbr.	Definition	Example
attribute	attr	a characteristic of the object	[memory] -> (attr) -> [fast].
identifier	idntf	identifier assigned to a device or value	[register] -> (idntf) -> [SCCR].
name	name	a name of an object	[counter] - -> (name) -> [program-counter],.
agent	agnt	the agent of an action or state	[reset] -> (agnt) -> [CPU].
instrument	inst	the instrument of an action	[transfer] -> (inst) -> [carrier].
operand	opnd	an operand of an action	[load] -> (opnd) -> [character].
source	src	a source object of an action	[read] -> (src) -> [register].
destination	dest	a destination object of an action	[write] -> (dest) -> [memory].
purpose	purp	a purpose of an action	[use] -> (purp) -> [read].
location	loc	the location of an object	[bit] -> (loc) -> [cache].
quantity	quant	the number of objects	[processor] -> (quant) -> [two].
size	size	the size of an object	[register] - -> (size) -> [length : @8bits],.
structure	struct	the structure of an object	[rom] -> (struct) -> [array].
part	part	a part of an object	[register] -> (part) -> [flip-flop].
attach	--	device is attached to a device	[device] ->(--) ->[device].

3 considers coreference of action or state concepts and their related agent, instrument, source, destination, attribute and operand concepts. Rule 4 considers the problem of possible coreferences of a concept in one conceptual graph with multiple concepts in another conceptual graph. These rules are applied in the same sequence in which they are enumerated above.

Following are some of the notations used in describing the rules. We let $c1 \approx c2$ mean that $c1$ and $c2$ are coreferent and $c1 \not\approx c2$ mean that $c1$ and $c2$ are not coreferent.

Let $c1 = c2$ mean that type($c1$) = type($c2$) and referent($c1$) = referent($c2$).

Also *relation*($c1$) represents the concept $c2$ such that $c1$ -> (*relation*) -> $c2$.

RG is the requirement graph being considered and SG is the aggregate specification graph being considered. For any conceptual graph G concepts(G) and relations(G) are the sets of concept nodes and relation nodes respectively.

Rule 1   a). idntf($c1$) = idntf($c2$) $\Rightarrow$ $c1 \approx c2$.

Two concepts are coreferent if they have the same identifier.
An identifier is associated with a unique entity. So if two concepts share an identifier, they are references to the same entity and are therefore coreferent. Since a name is not associated with a unique entity, concepts with the same name may not be coreferent.

b). idntf($c1$) $\neq$ idntf($c2$) $\Rightarrow$ $c1 \not\approx c2$.

Two concepts with different identifiers cannot be coreferent.
Since each entity has at most one identifier, two concepts with different identifiers cannot refer to the same entity.

Rules 1 a) and b) can be combined as follows

$c1 \approx c2$ $\Leftrightarrow$

$((\exists \, \text{idntf}(c1) \wedge \exists \, \text{idntf}(c2))$ $\Rightarrow$ $(\text{idntf}(c1) \approx \text{idntf}(c2))$

c). name($c1$) $\neq$ name($c2$) $\Rightarrow$ $c1 \not\approx c2$.

Two concepts with different names are not coreferent.
While many entities may be described by the same name, each entity has at most one name.

Rule 2   a). (type($r1$) $\leq$ *attr*) $\wedge$ ($r1(c1) \not\approx r1(c2)$) $\Rightarrow$ ($c1 \not\approx c2$).

Two concepts are not coreferent if the attributes of one are incompatible with those of the other.
When the attribute type relation being considered is *quant*, then an individual concept without an explicit quantity attribute attached to it is assumed to have a quant attribute of value 1.

b). ($c1 \not\approx c2$) $\Rightarrow$

$(\forall \, r \, , (\text{type}(r) \leq attr \wedge \exists \, r(c1) \wedge \exists \, r(c1))$ $\Rightarrow$ $(r(c1) \not\approx r(c2))$

Two attribute type concepts are not coreferent if the object concepts they modify are not coreferent.
Each attribute gives information about a specific entity. So attributes of distinct entities are themselves distinct entities.

Note that this rule requires comparing the coherences of each pair of attribute concepts being considered.

Rule 3 a). $((\text{type}(c1) \leq action \wedge \text{type}(c2) \leq action) \vee (\text{type}(c1) \leq state \wedge \text{type}(c2) \leq state)$

$\wedge \, (\text{agnt}(c1) \, !\approx \text{agnt}(c2) \vee \text{inst}(c1) \, !\approx \text{inst}(c2))) \Rightarrow (c1 \, !\approx c2)$

Two action or state concepts are not coreferent if their agents or instruments are not coreferent.

Recall that actions or states have at most one agent or instrument.

b).   i) Two action concepts sharing the type label "transfer", "shift", "read" or "send" are not coreferent if their sources are not coreferent.

ii) Two action concepts sharing the type label "transfer", "shift", "write" or "receive" are not coreferent if their destinations are not coreferent

iii) Two action concepts sharing the type labels "transfer", "shift", "read", "write", "send" or "receive" are not coreferent if their operands are not coreferent.

The schemas for the above action concepts indicate that the sources, destinations and/or operands are unique, so coreferent actions of these types should have coreferent sources, destinations and/or operands.

iv) Two state concepts sharing the type labels "is" or "contain" are not coreferent if their respective attributes are not coreferent.

The schemas for the above state concepts dictate that each such concept has a unique attribute attached.

c). $((\text{type}(c1) \leq action) \wedge (\text{type}(c2) \leq action) \wedge (c1 \, !\approx c2)) \Rightarrow$
$(\text{opnd}(c1) \, !\approx \text{opnd}(c2))$.

If two action concepts are not coreferent then their operands are not coreferent.

This rule is obtained by considering that while many operands may have the same value associated with them, each operand is associated with a unique action.

Note that applying this rule would require comparison of the coherences of the operand concepts being considered.

Rule 4. a) A concept from a requirement graph cannot be coreferent to more than one concept in the aggregate specification graph.

Note that distinct concepts in the specification graph are supposed to represent distinct entities or actions. All coreferent concepts in the aggregate graph should have already been joined by the integration algorithm.

b) If a concept in a requirement graph might be coreferent with more than one concept in the specification graph then it is coreferent with the concept with which it shares the greatest number of restrictive relations in common. Restrictive relations define an entity in greater detail and specialize the concept to which they are attached. Therefore, greater overlap among restrictive relations of two concepts increases the chances that they are coreferent.

c) $(c_1, c_2 \in \text{concepts}(RG) \wedge c_3 \in \text{concepts}(SG) \wedge c_1 \approx c_3) \Rightarrow (c_2 \,!\approx c_3)$.
Two concepts of a single requirement graph cannot be coreferent to the same concept in the specification graph.
This rule follows from the fact that each distinct concept in a requirement graph refers to a unique entity or action.

If, in the application of these rules, it is determined that the current requirement concept is not coreferent with the specification concept under consideration, the comparison is resumed with the next concept of the specification graph that has the same type label. If all the concepts in the specification graph have been considered and if no coreference has been detected, then the concept is classified as a definition and the next concept of the requirement graph is considered.
The schema-driven integrator uses a schematic database. This database consists of schemas which give design information about the various entities, actions or abstractions which occur in a typical system specification. Whenever the integrator comes across a new concept type in the requirement graph it searches the database for a schema for the concept type. If such a schema exists, it is joined to the graph on the given concept. The schema introduces generic concepts which may be specified as subsequent requirements graphs are processed. Concepts that are not specified may be considered omissions in the document. These must be specified either by the requirements author or the design engineers. The following integration algorithm is used to guide the integration process.

Integration Algorithm :

Step 0   The integration process begins by taking a null Aggregate Specification Graph (s-graph) and performing a maximal consistent join of a schema or a Requirement Graph (r-graph) resulting in a new s-graph. The next r-graph or schema is integrated with this s-graph by joining coreferent concepts to obtain a new s-graph. The steps in the algorithm are :-

Step 1. Search the r-graph for concepts with identifiers. Form a list of such concepts and consider each such concept for possible coreference with concepts in the s-graph. Join coreferent concepts.

Step 2. If no join is obtained in Step 1, repeat Step 1 but using name concepts instead of identifier concepts as a guide to the join. Note that every r-graph will have at least one concept with an identifier or a name. This is due to the fact that each requirement graph has at least one concept of type device or signal and all devices and signals have either an identifier or a name.

Arbitrarily select one of the concepts ROOT_R of the r-graph which has been joined to concept ROOT_S of the s-graph in Steps 1 or 2.

Step 3

a) Form an undirected spanning tree [9] of the r-graph with ROOT_R as the root.

b) If the tree has only one concept, go to Step 4.

c) For each concept CON_R in the r-graph which is adjacent to ROOT_R and has not yet been considered, repeat i) - iii)

i) Perform a breadth-first search around concept ROOT_S of the s-graph for a concept CON_S which can join with CON_R. i.e. a concept identical to or a generalization of CON_R. If a schema is being considered instead of a r-graph, then CON_S must be identical to or a specialization of CON_R. Use the set of rules to check whether the two concepts are coreferent. This would involve performing a temporary join on the concepts. If the join violates any of the rules, then it is undone.

ii) Join CON_R and CON_S.

iii) Using CON_R as the new ROOT_R and CON_S as the new ROOT_S, repeat Step 3.

Step 4. Repeat Steps 1-3 until all r-graphs are exhausted.

5. Implementation and Results

A design specification for a universal asynchronous receiver transmitter (UART) has been constructed from the Intel MC68HC11 Reference Manual [10]. Figure 5.1 a) shows a simplified block diagram of the UART. The topology of the conceptual graph of the simplified block diagram is shown in Figure 5.1 b). The specification consists of the simplified diagram and set of seventy-eight sentences.

To reduce labor and allow greater latitude in experimenting with conceptual graph notation, manual semantic analysis was performed on the specification set. The schema-driven integrator algorithm was applied to this set. Figure 5.2 a) shows the simplified graphical representation of a schema of the UART selected by the algorithm. The simplified graphical representation of an intermediate aggregate specification graph appears in Figure 5.2 b). This specification graph is obtained from the UART schema and 10 requirements graphs. Nodes formed by joining two or more concepts are shown as filled nodes. It was observed that the algorithm ensured complete elimination of ambiguity in performing the integration. However, this can be attributed, in part, to the manual development of the specification which avoided overly difficult constructs and complexity in text. Also, manual analysis eliminated the possibility of awkward or erroneous conceptual graphs which is a potential problem of imperfect automated analysis.

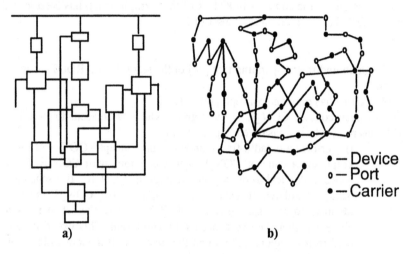

Figure 5.1.   a) **Simplified Block Diagram of the UART. and**
b) **Simplified Graphical Representation of Conceptual Graph for the Block Diagram of a UART.**

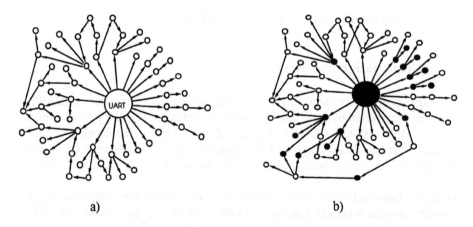

Figure 5.2   a) **Simplified Graphical Representation of a Schema of a UART**
b) **Simplified Graphical Representation of an Intermediate Aggregate Specification Graph**

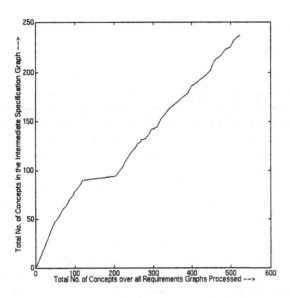

Figure 5.3 Plot of Total No. of Concepts in the Intermediate Specification Graph vs. The Total No. Of Concepts over all Requirement Graphs Processed

Figure 5.3 shows a plot of the number of concepts in the aggregate specification graph as a function of the sum of the concepts of the requirement graphs processed by the schema-driven integrator. The first few requirements graphs state the devices which compose the system and assign identifiers or names to these devices. So, as evident from the plot, there are very few joins till the total number of requirements graphs concepts reaches about 150. After this, we have requirements graphs describe simple interaction between the devices. Such graphs typically have 5 device concepts (coreferences) for every action or state concept (definitions). Since there are far more coreferences per graph than definitions, the total number of concepts in the specification graph increases slowly till the sum of the requirements graphs concepts reaches about 300. After this stage, more complex interactions are described which typically have two or more actions and at least one state (definitions) but few devices (coreferences). So the number of joins are comparatively fewer and the rate of increase in the number of concepts in the specification graph is closer to the rate of increase of the requirements graphs concepts.

Table 5.1 illustrates some of the results obtained from the set of 78 requirements graphs used. Some concepts of some of the sentences are shown in the leftmost column. A cell is marked "D" if the definition reference of the concept corresponding to the cell row number occurs in the sentence corresponding to the cell column number. If a coreference of the concept occurs, then the cell is marked "R".

Table 5.1 : A sampling of results

Concept		Sentence Number																			
NAME	TYPE	1	2	4	5	6	8	11	12	13	14	16	17	20	21	29	30	31	32	75	76
UART	device	D																			
SCI	id	D																			
shift-register	device		D																		
TSR	id		D						R	R	R		R								
TCL	id			D					R	R	R										
register	device				D																
SCCR	id				D															R	
register	device					D															
SCSR	id					D						R		R	R	R		R	R		
TDR	id						D		R	R		R	R	R	R						
register	device							D													
transfer	action								D												
data	data								D												
shift	action									D											
data	data									D											
load	action										D										
load	action											D									
TDRE	id											D		R	R						
value	value											D									
empty	state											D	R								
transfer	action												D								
load	action													D	R						
contains	state													D							
load	action														D						
read	action														D						
load	action															D					
IDLE	id															D					
RxD	id															D	R				
idle	state															D	R				
contains	state																D				
one	constant																D				
character	data																	D			
characteristic	length																	D			
character	data																		D		
characteristic	length																		D		
select	action																		D		
method	method																		D		
select	action																				D
method	method																				D

Except for rules 1 a) and 4 b), all the integration rules prevent joins. The table shows a large number of coreferences detected by rules 1 a) and 4 b) while several action, device and state concepts are classified correctly as definitions by the other rules.

6. Conclusions and Future Work

An algorithm and a set of rules have been developed to integrate conceptual graphs representing individual requirements in a system description into a single aggregate specification graph representing the entire system. The results indicate that the detection rules are quite reliable for a modest example.

The restricted sublanguage of English used for digital system description does not totally eliminate the problems caused by the inherent ambiguity of the English language. Questionable cases can be resolved through user interaction. Further studies with larger examples could provide enough experience to be able to decide when user interaction should occur. Extending the application domain to provide a general solution to the problem of coreference detection remains an important and unsolved problem.
A software package that implements the above algorithm will be developed.

7. References

[1]   IEEE, IEEE Standard VHDL Language Reference Manual, IEEE, NY, 1988.

[2]   J. F. Sowa, Conceptual Structures : Information Processing in Mind and Machine, Addison-Wesley, Reading, MA, 1984.

[3]   R. Greenwood and W. R. Cyre, "Conceptual Modeling of Digital Systems from Informal Descriptions", Proc. International Workshop on Modeling, Analysis and Simulation of Computer Telecommunications Systems, San Diego, CA, January 17-20, 1993.

[4]   Alex Honcharik, J. Armstrong and W. R. Cyre, "Mapping Conceptual Graphs to VHDL Descriptions", VHDL International User's Forum, Washington, DC, October 18-21, 1992.

[5]   M. L. Mugnier and M. Chein, "Polynomial Algorithms for Projection and Marching, Proc. 7th Annual Workshop on Conceptual Graphs, Las Cruces, NM, pp. 49-58, July, 1992.

[6]   Sidney Greenbaum et al., A Grammar of Contemporary English, Longman Group Limited, London, 1973.

[7]   S. Shankaranarayanan and W. R. Cyre, "Identification of Coreferences with Conceptual Graphs", Proc. Supp. 2nd International Conference on Conceptual Structures, College Park, NM, pp. 45-60, August 16-19, 1994.

[8]   W. R. Cyre, "Capture, Integration and Analysis of Digital System Requirements with conceptual Graphs", accepted for publication in IEEE Trans. on Knowledge and Data Engr., 1994.

[9]   J. A. Bondy and U. S. R. Murty, Graph Theory with Applications, New York : North Holland, 1976.

[10]   MOTOROLA, MC68HC11 Reference Manual, PRENTICE HALL, Englewood Cliffs, N. J. 07632, 1989.

[11]   J. F. Sowa and E. C. Way, "Implementing a Semantic Interpreter Using Conceptual Graphs," IBM J. Research and Development, Vol. 30, pp. 57-69, January, 1986.

Service Trading Using Conceptual Structures

A. Puder, S. Markwitz and F. Gudermann

Department of Computer Science
University of Frankfurt
D–60054 Frankfurt, Germany
{puder,markwitz,florian}@informatik.uni-frankfurt.de

Abstract. An open distributed environment can be perceived as a service market where services are freely offered and requested. Any infrastructure which seeks to provide appropriate mechanisms for such environments has to include some mediator functionality to bring together matching service requests and service offers. The matching algorithm that the mediator must perform commonly builds upon an IDL–based type definition for service specification. We propose a type specification notation based upon conceptual graphs to support the cognitive domain of application users. In our framework the trader implements a matching algorithm as well as a learning algorithm which are tailored to service trading in open environments.

Keywords: Open distributed environments, type graphs, trading, service matching, service knowledge base, conceptual graphs.

1 Introduction

An open distributed environment can be perceived as a service market where services are offered and requested. A crucial aspect of these environments is the way services are *typed*. A type specification abstracts from specific internal details as it decouples the service usage from its implementation. Service types play an integral role during the mediation process. A software component known as a *trader* matches service offers and service requests based on their types.

There are several ongoing standardization efforts to develop an appropriate infrastructure supporting these environments (for example [5, 4, 6]). All these infrastructures focus on *interface definition languages* (IDL for short) as the main technique of service typing. An IDL is a low level specification of a precisely defined programming interface describing function names and input/output parameters in terms of signatures. These kind of types are meant to be handled by programmers and therefore are not suitable for application users.

With the emergence of high level services as found in the World Wide Web (see [1]) the need for an appropriate typing mechanism becomes evident. An application user (i.e. *not* a programmer) must deal with service types to lookup matching service providers. So far there exist only simple keyword based traders which allow only for a limited expressiveness of service types. We have chosen to develop a type specification based on conceptual graphs which are particularly useful to support the cognitive domain of the users.

The paper is organized as follows: in section 2 we introduce concepts from the field of open distributed environments necessary for the scope of this paper. We present definitions for the terms *object graph* and *trader object* to motivate the need for type specifications of *User Types*. In section 3.1 we identify the requirements of a *learning model* and a *matching model* which will then be defined in sections 3.2 and 3.3 respectively. In these sections we also present appropriate algorithms. Finally a summary in section 4 concludes this paper.

2 Environment

The goal of this section is to motivate the need for an abstract type specification notation based on conceptual graphs. The starting point of our discussion will be the object graph model which serves as a suitable abstraction for an open service market where services are freely offered and requested. A crucial aspect of the object graph model is the typing of objects which is essential for supporting the matching process of service requests and offers. Finally we distinguish two levels of abstraction with respect to who — the programmer or the application user — is exposed to the type specifications. The premise of this paper is that currently there exists no suitable type notation which supports the cognitive domain of an application user.

2.1 Object graph model

The environment, which serves as a basis for this paper, builds upon the classical definition of an object (see [2]): *an object has state, behavior and identity; the structure and behavior of similar objects are defined in their common class. The terms instance and object are interchangeable.* Using this definition, a problem domain may be decomposed as a set of interacting and co-operating objects. A snapshot of such an object-based computation may be visualized as a directed graph, where nodes represent objects and arcs represent references. A *reference* (or arc) is therefore a referral of an object's identity. The direction of the arc determines whose identity is known to whom. For an object to hold a reference to another object means to know about the existence of this particular instance, allowing *operation invocations* (also commonly called method invocations). Thus a directed arc between two nodes (objects) represents the ability to invoke operations along the direction of this arc (i.e. the *service provider* is at the arc head, and the *requester* is at the tail). Service providers are also called *server objects* and service requesters are called *client objects*. A client object may be controlled by a (human) *user* who eventually decides the course of actions. The directed graph will be called an *object graph*.

An important requirement of the object model is that an object encapsulates data and code. The role of a *type specification* is therefore crucial in the sense that it should provide enough information to describe an object's behavior, yet conceal any implementation specific details. We assume that both references and objects are typed. Implementation details of a server object are irrelevant to a client. From a client object's perspective (the one holding the reference) a reference guarantees a contract that the server object must fulfill. Polymorphism occurs when the type of the reference is a super–type of the object to which it points. The server object therefore is a specialization to what the client expects, if it can fulfill the contract[1].

2.2 Trading and the role of types

In this section we assume some kind of type notation which allows the typing of references and objects as discussed above. Furthermore the object graph will be seen in the context of an open, distributed environment. By *open* we mean an environment spanning several administrative domains. Thus, the object graph and its modifications are to be seen as an abstraction of a service market, where services are freely provided and requested by independent parties.

In a distributed environment the object graph will generally be partitioned. A client object has only a limited view on the object graph, as global knowledge of its structure is generally impossible to acquire. References between objects induce a "knows–about–relation", it is also clear that without appropriate support from an underlying infrastructure a client can't see beyond a transitive closure of the references it holds (i.e. the partition of the object graph in which the client is embedded).

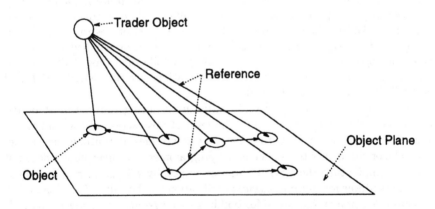

Fig. 1. The Trader Object possesses global knowledge of the object graph.

These considerations have led to system components like the ODP Trader (see [7]), which serve as a mediator between service requesters and service providers

---

[1] Polymorphism is often referred to as the "principle of substitutability" where an object of type A may be substituted by an object of type B without anything "bad" happening.

and therefore bridge the knowledge/visibility gap. The *trader* matches service requests with previously stored service offers and thereby helps to establish references in the object graph. The match process heavily depends on the precise definition of the type specification notation.

Within the object graph model introduced above, a trader can be modeled by an object as well (called the *trader object*, see figure 1). The trader object has a well known identity, so every object is given an implicit reference to this object (this fact is not shown in figure 1 to simplify the depiction). For a more detailed discussion see [9].

Conceptually the trader maintains a database of all service providers who have registered themselves previously. The database holds tuples each containing a *type specification* as one argument. The second argument represents a non–empty set of *addresses* of appropriate server objects providing the service described by the first argument. If a match of a request and a service offer succeeds, the trader uses the address to construct a reference which will be given to the client as the result.

2.3   System Types vs User Types

One can distinguish two different cases with respect to *when* a type specification is initially required: at *compile time* or at *runtime*. For compile time type notations there exists a wide range of notations based on interface signatures defined in some *interface definition language* (IDL for short). A type specification written in an IDL commonly lists a set of methods which are implemented by the server object. Special tools generate so–called *stubs* or *proxies* which eventually get linked to the client object. In terms of level of abstraction, an IDL is intended for programmers and is generally not seen or handled by users of the client object at runtime. In the following compile time types will be called *System Types*. Common examples for such notations are the CORBA–IDL and the DCE–IDL (see [5] and [4] resp).

In contrast a type specification notation used during runtime must be designed under different assumptions. The level of abstraction of the underlying objects is higher in the sense that the *user* determines at runtime the kind of service he or she wishes to use. Thus the programmer of a client object does not know at compile time the services the user eventually will use at runtime. Type notations suitable for runtime service specification will be called *User Types* (UT for short). A common technique for building such systems are *generic client objects* which are able to communicate with an a priori unknown server. Examples for such systems are the World Wide Web (WWW) or OpenDoc (see [1, 8]).

These systems have in common that different services can be provided at runtime without the need for a specific service type at compile time. Instead, a service provider has means to dynamically convey its particular user interface to the client via some sort of *graphical user interface* (GUI) description. The generic client is able to interpret these descriptions and to build and present an appropriate GUI for the end user. The user may interact with the generic client to invoke operations and to provide parameters embedded as widgets such as edit fields or check–boxes appearing in the GUI. These parameters are transfered to the service provider who takes appropriate actions to perform the request. For example, within the WWW system a GUI description is based on the *hypertext markup language* (HTML) which the generic client is able to translate into a visual presentation.

It is not immediately clear how a notation for User Type specifications should be defined. A type specification in this environment is more abstract than an IDL based specification. In particular it should support the cognitive domain of the users and not of the programmers. Our choice for User Type specification notation are *conceptual graphs* which in particular support the cognitive domain of the users. We do not compare conceptual graphs with other knowledge representation approaches. The reason for choosing conceptual graphs was their close relationship to natural language. We focus upon the non trivial mechanisms which the *trader* must perform during the mediation process in order to match service descriptions based on conceptual graphs. These mechanisms which the trader builds upon will be discussed thoroughly in the next section. The following table summarizes the differences between System Types and User Types.

	System Types	User Types (UT)
When initially required	At compile time.	At runtime.
Level of abstraction	Low. Supports fine grained objects manipulated by *programmers*.	High. Supports cognitive domain of *users* independent of implementation specific details.
Level of detail	High. Strict defined object interface based on signatures.	Possibly low. Vague and unprecise information due to the world of discourse.
Trading	Two phase trading: first service lookup then service usage. The traded service is always the desired one.	Multi phase trading: user may decide to switch back and forth between service lookup and service usage in order to obtain the desired service.
Supported kind of objects	Fine grained IDL–based typing of objects with a priori known service types.	Generic clients which handle a priori unknown service types (WWW, OLE2, OpenDoc).
Example notations	CORBA–IDL, DCE–IDL.	-

Table 1: Differences between System Types and User Types.

3 Trading through learning and matching

The basic interface that the trader offers its clients is that of a *service export* and a *service import* (see figure 2). A provider first has to export the type of service it wishes to advertise via the trader. The trader maintains a knowledge base of all registered service providers and therefore is able to map service types to appropriate providers. A requester on the other hands queries the trader for a suitable provider via a service import. An approximate portrayal of a service type is represented on the implementation level as a *conceptual graph*. Both import and export operations to the trader are formulated using conceptual graphs. In the following the service description passed as a parameter for an export operation will be called a *type graph* and the description passed as an argument for an import operation a *query graph* respectively.

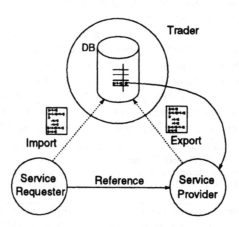

Fig. 2. Architectural overview: export and import of services.

The trader has to implement the import and the export operations. Translated to an AI context, a service export can be seen as a step of a *learning* process. Successive exports increase the quality of service descriptions in terms of decreasing semantic distance to a User Type. A service import on the other hand is implemented within the trader as a *match* process of conceptual graphs. The trader has to cope with the problem that the size (in terms of nodes) of a type graph may grow considerably due to the diversity of service descriptions. To handle this situation we have devised a *learning model* which represents a formal specification of the export operation described in section 3.3. Furthermore there is a need for a *matching model* which defines a metric expressing the degree of similarity between a query and a type graph and which serves as a basis for a formal specification of the import operation. This match is the major task that the trader must perform. The practical relevance of our proposal mainly rests upon the quality of this matching process. A step by step development of the match algorithm will be discussed in detail in section 3.2. The basic idea lies in the determination of an *akin index* which is a measure of the semantic difference of a query graph and a type graph. The best match (i.e. most appropriate service) is then decided by means of an *akin taxonomy* defined over all type graphs contained in the trader's knowledge base.

3.1 Requirements of the matching– and learning model

There is an immediate relation between the matching model and the learning model. Prior to the match of a query graph and a type graph (defined by the matching model), the trader must build up an approximation of a service type obtained by a sequence of learning phases (defined by the learning model). In section 3.2 we show that existing definitions of conceptual graphs do not satisfy the special requirements of a trader. One implication will be that weights need to be assigned to the nodes of a type graph. We identify the following requirements which a learning model must satisfy:

L1: The initial type graph describing a UT must be a "good" first approximation.

L2: One execution of a learning algorithm must either extend the type graph or adapt it in terms of a better approximation to the UT. The weights assigned to the nodes of a type graph should converge asymptotically to a final value.

L3: Eventually a sufficient approximation of the UT will be reached.

The semantic distance of a query graph and a type graph will by characterized by an akin index. We identify the following requirements for a matching model:

M1: The akin index denotes the quality of identical semantic context within a type graph and a query graph. A high akin index reflects major identical components and vice versa.

M2: The size of a type graph has no influence on the akin index. Increasing the size of a type graph merely reflects a more detailed description of a service (i.e. a closer approximation to the UT).

M3: The difference in akin indices is a true measure of the real semantic difference.

M4: It is possible to assign weights to the nodes of a query graph which will be taken into account by the match algorithm.

The matching model builds upon the learning model. The learning algorithm has to be repeated due to missing details or wrong weights (i.e. the approximation to a UT is insufficient), if the evaluation of the akin index does not reflect the real semantic distance. In order to properly motivate the way we assign weights to nodes in a conceptual graph we will first discuss the matching model prior to the learning model.

3.2 Fuzzy context matching

Practical experience has shown that *atomic conceptual graphs* (see [3]) are sufficient to express type graphs for approximation of UTs. The semantics of a simple conceptual graph can be defined through a mapping to a first order predicate formula (see ϕ operator defined in [11]). Using this mapping, example 1 shows a type graph and a query graph, as well as their semantics in terms of the existentially quantified first order formulas F_1 and F_2. The informal semantic of the type graph G_1 is that *there is a relational database which runs on AIX machines and possesses an SQL-interpreter*. The informal semantic of the query graph G_2 expresses the need for *something which supplies data security, runs on AIX machines and possesses an SQL-interpreter*[2].

Example 1. Let G_1 be a type graph and G_2 be a query graph.

G_1 : ```
[DATABASE]-
    ->(CHRC)->[RELATIONAL]
    ->(LOC)->[HOST:AIX]
    ->(PART)->[SQL-INTERPRETER]
```

$$F_1 : \exists x_1 \exists x_2 \exists x_3 \Big( DATABASE(x_1) \wedge RELATIONAL(x_2) \wedge HOST(AIX) \wedge$$
$$SQL-INTERPRETER(x_3) \wedge CHRC(x_1,x_2) \wedge LOC(x_1,AIX) \wedge PART(x_1,x_3) \Big)$$

$G_2$ :  ```
[SOMETHING]-
    ->(SUPP)->[DATA-SECURITY]
    ->(LOC)->[HOST:AIX]
    ->(PART)->[SQL-INTERPRETER]
```

$$F_2 : \exists x_1' \exists x_2' \exists x_3' \Big(SOMETHING(x_1') \wedge DATA-SECURITY(x_2') \wedge HOST(AIX) \wedge$$
$$SQL-INTERPRETER(x_3') \wedge SUPP(x_1',x_2') \wedge LOC(x_1',AIX) \wedge PART(x_1',x_3') \Big)$$

Every first order formula and therefore every conceptual graph has an intended semantic defined by an interpretation I over domain D. The interpretation maps every constant to a distinct element in D and all ground instances of every predicate symbol to *true* or *false*. Together with the usual semantics of the connectives, a first order formula can be assigned a truth value under an interpretation. Naturally the intended interpretation of a formula should be a model (i.e. evaluate the formula to *true*).

The akin index is obtained by determining the common subgraphs of both graphs. The predicate logic provides a proper framework for this operation. Given two conceptual graphs the common subgraphs are computed by mutually eliminating concept and relation nodes which are not contained in the other graph. Two nodes may be similar after a reduction specializing the nodes. This elimination process is sound due to the fact that the corresponding first order formulas only contain conjunctions. If the intended interpretation is a model for a formula then it is also a model for the formula obtained by removing one or more predicates. In example 2 the result of this elimination process is shown. The remaining common parts of the type graph and the query graph denote the *semantic intersection*. The akin index is computed based on the semantic intersection and the original query graph. This computation ignores the root concepts of those two graphs as the akin index should infer a metric of the similarity of two descriptions rather than of their root concept names.

---

[2] To keep the size of this and the following examples within reasonable size, the service descriptions are not further refined. One such possible refinement for the database example would be some kind of product information.

Example 2. Let G_3 be the semantic intersection of G_1 and G_2

G_3 : [SOMETHING]-   *(concurring context of G_1 and G_2)*
       ->(LOC)->[HOST:AIX]
       ->(PART)->[SQL-INTERPRETER]

$$F_3 : \exists x_1 \exists x_2 \Big((SOMETHING(x_1) \wedge HOST(AIX) \wedge LOC(x_1, AIX)) \wedge$$
$$(SOMETHING(x_1) \wedge SQL - INTERPRETER(x_2) \wedge PART(x_1, x_2)) \Big)$$

The similarity of the semantic intersection G_3 and the type graph G_1 can be computed by comparing their sizes in terms of number of nodes (see [13]). Employing this method it can be seen that (i) G_1 has been reduced in size to 66% to result in G_3 and (ii) to 66% of G_2 match with G_3 respectively. In example 3 a different type graph (labeled G_4) is given. Type graph G_4 also can be reduced to G_3 if it is to be compared with G_2. Now there is a 50% correspondence in size between G_3 and G_4 while there is still a 66% match between G_1 and G_3. The deviation obviously results from the higher level of detail of G_4. The measure presented in (i) violates the requirement **M2** of the matching model and is therefore not suitable.

Example 3. Let G_4 be a type graph.

G_4 : [DATABASE]-
       ->(LOC)->[HOST:AIX]
       ->(PART)->[SQL-INTERPRETER]
       ->(ATTR)->[OPTIMIZED-ACCESS]
       ->(SUPP)->[SCHEMA-EVOLUTION]

On the other hand the measure (ii) also is insufficient, as there is no justification that the akin index is solely based on the size of the semantic intersection. Summing up the arguments presented above, the mere size in terms of nodes of the semantic intersection is not sufficient as it disregards the context of the query graph within the type graph.

Our solution to this problem is the assignment of weights to the nodes in a conceptual graph. Naturally the weights reflect the importance of each node in a graph. In example 4 the notation is slightly extended to allow for a weight specification of each node. The nodes of the type graph G_1 and the semantical intersection G_3 are now assigned individual weights.

Example 4. Let G_1' and G_2' be conceptual graphs with associated weights.

G_1' : [DATABASE]-
       ->(CHRC,w1)->[RELATIONAL,w2]
       ->(LOC,w3)->[HOST:AIX,w4]
       ->(PART,w5)->[SQL-INTERPRETER,w6]

G_2' : [SOMETHING]-
       ->(SUPP,w1')->[DATA-SECURITY,w2']
       ->(LOC,w3')->[HOST:AIX,w4']
       ->(PART,w5')->[SQL-INTERPRETER,w6']

The root concept nodes of a graph do not get assigned a weight because they are ignored during the matching process as explained above. There are two distinct ways to assign weights to the nodes of a conceptual graph, depending on whether all the weights of the graph satisfy a global invariant or not. A global invariant might be the requirement that for each node the sum of all the weights of its successors must be equal to one (see [10]). The consequence of this invariant is that the absolute weight of a particular node may decrease if the graph is extended by further nodes, as shown in example 5. Therefore the akin index of this extended graph and the query graph decreases in the same sense due to the different weights in the semantic intersection. This clearly violates requirement M2 of the learning model.

Example 5. Let G_5 be a type graph and G_5' an extension assigned with weights.

```
G₅ :   [DATABASE]-
             ->(CHRC,1/2)->[RELATIONAL,1]
             ->(LOC,1/4)->[HOST:AIX,1]
             ->(PART,1/4)->[SQL-INTERPRETER,1]

G₅' :   [DATABASE]-
             ->(CHRC,2/5)->[RELATIONAL,1]
             ->(LOC,1/5)->[HOST:AIX,1]
             ->(PART,1/5)->[SQL-INTERPRETER,1]
             ->(SUPP,1/5)->[DATA-SECURITY,1]
```

This undesired effect of a decreasing akin index by an increased level of detail of the type graph can be eliminated by assigning weights which do not adhere to a global invariant. We therefore propose an independent assignment of weights to the nodes of a conceptual graph. In order to reach this independence, the first order formula of the semantic intersection has to be interpreted differently. All direct successors of a node represent alternative descriptions of this node. In the formula some conjunctive junctors have to be replaced by disjunctions. Example 6 shows how to interpret the semantic intersection.

Example 6. Let G_3' be the semantic intersection of G_1' and G_2' and F_3' the new interpretation.

```
G₃' :   [SOMETHING]-                    (concurring context)
             ->(LOC,w1'')->[HOST:AIX,w2'']
             ->(PART,w3'')->[SQL-INTERPRETER,w4'']
```

$$F_3' : \exists x_1 \exists x_2 \exists x_3 \Big((SOMETHING(x_1) \land HOST(AIX) \land LOC(x_1, AIX)) \lor$$
$$(SOMETHING(x_1) \land SQL - INTERPRETER(x_3) \land PART(x_1, x_3)) \Big)$$

Derived from our practical experiences we assign each node (except the root node) of a conceptual graph a natural number ranging from $1, \ldots, 100$. This scale is divided into three categories depending on the importance of the node for the graph it appears in: (I) $1, \ldots, 33$ minor importance, (II) $34, \ldots, 64$ indifferent, (III) $65, \ldots, 100$ major importance. The default value of a node is 50.

The new interpretation intuitively divides a conceptual graph as a set of alternating subgraphs which overlap each other (figure 3 depicts this idea). The semantic of each subgraph is given by the ϕ operator. The akin index is determined by matching all possible combinations of all subgraphs.

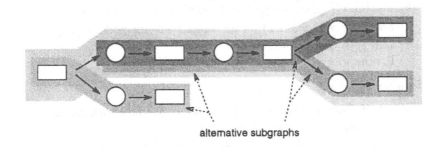

alternative subgraphs

Fig. 3. Alternative subgraphs of a conceptual graph.

The root node of each subgraph is explained by its successors and the weight of this subgraph therefore has to be reduced proportionally. This can be accomplished by a simple multiplication of weights. The weights of different subgraphs are computed by a *maximum operator* known from the Fuzzy Set Theory (see [14]). This operator satisfies the requirement of the independence of weights. Given a query graph, an increase in level of detail in a type graph can only result in a bigger semantic intersection. Together with the maximum operator the akin index can therefore only increase as well. The best alternative subgraph is chosen and the remaining graphs discarded. This scheme evaluates the quality of an alternative within the semantic intersection compared to the context of the original conceptual graph. Altogether the akin index is computed recursively. The complete match algorithm is given below.

Experiments have shown that two UTs may have a very close semantical relationship. To further differentiate between similar semantics we integrated the possibility to include counter examples with the keyword NOT. In this case the weight of the contradictory context is set to zero.

Algorithm 1: compute the akin index of a query graph and a type graph

1. **Input:** G_T type graph and G_Q query graph.
2. *Restrict$_C$*, the restrict–operation taken from the *canonical formation rules* (see [11]), determines the common subtype of the concept nodes. *Restrict$_R$* the restrict–operation for relation nodes respectively.
3. *weight* determines the weight of a node. Let the weight of the root nodes of G_T and G_Q be 100.
4. *index$_C$*(n_T, n_Q) determines the akin index of every pair of concept nodes from G_T and G_Q, starting with n_T and n_Q respectively. *index$_R$* is computed analogously.

5. Let n_T, n_Q be concept or relation nodes of G_T, G_Q. $index_C$ and $index_R$ are defined as follows:

$index_C(n_T, n_Q) =_{df}$

(a) $\frac{weight(G_T, n_T) \cdot 100}{weight(G_Q, n_Q)}$, if $Restrict_C(n_T, n_Q) \neq \perp$ or

(b) $\max(\max\{\frac{weight(G_T, n_T) \cdot 100}{weight(G_Q, n_Q)} \cdot index_R(n_T^+, n_Q^+)\}, 0) \cdot 1/100$, if $Res-trict_C(n_T, n_Q) = \perp$ and for all direct successor nodes n_T^+, n_Q^+ : $index_R(n_T^+, n_Q^+) \neq -1$ or

(c) 0, otherwise

$index_R(n_T, n_Q) =_{df}$

(a) -1, if $Restrict_R(n_T, n_Q) \neq \perp$ but n_T contradicts with n_Q and for all $n_T^+, n_Q^+ : index_C(n_T^+, n_Q^+) > 0$ or

(b) $\max\{\frac{weight(G_T, n_T) \cdot 100}{weight(G_Q, n_Q)} \cdot index_C(n_T^+, n_Q^+)\} \cdot 1/100$, if $Restrict_R(n_T, n_Q) \neq \perp$ or

(c) 0, otherwise

6. **Output**: $index(G_T, G_Q) =_{df} index_C(n_T, n_Q)$ determines the akin index of G_T and G_Q with n_T, n_Q the root concept nodes of G_T, G_Q.

To determine the best service provider, the maximum of all the akin indices resulting from the query graph and all the type graphs stored in the trader's knowledge has to be computed. The best possible match for a query graph G_Q with the smallest semantic distance is therefore given by $match(G_Q) =_{df} max\{index(G, G_Q) | G$ type graph of the knowledge base$\}$. The following example shows in detail a computation of an akin index according to algorithm 1.

Example 7. Determine the akin index of the semantic intersection of G_1' and G_2' with w1 = 50, w2 = w4 = w5 = w6 = 100, w3 = 60, w1'...w6' = 100.

$$
\begin{aligned}
index(G_1', G_2') \quad &= index_C(DATABASE, SOMETHING) \\
&= \max(\max\{100 \cdot 100/100 \cdot index_R(CHRC, SUPP), \\
&\quad 100 \cdot 100/100 \cdot index_R(CHRC, LOC), \\
&\quad 100 \cdot 100/100 \cdot index_R(CHRC, PART), \\
&\quad 100 \cdot 100/100 \cdot index_R(LOC, SUPP), \\
&\quad 100 \cdot 100/100 \cdot index_R(LOC, LOC), \\
&\quad 100 \cdot 100/100 \cdot index_R(LOC, PART), \\
&\quad 100 \cdot 100/100 \cdot index_R(PART, SUPP), \\
&\quad 100 \cdot 100/100 \cdot index_R(PART, LOC), \\
&\quad 100 \cdot 100/100 \cdot index_R(PART, PART)\}, 0) \\
&= \max(\max\{0, 0, 0, 0, 60, 0, 0, 0, 100\}, 0\} \\
&= \mathbf{100}, \text{ with}
\end{aligned}
$$

$index_R(CHRC, SUPP) = index_R(CHRC, LOC) = index_R(CHRC, PART) =$

$$index_R(LOC, SUPP) = index_R(LOC, PART) = index_R(PART, SUPP) =$$
$$index_R(PART, LOC) = 0$$
$$index_R(LOC, LOC) = \max\{60 \cdot 100/100 \cdot$$
$$index_C(HOST : AIX, HOST : AIX)\} = 60$$
$$index_R(PART, PART) = \max\{100 \cdot 100/100 \cdot$$
$$= index_C(SQL - INTERPRETER,$$
$$SQL - INTERPRETER)\} = 100$$

3.3 Fuzzy context learning

The assignment of weights to the nodes of a conceptual graph is central in our model for the computation of an akin index. We have devised a learning model tailored to the matching model introduced above. The learning model allows the incremental growth of type graphs which results in a better approximation to UTs. Within the framework of this model there needs to be a first type graph as initial approximation of a UT representing a valid semantic context (canonical, see [12]). The semantic distance between this first type graph to the UT may be high. Using the join-operation (as described in [11]) the type graph will be successively refined by further graphs. The semantic context remains valid. Example 8 demonstrates one application of the learning algorithm by combining G_6 and G_7 yielding G_8.

Example 8. G_6 is refined by joining it with G_7 resulting in G_8.

```
G6 : [DATABASE]-
         ->(CHRC,w1)->[RELATIONAL,w2]
         ->(LOC,w3)->[HOST:AIX,w4]

G7 : [DATABASE]-
         ->(LOC,w1')->[HOST:AIX,w2']
         ->(PART,w3')->[SQL-INTERPRETER,w4']

G8 : [DATABASE]-
         ->(CHRC,w1)->[RELATIONAL,w2]
         ->(LOC,w')->[HOST:AIX,w'']
         ->(PART,w3')->[SQL-INTERPRETER,w4']
```

The nodes of the example graphs can be attributed with weights, reflecting their relevance within the context of the graphs. The weight of those nodes which extend the type graph during a join operation remain unchanged. Synonym nodes or nodes resulting from a restrict operation require a re-computation of the weights. This step has to take into account how often the restrict operation already has been performed on this node in order to accomplish an asymptotic behavior of the learning process (requirement **L2** of the learning model). Therefore a counter is associated with every node and is being increased by one with every restrict during a join operation. The new weight is computed according to formula 1.

$$\text{new weight} = \frac{\text{counter} \cdot \text{old weight} + \text{weight of the example graph}}{\text{counter} + 1} \tag{1}$$

New weights only have a small influence on old weights with a high counter. They converge within smaller getting bounds. As a consequence examples joined with a type graph early in the course of time have a higher impact on the weights than at a later point of time. The influence of bad examples on type graphs decreases accordingly which results in a fault tolerant behavior of the learning model.

The converging of weights assigned to nodes implies a criterion to decide when the weight of a node has reached a fix point and therefore is not subject to further changes. Further examples have no more influence on this weight. The learning can be seen as partially finished if some counters reach a given bound. The best possible approximation is reached towards the end of the learning phase. The weights then reflect the importance of nodes which serve as a basis for the matching model.

Algorithm 2: adopted join

1. **Input**: Enter initial type graph.
2. Successive extension of the type graph through examples. This requires the re–computation of the type graph's weights according to formula 1, if those nodes are already contained in the type graph. Step 2 can be repeated an arbitrary number of times.
3. **Output**: The resulting type graph denotes a semantic approximation of a UT, where the weights of the nodes represent the importance for the context of embedded components of the graph.

4 Conclusion

Open distributed environments may be seen as a service market where services are freely offered and requested. The mediation of these services is done by a designated system component known as a *trader*. Current traders primarily base their matching algorithm of services upon IDL-based type notations. We have shown how conceptual graphs can be used for abstract type specification to support the cognitive domain of application users. This approach is particularly useful as the trader *learns* various ways of describing a service. We have developed formal *learning–* and *matching models* which serve as a basis of our implementation.

We have implemented the algorithms described in this paper as well as a *trading protocol* (see [9]). The complete source, using various C++ PD-class libraries and a Tcl/Tk–based GUI front end, are placed in the public domain and may be requested from the authors. Our implementation of an AI-based trader maintains a database of *uniform resource locators* (URL) of the World Wide Web.

Acknowledgements

We thank Wayne Brookes (University of Queensland, Australia) and an anonymous referee for their comments and discussions on this article.

References

1. Tim Berners-Lee et al. The World–Wide Web. *Communications of the Association for Computing Machinery*, 37(8):76–82, August 1994.
2. G. Booch. *Object Oriented Design with Applications*. BenjaminCummings Publishing Company, Inc, Redwood City, California, 1991.
3. Gerard Ellis. Compiled hierarchical retrieval. In T. E. Nagle, J. A. Nagle, L. L. Gerholz, and P. W. Eklund, editors, *Conceptual Structures, current research and practice*, chapter 31, pages 595–604. Ellis Horwood Limited, 1992.
4. Open Software Foundation. *Introduction to OSF DCE*. Prentice–Hall, Englewood Cliffs, New Jersey, 1992.
5. Object Management Group. *The Common Object Request Broker: Architecture and Specification Revision 1.1*. 1991.
6. ISO/IEC. Information Technology – Basic Reference Model of Open Distributed Processing – Part I. ISO/IEC COMMITTEE DRAFT ITU-T RECOMMENDATION X.901, 1993. ISO/IEC CD 10746–1.
7. ISO/IEC. *ODP–Trader, Document Title ISO/IEC JTC 1/SC 21 N 8192*. 1993.
8. Component Integration Laboratories. Shaping tomorrow's software (white paper). Technical report, cil.org:/pub/opendoc-interest/OD–overview.ps, 1994.
9. Arno Puder, Stefan Markwitz, Florian Gudermann, and Kurt Geihs. AI–based Trading in Open Distributed Environments. In *International Conference on Distributed Processing (ICODP'95)*. Chapman and Hall, 1995.
10. A. L. Ralescu and J. F. Baldwin. Concept learning from examples and counter examples. In B. R. Gaines and J. H. Boose, editors, *Machine Learning and Uncertain Reasoning, Knowledge–Based Systems*, pages 49–74. Academic Press, 1990.
11. John F. Sowa. *Conceptual Structures, Information Processing in Mind and Machine*. Addison–Wesley Publishing Company, 1984.
12. M. Wermelinger and J. Lopes. Basic Conceptual Structures Theory. In William M. Tepfenhart, Judith P. Dick, and John F. Sowa, editors, *Conceptual Structures: Current Practices. Second International Conference on Conceptual Structures, ICCS'94, College Park, Maryland, USA, August 1994*. Springer, 1994.
13. G. Yang, Y. Choi, and J. Oh. Cgma: A Novel Conceptual Graph Matching Algorithm. In H. D. Pfeiffer and T. E. Neagle, editors, *Conceptual Structures: Theory and Implementation, 7th Annual Workshop*, pages 252–261. Springer, 1992.
14. Hans Jürgen Zimmermann. *Fuzzy Set Theory and Its Applications*. Kluwer Academic Publishers, second, revised edition, 1992.

Sentence Generation from Conceptual Graphs

Nicolas Nicolov, Chris Mellish, Graeme Ritchie

Dept of Artificial Intelligence
University of Edinburgh
80 South Bridge
Edinburgh EH1 1HN
{nicolas,chrism,graeme}@aisb.ed.ac.uk

Abstract. This paper describes a technique for translating the semantic information encoded in a conceptual graph into an English language sentence. The use of a non-hierarchically structured semantic representation (conceptual graphs) allows us to investigate a more general version of the sentence generation problem where one is not pre-committed to a choice of the syntactically prominent elements in the initial semantics. We show clearly how the semantic structure is declaratively related to linguistically motivated syntactic representation. Our technique provides flexibility to address cases where the entire input cannot be precisely expressed in a single sentence.

1 Introduction

Natural language generation is the process of realising communicative intentions as text (or speech). The generation task is standardly broken down into the following processes: content determination (what is the meaning to be conveyed), sentence planning (chunking the meaning into sentence sized units, choosing words), surface realisation (determining the syntactic structure), morphology (inflection of words), synthesising speech or formatting the text output.

In this paper we address aspects of sentence planning (how content words are chosen but not how the semantics is chunked in units realisable as sentences) and surface realisation (how syntactic structures are computed). We thus discuss what in the literature is sometimes referred to as tactical generation, that is "how to say it"—as opposed to strategic generation—"what to say". We look at ways of realising a non-hierarchical semantic representation (conceptual graph) as a sentence, and explore the interactions between syntax and semantics.

This work improves on existing generation approaches in the following respects: *(i)* Unlike the majority of generators this one takes a non-hierarchical semantic representation as its input. This allows us to look at a more general version of the realisation problem which in turn has direct ramifications for the increased paraphrasing power of the generator; *(ii)* The generator can happen to convey more information than is originally specified in its semantic input. We have a principled way to account for such additions; *(iii)* We can make finer distinctions as to what counts as a valid rendition of the input semantics by building the corresponding semantics of the generated sentence and exploring how close this structure is to the original input semantics; *(iv)* We show how the semantics is systematically related to syntactic structures in a declarative framework. Alternative processing strategies using the same knowledge sources can therefore be envisaged.

Before giving a more detailed description of our proposals first we review some approaches to generation from semantic networks (Section 2). We proceed with some background about the grammatical framework we will employ—Tree Adjoining Grammars (Section 3) and after describing the knowledge sources available to the generator (Section 4) we present the generation algorithm (Section 5). This is followed by a step by step illustration of the generation of one sentence (Section 6). We then discuss further semantic aspects of the generation (Section 7) and the implementation (Section 8). We conclude with a discussion of some issues related to the proposed technique (Section 9). The actual generation algorithm is presented in the appendix.

2 Generation from Semantic Networks

The input for generation systems varies radically from system to system. Most generators expect their input to be cast in a tree-like notation which enables the actual systems to assume that nodes higher in the semantic structure are more prominent than lower nodes. The semantic representations used are variations of a predicate with its arguments. The predicate is realised as the main verb of the sentence and the arguments are realised as complements of the main verb—thus the control information is to a large extent encoded in the tree-like semantic structure. Unfortunately, such dominance relationships between nodes in the semantics often stem from language considerations and are not always preserved across languages. Moreover, if the semantic input comes from other applications, it is hard for these applications to determine the most prominent concepts because linguistic knowledge is crucial for this task. The tree-like semantics assumption leads to simplifications which reduce the paraphrasing power of the generator (especially in the context of multilingual generation). In contrast, the use of a non-hierarchical representation for the underlying semantics allows the input to contain as few language commitments as possible and makes it possible to address the generation strategy from an unbiased position. We have chosen conceptual graphs (CGs) to represent the input to our generator. This has the added advantage that the representation has well defined deductive mechanisms.

The use of semantic networks in generation is not new [15, 13]. It is surprising that although CGs were developed to express natural language semantics and in the seminal work [17] section 5.4 is entirely devoted to surface realisation there has been little work that has taken up this line of research.

Two main approaches have been employed for generation from semantic networks: *utterance path* and *incremental consumption*. An utterance path is the sequence of nodes and arcs that are traversed in the process of mapping a graph to a sentence. Generation is performed by finding a cyclic path in the graph which visits each node at least once. If a node is visited more than once, grammar rules determine when and how much of its content will be uttered [17]. Under the second approach, that of incremental consumption, generation is done by gradually relating (consuming) pieces of the input semantics to linguistic structure [4, 10]. Such covering of the semantic structure is more standard in generation research and is also the approach we have adopted. The borderline between the two paradigms is not clear-cut. Some researchers [16] are looking at finding an appropriate sequence of expansions of concepts and reductions of subparts of the semantic network until all concepts have realisations in the language. Others assume all concepts are expressible and try to substitute syntactic relations for conceptual relations [3].

Other work addressing surface realisation from CGs includes: generation from conceptual dependency graphs [19], generation using Lexical Conceptual Grammar [12], and generating from CGs using categorial grammar in the domain of technical documentation [18].

Our work improves on generation approaches from hierarchically structured input by looking at a more general version of the realisation problem. We also do not assume that the generator has to be either coherent or complete (see Section 9); and we show clearly how the semantic structure is declaratively related to linguistically motivated syntactic representation.

3 Background: Tree–Adjoining Grammars

Our generator uses a particular syntactic theory—Tree-Adjoining Grammar (TAG) which we briefly introduce because the generation strategy is influenced by the linguistic structures and the operations on them.

TAG postulates the existence of a finite set of *elementary trees* out of which bigger syntactic trees can be built. Elementary trees are divided into *initial trees* and *auxiliary trees*—initial trees represent minimal syntactic structures while auxiliary trees correspond to minimal recursive structures. Two operations are used to compose bigger syntactic structures in TAG: substitution and adjoining (Figure 1). Non-terminal nodes in the elementary trees can be annotated with

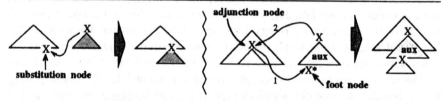

Fig. 1. Substitution and Adjoining in TAGs

information restricting the operations in which they can participate—some are marked for substitution, others for adjunction. Substitution simply attaches a substitution tree at the frontier non-terminal of another tree (left part of Figure 1). Adjoining takes an auxiliary tree (with a root non-terminal labeled X and a single non-terminal at the frontier labeled X*—known as the foot node) and another tree (either an initial tree or the result of a combination operation) with an internal node labeled **X** annotated to allow adjunction (adjunction node). The result of the adjunction operation is the attachment at the adjunction node in the initial tree of the auxiliary tree with the subtree dominated by the adjunction node of the original tree substituted at the foot node (right part of Figure 1). TAG is attractive for sentence generation because [7]:

1. On the syntactic level TAG trees factor out recursion of syntactic structures (in the auxiliary trees) and localise dependencies such as subcategorisation and filler-gap (in the initial trees). This makes it easy to describe syntactic phenomena.
2. TAG trees capture the function argument structure. This simplifies the mapping at the semantics-syntax interface level and allows for incremental processing.

4 Knowledge Sources

The generator assumes the existence of the following resources: a knowledge base, some mapping rules and a grammar module. The knowledge base is a more elaborate version (i.e. a specialisation) of the input CG. If the generator happens to introduce more semantic information by choosing a particular expression the knowledge base is the information where such additions can be checked for consistency (other generators have been given a large network of nodes some of which are marked to be expressed [16]). We return to how the knowledge base is actually used in Section 7. Mapping rules state how the semantics is related to the syntactic representation. We do not impose any intrinsic directionality on the mapping rules and view them as declarative statements. Although we will not use mapping rules in the direction of *syntax → semantics* we think of them as being free from control information so that mapping rules can be used with different generation strategies. In order to cut down on the number of mapping rules we impose the condition of minimality of the description of mapping rules—only those details from both semantic and syntactic structures that are relevant for the mapping are stated. That implies that the mapping rules will link partial semantic structures with partial syntactic structures. We also assume a grammar module that will have the syntactic (i.e. linguistic proper) knowledge of how to combine partial syntactic structures and how to transform a partial syntactic structure into a complete syntactic structure at the end of the processing.

4.1 Mapping rules

Mapping rules specify the *semantics ↔ syntax* relation. In our generator a mapping rule is represented as a partial syntactic tree where certain nodes in the tree are annotated with the semantics of the constituent the node represents. Partial syntactic trees are descriptions of trees (i.e. quasi-trees [20])—which apart from specifying that some nodes immediately dominate other nodes can contain

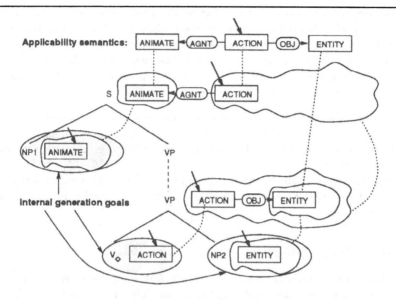

Fig. 2. A mapping rule for transitive constructions

information that a node just dominates another node. Graphically we will use a dashed line to indicate the (reflexive) dominance relation between nodes. A pair of nodes one of which dominates the other is called a quasi-pair (or quasi-node) and in the final tree it can correspond to a single node or to two separate nodes if there is additional material adjoined in between. The use of descriptions of trees (quasi-trees) allows us to view the adjoining operation as monotonic (preserving the information in the original tree) which would not be possible if trees were used instead. The nodes in the syntactic structure will be feature structures and we use unification to combine two syntactic nodes [8].

The semantic annotations of the syntactic nodes are either conceptual graphs or *instructions* indicating how to compute the semantics of the syntactic node from the semantics of the daughter syntactic nodes. Graphically we use dotted lines to show the coreference between graphs (or concepts). Each graph appearing in the rule has a single node ("the semantic head") which acts as a root (indicated by an arrow in Figure 2). This hierarchical structure is imposed by the rule, and is not part of the semantic input. Every mapping rule has associated applicability semantics which is used to license its application. The applicability semantics can be viewed as an evaluation of the semantic instruction associated with the top syntactic node in the tree description. This contains all the semantic material in the mapping rule and is different from the instruction itself. We want our mapping rules to be minimal in the sense of mentioning only those details in the semantic and syntactic structures which are relevant for the correspondence. So the syntactic and semantic structures have to be partial.

Figure 2 shows an example of a mapping rule. The applicability semantics of this mapping rule is: $\boxed{\text{ANIMATE}} \leftarrow (\text{AGNT}) - \boxed{\text{ACTION}} - (\text{OBJ}) \rightarrow \boxed{\text{ENTITY}}$. If this structure matches part of the input semantics (we explain more precisely what we mean by matching later on) then this rule can be triggered (if it is syntactically appropriate—see Section 5). The internal generation goals (shaded areas) express the following: (1) generate $\boxed{\text{ACTION}}$ as a verb and substitute (attach) the verb's syntactic structure at the $V\diamond$ node; (2) generate $\boxed{\text{ANIMATE}}$ as a noun phrase and substitute the newly built structure at $NP1$; and (3) generate $\boxed{\text{ENTITY}}$ as another noun phrase and substitute the newly built structure at $NP2$. The newly built structures are also mixed syntactic-semantic representations and they are incorporated in the mixed structure corresponding to the current status of the generated sentence. A mapping rule is a mixed representation similar in spirit to the *syntactic-semantic correspondence* in [21] where such mappings are used in the direction of understanding.

5   Sentence Generation

In this section we informally describe the generation algorithm. In Figure 3 and later in Figure 7, which illustrate some semantic aspects of the processing, we use a diagrammatic notation to describe the semantics which is actually encoded using conceptual graphs.

The input to the generator is an *Initial Graph* and a mixed structure, *Partial*, which contains a syntactic part (usually just one node but possibly something more complex) and a semantic part which takes the form of semantic annotations on the syntactic nodes in the syntactic part. Initially *Partial* represents the syntactic-semantic correspondences which are imposed on the generator. It has the format of a mixed structure like the representation used to express mapping rules (Figure 2). Later during the generation *Partial* is enriched and at any stage of processing it represents the current syntactic-semantic correspondences.

We have augmented the TAG formalism so that the semantic structures associated with syntactic nodes will be updated appropriately during the substitution and adjoining operations. This process is guided by the annotations for semantic heads (arrows). The stages of generation are: (1) building an initial skeletal structure; (2) attempting to consume as much as possible of the semantics uncovered in the previous stage; and (3) converting the partial syntactic structure into a complete syntactic tree.

5.1   Building a skeletal structure

Generation starts by first trying to find a mapping rule whose semantic structure matches[1] part of the initial graph and whose syntactic structure is compatible with the goal syntax (the syntactic part of *Partial*). If the initial goal has a more elaborate syntactic structure and requires parts of the semantics to be expressed as certain syntactic structures this has to be respected by the mapping rule.

---

[1] via the maximal join operation

Such an initial mapping rule will have a syntactic structure that will provide the skeleton syntax for the sentence. If Lexicalised TAGs [1] are used as the base syntactic formalism at this stage the mapping rule will introduce the head of the sentence structure—the main verb. If the rule has internal generation goals then these are explored recursively (possibly via an agenda—we will ignore here the issue of the order in which internal generation goals are executed). Because of the minimality of the mapping rule, the syntactic structure that is produced by this initial stage is very basic—for example only obligatory complements are considered. Any mapping rule can introduce additional semantics and such additions are checked against the underlying knowledge base. When applying a mapping rule the generator keeps track of how much of the initial semantic structure has been covered (i.e. consumed). So at the point when all internal generation goals of the first (skeletal) mapping rule have been exhausted the generator knows how much of the initial graph remains to be expressed.

5.2 Covering the remaining semantics

In the second stage the generator aims to find mapping rules (like the new mapping rule in Figure 3) in order to cover most of the remaining semantics. The choice of mapping rules is influenced by the following criteria:

Connectivity: The semantics of the mapping rule has to match (cover) part of the covered semantics and part of the remaining semantics.

Integration: It should be possible to incorporate the semantics of the mapping rule into the semantics of the current structure being built by the generator.

Realisability: It should be possible to incorporate the partial syntactic structure of the mapping rule into the current syntactic structure being built by the generator.

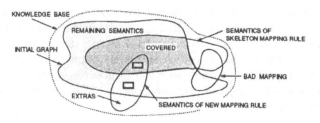

Fig. 3. Covering the remaining semantics with mapping rules

Note that the connectivity condition restricts the choice of mapping rules so that a rule that matches part of the remaining semantics and the extra semantics added by previous mapping rules cannot be chosen (e.g. the "bad mapping" in Figure 3). While in the stage of fleshing out the skeleton sentence structure (Section 5.1) most of the syntactic integration involves substitution, in the stage of covering the remaining semantics it is adjunction that is used. When incorporating semantic structures the semantic head has to be preserved—for example when adjoining the auxiliary tree for an adverbial construction the semantic head of the auxiliary construction has to be the same as the semantic head of the adjunction (quasi-) node. This explicit marking of the semantic head differs from [14] where the semantic head is a term with exactly the same structure as the input semantics.

5.3 Completing a derivation

In the preceding stages of building the skeletal sentence structure and covering the remaining semantics, the generator is mainly concerned with consuming the initial semantic structure. In those processes, parts of the semantics are mapped onto partial syntactic structures which are integrated and the result is still a partial syntactic structure. That is why a final step of "closing off" the derivation is needed. The generator tries to convert the partial syntactic structure into a complete syntactic tree. A morphological post-processor reads the leaves of the final syntactic tree and inflects the words.

Based on the above considerations we have created an algorithm for generating sentences from conceptual graphs. This algorithm is detailed more precisely in the appendix.

6 Example

In this section we illustrate how the algorithm works by means of a simple example. Suppose we start with an initial semantics as given in Figure 4.

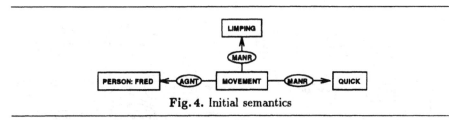

Fig. 4. Initial semantics

In the stage of building the skeletal structure the mapping rule *(i)* in Figure 5 is used. Its internal generation goals are to realise the instantiation of [ACTION] (which is [MOVEMENT]) as a verb and similarly [PERSON:FRED] as a noun phrase. The generation of the subject noun phrase is not discussed here. The main verb is generated using the terminal mapping rule[2] *(iii)* in Figure 5.[3] The skeletal structure thus generated is *Fred limp(ed)* (see *(i)* in Figure 6). An interesting point is that although the internal generation goal for the verb referred only

---

[2] Terminal mapping rules are mapping rules which have no internal generation goals and in which all terminal nodes of the syntactic structure are labelled with terminal symbols (lexemes).

[3] In Lexicalised TAGs the main verbs would be already present in the initial trees.

to the concept ⎡MOVEMENT⎤ in the initial semantics, all of the information suggested by the terminal mapping rule *(iii)* in Figure 5 is consumed. We will say more about how this is done in Section 7. At this stage the only concept that

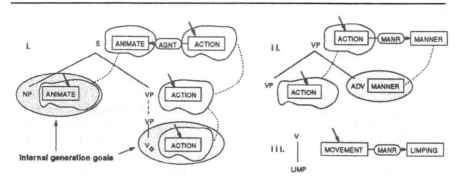

Fig. 5. Mapping rules

remains to be consumed is ⎡QUICK⎤. This is done in the stage of covering the remaining semantics when the mapping rule *(ii)* is used. This rule has an internal generation goal to generate the instantiation of ⎡MANNER⎤ as an adverb, which yields *quickly*. The structure suggested by this rule has to be integrated in the skeletal structure. On the syntactic side this is done using the adjunction operation. The final mixed syntactic-semantic structure is shown on the right in Figure 6. In the syntactic part of this structure we have no more quasi-nodes

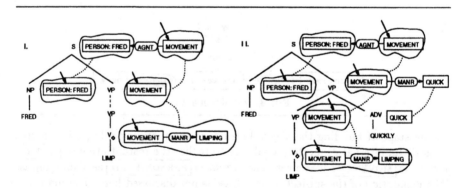

Fig. 6. Skeletal structure and final structure

where adjunction can occur. Also all of the input semantics has been consumed. The semantic annotations of the *S* and *VP* nodes are instructions about how the graphs/concepts of their daughters are to be combined. If we evaluate in a bottom up fashion the semantics of the *S* node, we will get the same result

as the input semantics in Figure 4. After morphological post-processing and lin-
earisation the result is *Fred limped quickly*. The same input semantics, of course,
can give rise to other sentences, for instance—*Fred hurried with a limp*.[4]

7 Matching the applicability semantics of mapping rules

Matching of the applicability semantics of mapping rules against other semantic
structures occurs in the following cases: when looking for a skeletal structure;
when exploring an internal generation goal; and when looking for mapping rules
in the phase of covering the remaining semantics. During the exploration of internal generation goals the applicability semantics of a mapping rule $(sem(R))$
is matched against the semantics of an internal generation goal $(GoalSem)$
(see Figure 9 in the appendix). We assume the following conditions hold:

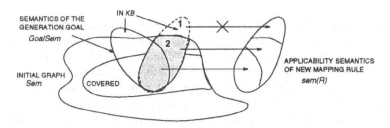

Fig. 7. Interactions involving the applicability semantics of a mapping rule

1. The applicability semantics of the mapping rule can be maximally joined
   with the goal semantics.
2. Any information introduced by the mapping rule that is more specialised
   than the goal semantics (additional concepts/relations, further type instantiation, etc.) must be in the knowledge base. If this additional information
   is within the input semantics, then information can propagate from the input semantics to the mapping rule (the shaded area 2 in Figure 7). If the
   mapping rule's semantic additions are merely in the knowledge base, then
   information cannot flow from the knowledge base to the mapping rule (area 1
   in Figure 7).

Similar conditions hold when in the phase of covering the remaining semantics the applicability semantics of a mapping rule is matched against the initial
semantics (see Figure 10 in the appendix). This way of matching allows the generator to convey only the information in the original semantics and what the
language forces one to convey even though more information might be known
about the particular situation.

---

[4] Our example is based on Iordanskaja *et al.*'s notion of maximal reductions of a
semantic net (see [6] p.300).

8 Implementation

Our implementation of the algorithm is in LIFE [2]. Conceptual graphs are represented as partially instantiated terms. Further instantiation of these terms allows their interpretation as conceptual graphs to simulate, for example, the addition of new concepts and relations to the graphs. The choice of this representation enables us to express operations such as *maximal_join* in a destructive way. That is, rather than viewing the *maximal_join* operation as producing a separate third graph, both graphs' term representations are further instantiated so that their interpretation after the operation is the same graph (while their representations can be different terms). This facilitates the propagation of the instantiation of the applicability semantics of a mapping rule to the internal generation goals. This is an implementational convenience.

9 Discussion

During generation it is necessary to find appropriate mapping rules. However, at each stage a number of rules might be applicable. Due to possible interactions between some rules the generator may have to explore different choices before actually being able to produce a sentence. Thus, generation is in essence a search problem. In order to guide the search a number of heuristics can be used. In [11] the number of matching nodes has been used to rate different matches, which is similar to finding maximal reductions in [6]. Alternatively a notion of semantic distance [5] might be employed.

We use instructions showing how the semantics of a mother syntactic node is computed because we want to be able to correctly update the semantics of nodes higher than the place where substitution or adjunction has taken place—i.e. we want to be able to propagate the substitution or adjunction semantics up the mixed structure whose backbone is the syntactic tree.

We also use a notion of headed conceptual graphs, i.e. graphs that have a certain node chosen as the semantic head. The initial semantics need not be marked for its semantic head. This allows the generator to choose an appropriate (for the natural language) perspective. The notion of semantic head and their connectivity is a way to introduce a hierarchical view on the semantic structure which is dependent on the language. When matching two conceptual graphs we require that their heads be the same. This reduces the search space and speeds up the generation process.

Our generator is not coherent or complete (i.e. producing sentences with more general/specific semantics than the input semantics). We try to generate sentences whose semantics is as close as possible to the input in the sense that they introduce little extra material and leave uncovered a small part of the input semantics. We keep track of more structures as the generation proceeds and are in a position to make finer distinctions than was done in previous research. The generator never produces sentences with semantics which is more specific than the contents of the knowledge base which gives some degree of coherence. Our generation technique provides flexibility to address cases where the entire input cannot be expressed in a single sentence by first generating a "best match" sentence and allowing the remaining semantics to be generated in a follow-up sentence.

In a way, the algorithm described above comes close to semantic head-driven generation [14]. It can also be seen as a version of syntax-driven generation [9]— our use of Lexicalised TAG means that the algorithm in effect looks first for a syntactic head.

The algorithm has to be checked against more linguistic data and we intend to do more work using different generation strategies using knowledge sources free from control information. A declarative definition of "derivation" in this framework would enable better comparisons with relevant generation work.

10 Conclusion

We have presented an algorithm for generation of sentences from conceptual graphs. The use of a non-hierarchical representation for the semantics increases the paraphrasing power of the generator (i.e. more sentences can be produced) and enables the production of sentences with radically different syntactic structure due to alternative ways of grouping concepts into words. In generation from hierarchically structured representations this can be done only by performing transformations on the input semantics because the input already contains certain language commitments. Thus, a non-hierarchical semantic representation is particularly useful for multilingual generation. The use of a syntactic theory (Tree-Adjoining Grammars) allows for the production of linguistically motivated syntactic structures which will pay off as better coverage of the language and overall maintainability of the generator. The syntactic theory also affects the processing—we have augmented the syntactic operations to account for the integration of the semantics. As a result of the way the semantics is 'consumed', our generator is not constrained to produce sentences with semantics either more specific or more general than the input semantics. We have deliberately aimed at a generation architecture which guarantees that the sentence's semantics covers as much as possible from the input semantics and leaves out as little as possible. The generation architecture makes explicit the decisions that have to be taken and allows for experiments with different generation strategies using the same declarative knowledge sources.

References

1. A. Abeillé, K. B. Cote, and Y. Schabes. A lexicalised tree adjoining grammar for english. Technical report, Dept. of CS, University of Pennsylvania, 1990. MS-CIS-90-24.
2. H. Aït-Kaci and A. Podelski. Towards a meaning of LIFE. *Journal of Logic Programming*, 16(3&4):195–234, July-August 1993.

3. F. Antonacci et al. Analysis and Generation of Italian Sentences. In T. Nagle, J. Nagle, L. Gerholz, and P. Eklund, editors, *Conceptual Structures: Current research and Practice*, pages 437–460. Ellis Horwood, 1992.

4. M. Boyer and G. Lapalme. Generating paraphrases from meaning-text semantic networks. *Computational Intelligence*, 1(1):103–117, 1985.

5. N. Foo et al. Semantic distance in conceptual graphs. In J. Nagle and T. Nagle, editors, *Fourth Annual Workshop on Conceptual Structures*, 1989.

6. L. Iordanskaja et al. Lexical Selection and Paraphrase in a Meaning-Text Generation Model. In C. Paris, W. Swartout, and W. Mann, editors, *Natural Language Generation in Artificial Intelligence and Computational Linguistics*, pages 293–312. Kluwer Academic, 1991.

7. A. Joshi. The Relevance of Tree Adjoining Grammar to Generation. In G. Kempen, editor, *Natural Language Generation*, pages 233–252. Kluwer Academic, 1987.

8. M. Kay. Unification grammar. Technical report, Xerox Palo Alto Research Center, Palo Alto, California, 1983.

9. E. König. Syntactic head-driven generation. In *COLING'94*, 475-481, Kyoto, 1994.

10. J.-F. Nogier. *Génération automatique de langage et graphs conceptuels*. Hermes, Paris, 1991.

11. J.-F. Nogier and M. Zock. Lexical Choice as Pattern Matching. In T. Nagle, J. Nagle, L. Gerholz, and P. Eklund, editors, *Conceptual Structures: Current research and Practice*, pages 413–436. Ellis Horwood, 1992.

12. J. Oh, S. Graham, W.-J. Hsin, G.-C. Yang, Y. B. Choi, K.-S. Choi, and S.-H. Myaeng. NLP: Natural Language Parsers and Generators. In G. Ellis and R. Levinson, editors, *Proc. of 1st Int. Workshop on PEIRCE: A Conceptual Graph Workbench*, pages 48–55, 1992.

13. S. Shapiro. Generalized augmented transition network grammars for generation from semantic networks. *American J. of Computational Linguistics*, 2(8):12–25, 1982.

14. S. Shieber et al. A semantic head-driven generation algorithm for unification-based formalisms. *Computational Linguistics*, 16(1):30–42, 1990.

15. R. Simmons and J. Slocum. Generating English Discourse from Semantic Networks. *CACM*, 15(10):891–905, 1972.

16. M. Smith, R. Garigliano, and R. Morgan. Generation in the LOLITA system: an engineering approach. In *Proc. of 7th Int. Workshop on Natural Language Generation*, pages 241–244, 1994.

17. J. Sowa. *Conceptual Structures: Information Processing in Mind and Machine*. Addison-Wesley, 1984.

18. S. Svenberg. Representing Conceptual and Linguistic Knowledge for Multilingual Generation in a Technical Domain. In *Proc. of 7th Int. Workshop on Natural Language Generation*, pages 245–248, 1994.

19. A. van Rijn. *Natural Language Communication between Man and Machine*. PhD thesis, Technical University Delft, 1991.

20. K. Vijay-Shanker. Using Descriptions of Trees in a Tree Adjoining Grammar. *Computational Linguistics*, 18(4):481–517, 1992.

21. M. Willems. Pragmatic semantics by conceptual graphs. In W. Tepfenhart, J. Dick, and J. Sowa, editors, *Proc. 2nd Int. Conference on Conceptual Structures*, pages 31–44. Springer-Verlag, 1994.

Appendix: Generation Algorithm

Sem is the initial semantics. It remains unchanged throughout the generation process. *KB* ($KB \leq Sem$) is the knowledge base. Whenever the generator needs to say more information than is encoded in *Sem* then this is checked against the underlying knowledge base.

Partial is a tree description (quasi-tree [20]) whose nodes are annotated with corresponding semantics. *Partial* can be updated—at any stage it contains the (mixed syntactic-semantic) structure generated so far. *Sem*, *KB* and *Partial* are inputs to the algorithm.

Cover (a subgraph of *Sem*) represents how much of *Sem* has already been covered.

BuiltSem reflects the semantic structure of the string being produced (which is not necessarily identical with *Sem* or *Cover*).

All the above data structures: *Sem*, *KB*, *Cover*, *BuiltSem* and *Partial* are global:

type: *Sem, KB, Cover, BuiltSem : cg_graph*

      Partial : semantically annotated quasi_tree

$syn(MS)$ and $sem(MS)$ are selectors—they give the syntactic/semantic structure of a mixed structure MS.

---

generate() : syn_tree   % inputs are *Sem*, *KB* and *Partial*

begin

    Cover := empty_graph

    BuiltSem := empty_graph

    generate2(Sem, Partial)

    cover_remaining()

    close_derivation()

    return(*syn(Partial)*)

end

Fig. 8. Generation algorithm—top level

Figure 9 describes the stage of finding an initial skeletal mapping and the stage of recursively exploring its embedded generation goals. A mapping rule R is simply an elementary quasi-tree whose nodes are annotated with appropriate semantics—a mixed structure. $syn(R)$ of a mapping rule is only the syntax part of the mapping rule. $sem(R)$ is the semantics which acts as applicability condition for the mapping rule. $sem(R)$ can be viewed as evaluated semantics for the top syntactic node of the mapping rule.

The function $merge(CG1, CG2)$ produces a join of $CG1$ and $CG2$ on a projection consisting of the concepts and relations that are the same in both graphs.

IntStr in the internal generation goals is a usually a syntactic node (thus the generation goal will have to generate the *IntSem* as a syntactic sub-tree conforming to this syntactic node). In general, however, *IntStr* can be more complex—it is a semantically annotated quasi-tree (mixed structure).

During the matching of graphs indices of concepts from $sem(R)$ become the same as indices of concepts in *Sem* (or *KB*). This allows for the correct integration of the mapping rule's mixed structure in *Partial*—we know which concepts in *Partial* correspond to which concepts in the original semantics and can make sure that the integration of additional structures into *Partial* leads to the top syntactic node having the correct associated semantics.

---

generate2(GoalSem, GoalStr)
type: *GoalSem* : *cg_graph*
      GoalStr : *semantically annotated quasi_tree*
begin
    Find a mapping rule *R* s.t. :
        syn(R) is compatible with *syn(GoalStr)* **and**
        sem(R) matches *GoalSem*   % within *Sem* and *KB*
      BuiltSem := *merge(BuiltSem, sem(R))*   % join on the same concepts
      Cover := *merge(Cover, max_projection(Sem, sem(R)))* % *Sem* ≤ *Cover*
      Partial := integration of the mixed structure of *R* into *Partial*
      for each internal generation goal ⟨*IntSem, IntStr*⟩ ∈ mapping rule *R* **do**
          call *generate2(IntSem, IntStr)*
end

Fig. 9. Processing internal generation goals

---

cover_remaining()
begin
    while *Sem* ≤ *Cover* (*Sem* is more specific than *Cover*, i.e. there is remaining semantics) **and**
      there is a mapping rule *R* s.t. :
        sem(R) matches at least one concept in *Cover* **and**
        sem(R) matches at least one concept or relation in *Sem* but not in *Cover*
    do apply mapping rule *R*   % Integrate *R* into *Partial* and update *Cover* and *BuiltSem*
end

Fig. 10. Covering the remaining semantics

Applying a mapping rule in Figure 10 means finding a syntactic node in *Partial* such that: (i) syntactically we can adjoin the auxiliary tree suggested by the mapping rule at the found node; and (ii) semantically the mapping rule has the same head as the semantics associated with the found adjunction node. *Cover* and *BuiltSem* will need to be updated.

---

close_derivation()
begin
    For each quasi-node in *syn(Partial)*
      try to unify the quasi-root and quasi-foot
      If the quasi-root and quasi-foot do not unify
      then find an appropriate auxiliary tree and adjoin it at this quasi-node
end

Fig. 11. Closing a derivation

Discourse Spaces:
a Pragmatic Interpretation of Contexts

Bernard Moulin, Professor[1]
Laval university, Computer Science Department
Pavillon Pouliot, Ste Foy, QC G1K 7P4 Canada
Ph: (418) 656 5580, fax: (418) 656 2324, Email: moulin@ift.ulaval.ca

Abstract

In human interactions the context in which verbal and non-verbal acts are performed has a prime importance. In this paper we propose an approach emphasizing the cognitive dimension of contexts. Our basic assumption is that a human agent naturally uses contextual information because she cannot manipulate or communicate knowledge without positioning herself relatively to that knowledge. We introduce the notion of *discourse space (DS)* a pragmatic form of context which is used to structure knowledge contained in a discourse. Discourse spaces are created, updated or evoked by an agent who tries to generate or understand a discourse. We extend the basic conceptual graph framework with notions like temporal objects, temporal localizations and different kinds of discourse spaces: situations, definitional and focal DSs. Then, we present agent-related DSs (narrator's and agent's perspectives, agent's attitudes) and inference-related discourse spaces (conditional, alternative, generalized, hypothetical and counterfactual DSs). We also show how these DSs are interleaved in the representation of a discourse.

1. Introduction

In human interactions the context in which verbal and non-verbal acts are performed is of particular importance. The same sentence can take different meanings in different contexts. For instance, the sentence "I love you" is interpreted differently depending on who says it and in which circumstances: John can say that he loves Mary and an actor can say that sentence to an actress in a movie, etc. When communicating some information, a person usually presupposes that her interlocutor is aware of the context to which that information must be related. If this contextual relationship is not obvious or readily found, the addressee agent asks the originator agent to indicate explicitly which is the relevant context for that information. The notion of context is fundamentally pragmatic [21] since it involves the relationships established between locutors and the speech acts [32] they perform. The term *context* has been used to refer to a variety of different notions in cognitive sciences and in artificial intelligence. In AI contexts have been introduced as means of partitioning knowledge into manageable sets [15], or they have been considered as logical constructs that should facilitate reasoning activities [14, 33]. Both views are legitimate, but they fail to grasp the cognitive dimension of contexts. Our basic assumption is that a human agent naturally uses contextual information because she cannot manipulate or communicate knowledge without positioning herself relatively to that knowledge.

In linguistics and cognitive psychology, the notion of context has been recognized as an important one when it comes to language comprehension and generation.

---

[1] This research is supported by the Natural Sciences and Engineering Research Council of Canada (grant OGP 05518) and FCAR. Many thanks to Claire Girard who has provided thoughtful suggestions that helped me clarify this work on discourse spaces.

Langacker [20] claimed that all linguistic units are context-dependent.

Fauconnier [12] suggested that understanding a discourse "leads to the study of domains that we set up as we talk or listen, and that we structure with elements, roles, strategies and relations. These domains, or *mental spaces*, are used by locutors to evoke and accumulate information during the processing of discourses, primarily in response to the occurrence of linguistic elements called *space builders* such as "Mary thinks that ----", "In 1881, ----", and so forth. For instance, processing the sentence "John thinks that Mary will go to Jamaica" involves either creating or locating a belief space for John and representing in it the information that Mary will go to Jamaica. Using such an approach, Fauconnier provided a "simple and uniform account of a wide variety of problems: referential opacity and transparency, specificity of reference, definite reference in discourse, the interaction of quantifiers with modalities, the projection problem for presuppositions, the semantic processing of counterfactuals and the use of comparatives in modal contexts" [8].

A precursory idea of context can be traced back to Peirce's *existential graphs* [31]. Existential graphs use a logical form of context called a *cut* which shows in a topological way the scope of a negative context (logical negation) on a sheet of paper (the *sheet of assertion*). Sowa [34] introduced *conceptual graphs (CGs)* as an extension of existential graphs with elements borrowed from linguistics and artificial intelligence. Sowa's first formal definition of a context was presented as a second-order concept, called a PROPOSITION, whose referent contains one or several conceptual graphs. The equivalence between CGs and Peirce's existential graphs was emphasized. Influenced by Barwise and Perry's work [2] Sowa provided another interpretation of a context as a SITUATION which is described by a PROPOSITION which itself is stated by a GRAPH[2]. Recently, Sowa [35] informally introduced another interpretation of contexts. Using the example of a concept [BIRTHDAY-PARTY], he says: "The concept box with the label BIRTHDAY-PARTY says that there exists a birthday party, but it does not specify any details of what happened... To see the details of the party, it is necessary to open up the box and look inside". Here, the so-called notion of context is a kind of "focalization device" that has no apparent common ground with the preceding interpretations of PROPOSITION and SITUATION, or with EGs. Esch [10,11] proposed to consider a context as a "white box" view of the concept whose detailed description is provided by the context. Esch defined the χ operation that is used to expand a concept type or individual, given its definition.

Since the early nineties [24], we have been advocating a pragmatic approach for the representation of temporal knowledge in discourses, emphasizing the importance of explicitly introducing pragmatic constructs in our representations [25, 26]. On the basis of cognitive assumptions inspired by cognitive psychologists and linguists [17, 20] we extend that approach to encompass a variety of different context types interpreted as mental spaces [12, 9]. We suppose that an agent creates and manipulates a mental model representing elements contained in the real world or in imaginary worlds to which she can mentally access. Such a mental model provides the cognitive basis on which an agent draws the information that will be included in her discourse. Similarly, when understanding a discourse, an agent evokes, creates or modifies a mental model structuring the elements that she has extracted from the discourse. Because of the pre-eminence of spatial and temporal information in human

---

[2] Dick [7] reviews how the notion of context evolved in Sowa's notation.

cognitive experience, we assume that an agent's mental model is structured in terms of mental spaces. A mental space is a topological entity containing elements related together by relations that are compatible with the space topological properties. The structure of a discourse reflects the cognitive operations[3] that an agent applies when creating and manipulating her mental spaces and the elements contained in them. Locutors use those cognitive operations to structure the contents of their discourses.

We use the term *mental space* to denote the constructs that structure the knowledge in an agent's mental model. In this paper we will not analyze mental spaces but their counterparts, called *discourse spaces* (DS), that are used to structure the content of a discourse. Mental spaces are manipulated by an agent's reasoning mechanisms while discourse spaces are generated by an agent in order to structure the information that she will transmit when performing speech acts. Discourse spaces are also created, updated or evoked by an agent who tries to understand a discourse. In section 2 we introduce the main constituents of our model: concepts and conceptual relations, temporal objects and localizations, basic discourse spaces (situations, definitional and focal discourse spaces). In section 3, we present agent-related DSs (narrator's and agent's perspectives, agent's attitudes). In section 4 we discuss inference-related discourse spaces: conditional, alternative, generalized, hypothetical and counterfactual DSs.

2. A Discourse Representation Model

The model that we propose to represent the content of a discourse integrates the notion of discourse space within a knowledge representation framework based on conceptual graphs [34]. Since the notion of space is a topological one, we base our presentation on an analogy with biology. We consider the following analog notions: concepts viewed as atoms; conceptual relations as atomic links associating atoms; conceptual graphs as molecules and spaces as organelles in a cell.

A *space* is a topological entity composed of an envelope and an interior. The envelope is a region delimiting the interior of the space: it is the "interface" between the interior and the exterior of the space. A space's envelope and interior are regions containing other spaces and/or elements (concepts and/or conceptual graphs) used by an agent to describe the real or imaginary worlds in which she is involved. Graphically we represent a space as a double rectangle. In the upper part we specify the space's description and the elements contained in its envelope. In the lower part we specify the elements contained in the space's interior:

Space-desc	envelop
	interior

} A space

---

[3] Here is a list of the main cognitive operations that locutors use when building or understanding a discourse: describe spatio-temporal situations and the elements they contain; identify spatial and temporal localizations; position oneself and spatio-temporal situations with respect to spatial and temporal localizations; describe the setting in which an agent experiences attitudes and/or performs speech acts; evoke hypothetical (or counterfactual) situations and their expected consequences; evoke conditional or generalized situations and their expected consequences; use a concept to evoke a set of concepts or of situations based on locutor's knowledge of the concept interpretation; adopt a certain point of view to report a given situation.

2.1 Fundamental Constituents

Concepts can represent any concrete or abstract object or any agent contained in any agent's mental or discourse space. *Conceptual relations* establish links between concepts. Concepts and conceptual relations are represented according to Sowa's conventions [34]. *Conceptual graphs* (CG) are non-temporal structures composed of concepts and conceptual relations. They are the molecular components in our model. According to Barwise and Perry [2] "Reality consists of situations - individuals having properties and standing in relation at various spatio-temporal locations". In our approach a situation is viewed as an elementary space that contains one or several knowledge molecules in the form of conceptual graphs.

Agents create discourses in which they describe situations, express mental states or judgments. In contrast with objects, agents are intentional and can exercise a certain control on objects and/or situations. An agent can play several *roles* (such as "king of France" or "president") according to the situations in which she is involved. A role has the same cognitive and representational properties as an agent. Agents (and roles) can appear as concepts in situations. The architecture of a discourse is a set of related spaces, called *discourse spaces* structuring the discourse content. The discourse architecture can be thought of as a "cognitive map" built on the basis of the concepts transferred by locutors when they perform their speech acts. When analysing the surface form of a discourse, we find various syntactic and lexical constructs such as cue phrases, adverbial locutions and verb tenses emphasizing the presence of discourse spaces. Formally, a *discourse space* is a triple < DSdes, DSenv, DSint >

where DSdes is the discourse space description, which is composed of the discourse space type and of an identifier; DSenv and DSint are the envelope and the interior of the discourse space respectively. A discourse space's envelope and interior can contain situations and/or other discourse spaces.

The analysis of the cognitive operations that human agents use when producing or understanding a discourse leads to the identification of different kinds of discourse spaces (DS) which are discussed in the following sections.

2.2 Situations, Temporal and Focal Spaces

An elementary situation is associated with a conceptual graph, a time interval and a spatial region. Time plays a particular role in natural language since it is used to relate situations together in a discourse. Situations are usually temporally located relatively to an *absolute time coordinate system* (given by dates, hours, etc.) or to an agent (using verb tenses) or to other situations (using temporal conjunctions). Temporal locations are specified using time intervals located on an absolute time axis. We defined relations between time intervals [24] which are extensions of Allen's relations [1]. An elementary situation can be viewed as a *temporal discourse space* whose envelope is characterized by a time interval and whose interior contains the situation description in the form of a conceptual graph.

More precisely, an *elementary situation* is a triple <SD, SEN, SPC> where:

- The situation description SD is a couple [situation-type, situation-descriptor] used to identify the temporal situation. The situation type is used to semantically distinguish different kinds of temporal situations : events, states, processes etc. The situation descriptor is used for referential purposes.

- The situation's envelope SEN is composed of several parameters <STI, TV, DOP>. The time interval STI is a mandatory parameter that provides the temporal information associated with the situation (begin- and end- times, duration, duration scale, etc.). The parameter TV provides the truth value associated with the situation (if TV=+1, the situation is described by its propositional content; if TV=-1, the situation is described by the negation of its propositional content). The deontic operator parameter DOP is optional and is used to specify deontic modalities (obligation, permission, interdiction) applying on the situation, if any.
- The situation's propositional content SPC is the interior of the corresponding temporal space and contains a non-temporal knowledge structure described by a conceptual graph.

Graphically, we represent a temporal situation using a rectangle decomposed into two parts. The rectangle represents the corresponding time interval and symbolizes the temporal space's envelope. In the upper part of the rectangle we indicate the situation description SD, as well as the relevant parameters of the situation's envelope. In the lower part of the rectangle we represent the situation's propositional content SPC. For instance, the sentence "John bought a blue car in Montreal, on February 12 1994" is represented by EVENT: ev1 in figure 1.

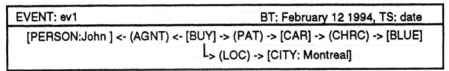

Figure 1: A situation describing the purchase of a car

Because their envelopes are characterized by time intervals, two situations can be related together by different kinds of relations: temporal relations relating their time intervals; rhetorical relations [22] such as CAUSE, RESULT, JUSTIF indicating that one situation is related to another situation by a causality or justification link. The situation type reflects the way an agent perceives the situation. For quite a long time, linguists have studied the *aspectual properties of verbs* [5, 3] in order to explain the difference between sentences like "John opened the door" and "John was opening the door, when ..." Different ways of categorizing temporal situations have been proposed such as Vendler's four categories [37]: events, states, achievements and accomplishments. The situation type is an important parameter of a temporal DS because it is needed to determine verb tenses [27].

Usually a temporal situation is considered as a whole. However, the semantic interpretation of certain situations can require to detail the content of the corresponding space's interior. That is the case of causative and inchoactive verbs. For example, a semantic interpretation of the sentence "Peter closes the door tight" is "Peter does something that causes the door to move in someway that the resulting state is the door being closed tight". Such situations are modelled using a focal DS.

A *Focal DS* is a discourse space whose envelope contains a situation S and whose interior contains a set of situations describing S in more details, those situations are usually related by temporal or rhetorical relations. In figure 2 we have an example of a focal DS. "Peter closes the door tight" is the situation that specifies the envelope of the focal DS and corresponds to the initial sentence. "Peter acted" (event ev1.1), "the door moved" (event ev1.2) and "the door is tightly closed" (state st1.3) are the situations appearing within the focal DS's interior. They are related by the causality relations CAUSE and RESULT.

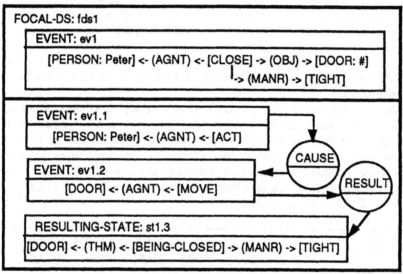

Figure 2 : "Peter closed the door tight"

Usually in a discourse, there is no explicit information showing that a locutor is using a focal DS when uttering a sentence. The need to use the focal DS associated with a concept appears whenever a locutor refers to a situation contained within the focal DS's interior. That is the case in our example. The adverb "tight" does not apply to the verb "close" but to the resulting state st1.3 appearing in the interior of focal DS fds1. In fact, situation ev1 appearing in the envelope of a focal DS is close to the sentence surface form in the discourse. It should be considered as a convenient approximation. This approximation may be reasonable for certain analyses of the sentence. The exact semantic interpretation appears in the focal DS's interior. Kocura [18] presented a detailed study of inchoactive and causative situations using conceptual graphs. A complementary study was done by [38]. These results can be applied in our framework to refine the descriptions of situations contained in focal DS's interiors. The detailed description of causative and inchoactive situations can lead to pretty complex graphs as some examples in [18] suggest. Using focal DSs provides a means to structure those descriptions. We assume that focal DSs are recorded in a lexicon that an agent can use when producing or interpreting a discourse.

2.3 Temporal Objects, Definitional DS and Temporal Localizations

By their nature, some entities are inherently associated with some sort of space. Material objects (i.e. an apple, a book, a car, etc.), animals (a tiger, a fly) and agents (John, Mary, etc.) are contained in spatial regions. Temporal objects (such as a specific day, year or minute) are represented by concepts associated with spaces having temporal topological properties. A situation is a spatio-temporal object that can be specified either by a concept (i.e. [PURCHASE: #]) or by one or several conceptual graphs describing it (see figure 1). This is analog to the duality observed by Esch [11] between concepts and contexts. Let us consider the case of a temporal object. It is a concept characterized by a time interval defining a temporal DS. "Day", "week", "month", "year" etc. are typical examples of temporal objects.

A *temporal object* is characterized by a triple <OD, OTI, OPC> where:
   - The object description OD is a couple <object-descriptor, object-definition>; the object-descriptor is used for reference purposes; the object-definition corresponds to the concept that represents the temporal object; it is specified using the linear form of Sowa's conceptual graphs.
   - The object time interval OTI aggregates the temporal information associated with the object.
   - The object propositional content OPC contains the description of other temporal objects or situations that semantically characterize the object.

The temporal object can also be viewed as a temporal DS whose envelope is characterized by the object definition and its time interval OTI. The temporal object propositional content corresponds to situations or objects contained in the temporal space's interior. Graphically, we represent a temporal object with a rectangle decomposed into two parts. The rectangle represents the corresponding time interval. In the upper part of the rectangle we indicate the object description OD, as well as the relevant parameters of the object time interval OTI. In the lower part of the rectangle we represent the object's propositional content in the form of temporal objects or situations related to the embedding rectangle by a relation ("part of" or a "temporal relation"). For example "the end of a journey of ten days" is represented in figure 3. Note that no time interval parameter is known for the temporal object "end", while duration parameters are known for the object "journey"[4].

Figure 3b displays another example where the temporal object [YEAR] is characterized by a situation [EVENT: ev2].

Some entities (usually corresponding to objects) can be denoted by n concepts linked by conceptual relations. However, those n concepts must not be considered independently, especially in the case of quantification. Consider for example "Tremblay is a wise chairman": "wise chairman" must be considered as a whole. A classical way of dealing with this problem is to introduce lambda-abstractions [34]. For example, "Tremblay is a wise chairman" is represented by the following conceptual graph (a definition of the concept WISE-CHAIRMAN):

   [PERSON: Tremblay *t]   [[CHAIRMAN: λ] -> (CHRC) -> [WISE]: *t]

where *t is a coreference variable used by Sowa to express the copula "be", and λ is the mark of the concept representing the formal parameter in the lambda abstraction.

It is convenient to use lambda abstractions in simple cases but this notation has some limits. Specifically, Sowa's notation makes it difficult to express (temporal) relations linking the conceptual graph contained in the lambda abstraction with conceptual graphs situated outside it, as for example when using temporal relations [26].

We introduce the notion of a *definitional discourse space* which is used to provide the definition of a concept when it is appropriate. A definitional DS is a

---

[4]   A time interval is characterized by a list of parameters: begin-time BT and end-time ET (lower and upper bounds of the time interval on TR); the time scale TS (unit used to measure the begin- and end- times on TR); the time interval duration DU and the duration scale DS.

discourse space whose envelope contains the concept to be defined, and in the definitional DS's interior we find the conceptual graph which defines this concept.

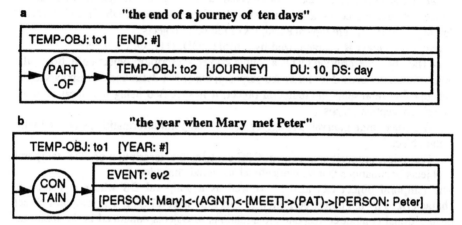

Figure 3: Two examples of temporal objects

Figure 4a illustrates the representation of "Tremblay is a wise chairman"[5]. Notice that the symbol λ marks the concept [CHAIRMAN] considered as an entry point in the conceptual graph [16]: this means that conceptual relations pointing towards the conceptual graph or exiting from it should be attached to the concepts marked by λ. The parameter "implicit" associated with the concept [WISE-CHAIRMAN] indicates that it is not expressed in the text. This notation is equivalent to Sowa's lambda abstraction, but it provides a richer expressive power.

Let us consider the case of apposition in sentences such as in the following example: "Last year, two friends, John and his coach Fred, passed away after John's car crashed on Road R375". It is represented in figure 4b. The concept [FRIEND: $\{\lambda1, \lambda2\}$ @2] in the definitional DS dds2 corresponds to "two friends" in the sentence, and the conceptual graph contained within the interior of dds2 corresponds to the apposition "John and his coach Fred". Note that both [PERSON: John $\lambda1$] and [COACH: Fred $\lambda2$] are marked by a λ mark. Definitional DSs extend the notion of "white box context" proposed by Esch [10, 11].

A temporal localization is a discourse space which sets a secondary time coordinate system. It can be thought of as a temporal capsule within which temporal situations or DSs can be localized. A *temporal localization* is a triple <TLD,TLENV, TLINT> where TLD identifies the temporal localization description. TLENV and TLINT are called the temporal localization's envelope and interior respectively. The *temporal localization description* TLD is a couple [localization-descriptor, localization-description]. The localization-descriptor is used for reference purposes. The localization-description is specified by a conceptual graph that usually describes a temporal object.

---

[5] We represent the copula BE by a concept instead of a coreference link: This follows Desclès' suggestion to consider BE as an archi-operator [6]. This representation is needed in order to distinguish different tenses such as "Tremblay will be a wise chairman" and "Tremblay was a wise chairman".

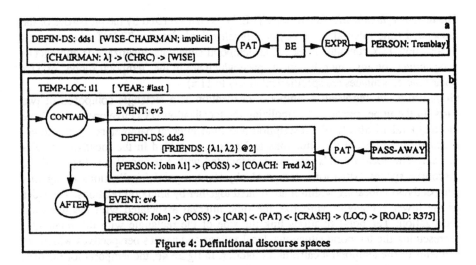

Figure 4: Definitional discourse spaces

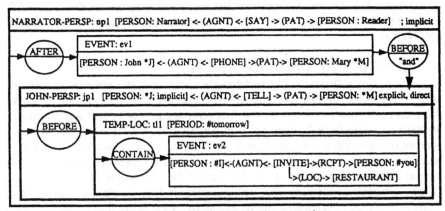

Figure 5 : Narrator's and agent's perspectives

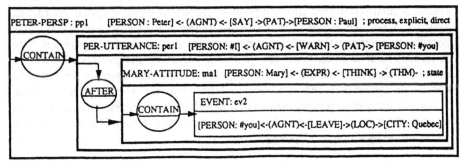

Figure 6: Peter says to Paul: "I warn you that Mary thinks that you left Quebec-city"

The *temporal localization's envelope* TLENV is characterized by the properties of the associated time interval. The *temporal localization's interior* TLINT of a temporal localization TLx contains the set of all temporal situations, localizations and agent's perspectives, or any other discourse space which are in the "temporal scope" of TLx (their time coordinates are specified relatively to TLx).

Graphically, we represent a temporal localization as a rectangle composed of two parts. In the upper rectangle symbolizing the localization DS's envelope, we indicate the temporal localization's description and time interval parameters. In the lower rectangle we indicate the situations and spaces contained in the localization DS's interior. The localization's interior is associated with a temporal topology and the discourse spaces (situations, temporal objects, temporal localizations or agent's perspectives) appearing within it are related together by temporal relations. The sides of the lower rectangle symbolize the time interval associated with the temporal localization. Temporal relations link the side of the rectangle with the rectangles representing the temporal situations, localizations and agent's perspectives which are elements of the temporal localization's interior. In figure 4b, the temporal localization tl1 corresponds to the sentence portion "last year". The temporal relation CONTAIN relates the time intervals of the temporal localization tl1 and of the situation EVENT:ev3.

3. Agent-Related Discourse Spaces

When producing a discourse, an agent considers her own spatio-temporal localization as the main coordinate system ("here and now"), which sets a reference for localizing other temporal situations, objects as well as other agents. Each agent acts physically, mentally or illocutionarily from her own perspective. In a discourse the narrator's perspective sets the main time coordinate system. When reporting the words of another agent, the discourse perspective changes from the narrator's to the agent's. In natural language it is important to characterize the time when an agent uttered a sentence, especially for verb tense determination. In [24] we claimed that much of temporal linguistic phenomena can be explained by explicitly introducing temporal coordinate systems in the semantic representation of discourses and by specifying the temporal relations which link temporal situations to these temporal coordinate systems. When we analyze discourses, we notice that locutors use several time coordinate systems: the *official time coordinate system* which provides time references such as dates, hours and years; the *narrator's perspective* which is the narrator's time coordinate system; the *agent's perspective* which is the temporal coordinate system attached to a person whose words are reported in a discourse. Narrator's or agents' perspectives can be thought of as discourse spaces explicitly situating the narrator or agent within the discourse cognitive map. These DSs are also characterized by a temporal topology.

A *narrator's perspective* is a triple <NPD, NPENV, NPINT> where NPD identifies the perspective description. NPENV and NPINT are called the perspective's envelope and interior respectively. The *narrator's perspective description* NPD is a couple [perspective-descriptor, perspective-description]. The perspective-descriptor is used for reference purposes. The perspective-description is specified by a conceptual graph that describes the narrator's action when generating the discourse:

[PERSON: narrator]<- (AGNT) <- [SAY] -> (PAT) ->[PERSON: reader].

Usually, these parameters are not explicitly stated in a discourse.

The *envelope* NPENV of a narrator's perspective specifies the characteristics of the time interval associated with the time coordinate system origin corresponding to the narrator's perspective (the narrator's "now").

The *interior* NPINT of a narrator's perspective NPx contains the set of all temporal situations, temporal localizations and perspectives, or any other DS which is in the "temporal scope" of NPx. The narrator's perspective NPx sets the time coordinate system origin. The time coordinates of the temporal DSs contained within the temporal scope of NPx are evaluated according to NPx.

Graphically, we represent a narrator's perspective as a rectangle composed of two parts. In the upper rectangle symbolizing the perspective DS's envelope, we indicate the narrator's perspective description and time interval parameters. In the lower rectangle we indicate the elements contained in the perspective DS's interior. The lower rectangle corresponds to the temporal scope of the perspective. The sides of the lower rectangle symbolize the time interval associated with the time coordinate system origin. Temporal relations link the side of the rectangle with the rectangles representing the temporal situations, localizations and perspectives which are contained in the perspective's interior.

The sentence *John phoned Mary and told her: "Tomorrow I invite you to a restaurant"* is represented in figure 5 where the outer rectangle corresponds to the narrator's perspective np1. In the interior of np1, we have a situation (event: ev1) and an agent's perspective (JOHN-PERSP: jp1). Considering the temporal relation AFTER holding between the time intervals of np1 and ev1, and the type of situation ev1 (event), we can determine that the verb tense is preterite: "phoned" [27].

An *agent's perspective* has the same characteristics as a narrator's perspective. It is specified by a triple <APD, APENV, APINT> where APD identifies the perspective description, APENV and APINT are the envelope and interior of the agent's perspective respectively. The graphical conventions used to represent the agent's perspective are the same as those used for a narrator's perspective. The only difference lies in the perspective header which is characterized by the agent's name, as in the perspective JOHN-PERSP in figure 5. The interior of an agent's perspective may contain other discourse spaces, temporal situations, localizations as well as other agent's perspectives. In figure 5, JOHN-PERSP's envelope contains the perspective description as a conceptual graph: [PERSON: *J] <- (AGNT) <- [TELL] -> (PAT) -> [PERSON: *M]. This corresponds to the sentence portion: "he told her"[6]. In the interior of perspective jp1 there is a temporal localization tl1 (corresponding to "tomorrow" in the text) which contains an event ev2[7].

---

[6] *Anaphoric references* (like he, she, it, them) are expressed according to Sowa's conventions [34]. *Indexicals* referring to the narrator and to her locutor (first and second persons of singular or plural in a text) must be resolved with respect to agent's perspectives [24, 25, 26]. See also [28].

[7] Note that the temporal relation linking the time intervals of jp1 and tl1 is BEFORE, and the temporal relation holding between the time intervals of tl1 and ev2 is CONTAIN. Hence, the verb tense corresponding to INVITE is the future ("I will invite you to a restaurant"), relatively to the time of John's utterance specified by John's perspective jp1.

The agent's and narrator's perspectives provide means to represent within a temporal model of discourse the notions of "point of view" [19, 29] and of mental spaces [12, 9]. Agent's perspectives also provide a formal specification of the notion of context of utterance in speech act theory [32].

Attitude situations are situations which express how agents relate to other situations through perception or cognition. Typical verbs of attitude are: to see, to hear, to believe, to want, to desire, etc. Attitude situations are obviously associated with discourse spaces. Some of them were mentionned as typical mental spaces in Fauconnier's theory [12]. We specify them in a similar way as agent's perspectives. The only difference is that an attitude description specifies an agent's attitude instead of describing the communicative act that an agent performs in an agent's perspective. An *agent's attitude* is a triple <AAD, AAENV, AAINT> where AAD identifies the agent attitude description, AAENV and AAINT are the envelope and interior of the corresponding DS. The properties of AAENV and AAINT are similar to those of an agent's perspective.

Graphically, an agent's attitude is represented in a similar way as an agent's perspective. Instead of using the structure type AGENT-PERSPECTIVE we use the type AGENT-ATTITUDE. An example of an attitude situation MARY-ATTITUDE: ma1 is given in figure 6: "Mary thinks that you left Quebec city".

Some verbs such as "promise", "assure", "warn", "supplicate", can be used performatively in performative sentences: they are called *performative verbs*. They name possible *illocutionary forces* of utterances [32]. A performative situation is associated with a discourse space (called a *performative utterance*) whose envelope contains the situation expliciting the illocutionary force and whose interior contains the situations corresponding to the speech act. Figure 6 provides the representation of the sentence: *Peter says to Paul: "I warn you that Mary thinks that you left Quebec city"*. The envelope of the performative utterance per1 contains the graph corresponding to the sentence portion "I warn you that". The interior of DS per1 contains the graph corresponding to the sentence portion "Mary thinks that you left Quebec city" which is itself an agent's attitude.

4. Inference-Related Discourse Spaces

An agent involved in a discourse has the ability to reason on knowledge using different kinds of reasoning approaches such as temporal, modal and hypothetical reasonings. We will not detail here these different kinds of reasoning mechanisms [4] but we will focus on the main kinds of discourse spaces that an agent can create or evoke in order to communicate inferential knowledge to other agents: conditional, alternative and generalized DSs are used to support inference mechanisms; hypothetical and counterfactual DSs are used to support hypothetical reasoning. Conditional, alternative and generalized discourse spaces are used to communicate knowledge corresponding to logical expressions that an agent can use to reason. These discourse spaces are interpreted in terms of Peirce's existential graphs.

A *conditional discourse space* is a DS whose envelope contains a set S1 of DSs (usually temporal DSs) and whose interior contains another set S2 of DSs whose existence is dependant on the existence of the discourse spaces of S1. A conditional DS is equivalent to Peirce's notion of *scroll* noted in terms of existential graphs ¬[S1 ¬[S2]]. The envelope and the interior of the conditional DS correspond to the outer and inner parts of the scroll respectively. The DSs that are usually found in a conditional DS's envelope and interior are situations and agent-related DSs (agent's perspectives, attitudes or judgements) eventually associated with temporal localizations. Conditional DSs correspond to rules known or believed by an agent.

A conditional discourse space is a triple <CDSD, CDSENV, CDSINT> where CDSD is the conditional DS descriptor. CDSENV and CDSINT are called the envelope and interior of the conditional DS respectively. Graphically, a conditional DS is represented as a rectangle composed of two parts. In the upper rectangle symbolizing the conditional DS's envelope, we indicate the conditional DS's descriptor and the DSs that are contained in its envelope. In the lower rectangle we indicate the DSs contained in its interior. Note that we use dashed lines to draw the rectangles symbolizing inference-related DSs in order to distinguish them from temporal DSs.

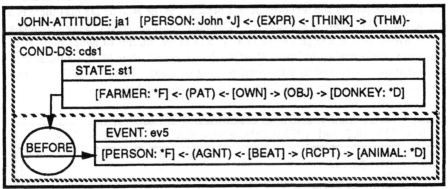

Figure 7: John thinks that if a farmer owns a donkey, he beats it

For example figure 7 contains the representation of the sentence: "John thinks that if a farmer owns a donkey, he beats it". In the attitude DS ja1 there is an conditional DS cds1. In the envelope of cds1 there is a state st1 corresponding to the situation "if a farmer owns a donkey" and in the interior of cds1 there is an event ev5 corresponding to "he beats it". Note that these situations are generic situations in the sense that their time intervals are not instantiated: that is the reason why they are not linked by temporal relations to the attitude DS's rectangle.

Statements like "Either situation P or situation Q" set *alternative discourse spaces*. This is a special case of a conditional DS since there is no precedence relation between situation P and Q. In an alternative discourse space the envelope and the interior have equivalent status: one situation is the envelope of the other considered as the DS's interior, and vice-versa. This view conforms to Peirce's representation of alternatives using existential graphs. We often use universal quantifiers to express general rules. In our framework a *universally quantified situation* is expressed using a conditional discourse space. In fact, figure 7 can also represent the sentence "John thinks that every farmer who owns a donkey beats it".

Conditional DSs must be distinguished from hypothetical DSs. Conditional DSs represent general laws or rules that apply in discourse spaces, whereas an hypothetical DS is created by an agent raising some hypothesis and drawing inferences from it. Consider for example the following sentence:

Mary thinks: "If I get some money, I will offer John a present".

Agents can raise hypotheses in various circumstances, with different degrees of certainty. Hypothetical discourse spaces can be created as the result of certain kinds of speech acts such as promises.

A *hypothetical discourse space* is always created in the interior of an agent's attitude DS (usually a belief DS). The envelope of an hypothetical DS contains situations that correspond to the hypothesis raised by the agent. The interior of an hypothetical DS contains situations corresponding to the agent's expectations if the hypotheses are realized. An hypothetical DS is represented in the same way as a conditional DS: the only difference lays in the discourse space type which is set as "HYPOTH-DS". An hypothetical DS can contain other hypothetical DSs in its interior.

For example figure 8 contains the representation of the sentence: "Mary thinks that if she gets some money she will offer John a present". In the attitude DS ma1 there is a hypothetical DS hyds1. In the envelope of hyds1 there is an event ev2 corresponding to the hypothesis "if she gets some money" and in the interior of hyds1 there is an event ev3 corresponding to "she will offer John a present". Here Mary's attitude ma1 and event ev2 are related by a temporal relation BEFORE. This is an indication that ev2 corresponds to an instantiated situation used as an hypothesis.

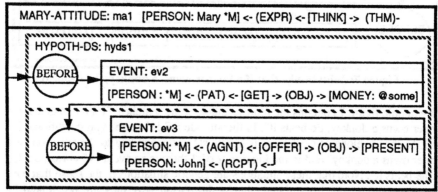

Figure 8: Mary thinks that if she gets some money she will offer John a present

Counterfactuals open discourse spaces under an hypothesis that the agent knows to be false in some reference space (usually reality). For example Mary can think:"If John had won the lottery, he would be rich. He would have moved to Tahiti". But Mary knows that John has not won the lottery.

A *counterfactual discourse space* is a DS whose envelope contains the hypothetical situation and whose interior contains the situations that would hold if the hypotheses were true. We use the same notation as for hypothetical discourse spaces changing the discourse space type for COUNTERFACT-DS.

5. Interrelations Between Discourse Spaces

As we have seen in several preceding examples, different kinds of discourse spaces can be found in other DSs' interiors. This illustrates the complexity and conciseness of natural language expressions. In general, the architecture of a discourse is composed of a nesting of temporal DSs: a narrator's perspective along with any combination of temporal localizations, agent's perspectives, attitudes or performative utterances. Within the interior of agent's perspectives or attitudes we can find any kind of temporal DS (perspectives, localizations, attitudes or situations) or of inference DS (conditional, generalized, alternative, hypothetical or counterfactual).

Temporal DSs are associated with a temporal topology: the situations and discourse spaces included in their interiors are associated with time intervals which are related by temporal relations. Inference spaces (conditional, alternative and generalized DSs) have other properties which are used to draw inferences (they are based on Peirce's inference rules for existential graphs). Hypothetical and counterfactual DSs are used to factor knowledge (situations and DSs) under some hypotheses. Definitional and focal DSs are used to refine a concept or a situation by replacing it by its definition. Different kinds of reasoning activities can take place on the basis of this various kinds of DSs.

6. Conclusion

In this paper we presented a pragmatic interpretation of the notion of context which can be used to model the architecture of discourses. This is an adaptation of Fauconnier's notion of mental space in the framework of CG theory. Our basic assumption is that, when producing or understanding a discourse, an agent evokes, creates or modifies a mental model composed of discourse spaces (DS) that structure the elements contained in the discourse. We introduced the main categories of discourse spaces that are useful to model the contents of a discourse: situations, definitional and focal discourse spaces, agent-related DSs (narrator's and agent's perspectives, agent's attitudes) and inference-related discourse spaces (conditional, alternative, generalized, hypothetical and counterfactual DSs).

Discourse spaces also play an important role in the resolution of reference and indexical expressions [25, 26, 27]. This approach can be extended to model other linguistic phenomena like metonymy, referential opacity and transparency [28].

References

1. Allen, J. F. (1983). Maintaining Knowledge about Temporal Intervals. *CACM*, vol 26 n11.
2. Barwise, J. & Perry, J. (1983). *Situations and Attitudes*. The MIT Press.
3. Binnick, R.I. (1991). *Time and the Verb: A Guide to Tense and Aspect*. Oxford U. Press.
4. Boury-Brisset, A.-C. & Moulin, B. (1994). Mise en oeuvre de raisonnements multiples dans un système multi-agent. Actes des Journées Intelligence Artificielle Distribuée et Systèmes Multi-agents. Voiron, France. Mai 1994, 169-182.
5. Comrie, B. (1976). *Aspect*. Cambridge University Press.
6. Descles, J-P. (1989). State, Event, Process and Topology. *General Linguistics*, vol 29 n3 161-199, Pensylvania State University Press.
7. Dick, J. P. (1994). Using contexts to represent text. In [36].
8. Dinsmore, J. (1987). Mental spaces from a functional perspective, *Cognitive Science*, vol 11, 1-21.
9. Dinsmore, J. (1991). *Partitioned Representations*. Berlin: Kluwer.
10. Esch, J. (1993). Contexts as white box concepts. In proceedings of the First Int. Conf. on Conceptual Structures. Mineau, G. & Moulin, B. (Eds.). Quebec 1993. 17-29.
11. Esch, J. (1994). Contexts and concepts: abstraction duals. In [36].

12. Fauconnier, G. (1985). *Mental Spaces*. The MIT Press.

13. Findler, N. V. Edt. (1979). *Associative Networks*. New York: Academic Press.

14. Guha, R. V. (1991). Contexts: a Formalization and some applications. MCC tech.report ACT-CYC-423-91.

15. Hendrix, G. (1979). Encoding knowledge in partitioned networks. In [13]. 51-92.

16. Janas, J. M. & Schwind, C. B. (1979). Extensional semantic networks: their representation, application and generation. In [12]. 267-302.

17. Johnson-Laird, P. N. (1983). *Mental Models*. Harvard University Press.

18. Kocura, P. (1993). Towards a semantic of inchoactive and causation events in conceptual graphs. In [30].

19. Kuroda, S. Y. (1973). Where epistemology, style and grammar meet. In S. Anderson & P. Kiparsky (Eds.). *A Festschrift for Morris Halle*. Holt, Rinehart & Winston.

20. Langacker R. W. (1987), *Foundations of Cognitive Grammar*, Stanford U. Press.

21. Levinson, S. C. (1983). *Pragmatics*. Cambridge University Press.

22. Mann, W.C., & Thompson, S.A. (1987). Rhetorical structure theory: description and construction of text structures. In Kempen G. (Ed.), *Natural Language Generation*. Dordrecht: Martin Nijhoff. Pub.

23. Mineau, G. W., Moulin, B. & Sowa, J. F. (Eds.). *Conceptual Graphs for Knowledge Representation*. Lecture Notes in A. I., 699. Springer Verlag.

24. Moulin, B. (1992). A conceptual graph approach for representing temporal information in discourse. *Knowledge-Based Systems*. vol5 n3, 183-192.

25. Moulin, B. (1993a). Representing temporal knowledge in discourse: an approach extending the conceptual graph theory. In [30].

26. Moulin, B. (1993b). The representation of linguistic information in an approach used for modelling temporal knowledge in discourses. In [23].

27. Moulin, B. & Dumas, S. (1994).The temporal structure of a discourse and verb tense determination. In [36].

28. Moulin, B. (1995). A pragmatic representational approach of context and reference in discourses. In Proceedings of the International Conference on Conceptual Structures ICCS'95, G. Ellis, R.A. Levinson, W. Rich, J. Sowa (Edts.). Springer Verlag.

29. Palacas, A. L. (1993). Attribution semantics: linguistic worlds and point of view. *Discourse Processes* 16, 239-277.

30. Pfeiffer, H. & Nagle, T. (Eds.) (1993). *Conceptual Graph: Theory and Implementation*. Lecture Notes in Artificial Intelligence 754. Springer Verlag.

31. Roberts, D.D. (1973). *The Existential Graphs of Charles S. Peirce*. Mouton.

32. Searle, J. R. & Vanderveken D. (1985). *Foundations of Illocutionary Logic*. Cambridge Univ. Press.

33. Shoham, Y. (1991). Varieties of contexts. In V. Lifschitz (Ed.), *Artificial Intelligence and Mathematial Theory of Computation*.Academic Press.

34. Sowa, J. F. (1984). *Conceptual Structures*. Reading Mas: Addison Wesley.

35. Sowa, J. F. (1993). Logical foundations for representing object-oriented systems. *Journal of Experimental and Theoretical Artificial intelligence*.Vol.5 n.2, 237-261.

36. Tepfenhart, W. M., Dick, J. P., & Sowa, J. F. (Edts.) (1994). *Conceptual Structures: Current Practices*. Lecture Notes in A.I. 835. Springer Verlag.

37. Vendler Z. (1967). Verbs and times. In Z. Vendler ed., *Linguistics and Philosophy*, Ithaca: Cornell Univ. Press.

38. Willems, M. (1993). A conceptual semantics ontology for conceptual graphs. In [23].

A Pragmatic Representational Approach of Context and Reference in Discourses

Bernard Moulin, Professor[1]
Laval university, Computer Science Department
Pavillon Pouliot, Ste Foy, QC G1K 7P4 Canada
ph: (418) 656 5580, fax: (418) 656 2324, Email: moulin@ift.ulaval.ca

Abstract
Apart from general rules or laws, knowledge must be considered with respect to the agents that manipulate or communicate it. We introduced the notion of *discourse space* which refers to constructs which are created, updated or evoked by an agent who tries to generate or understand a discourse. Our approach which is based on a generalization of Sowa's conceptual graph theory is used to explain a variety of referential mechanisms such as indexicality, metonymy, referential opacity and transparency.

1. Introduction

Any verbal or non-verbal action that an agent performs is done in a given context that has temporal, spatial and social characteristics. When an agent communicates some information, she often presupposes that her interlocutor is aware of the context to which that information must be related. If this contextual relationship is not obvious or readily found, the addressee agent might ask the originator agent to indicate explicitly what is the relevant context for that information. The notion of context has recently received much attention from the knowledge representation community; this interest being fostered by an increasing awareness that knowledge should be relativized or situated with respect to the agents that provide it as well as to the place, time and social settings in which it is produced. In artificial intelligence, contexts have been usually considered as useful structures that are used to partition knowledge and to facilitate reasoning activities. Our basic assumption is that a human agent naturally uses contextual information because she cannot manipulate or communicate knowledge without positioning herself relative to that knowledge. Apart from general rules or laws, knowledge must be considered with respect to the agents that manipulate or communicate it. Such a pragmatic approach is necessary not only to understand how knowledge is structured in a discourse, but also to explain referential mechanisms such as indexicality, metonymy, referential opacity and transparency.

In this paper we propose a pragmatic approach using discourse spaces to represent contexts within the conceptual graph framework. We show how this notion is an important one when it comes to explain referential phenomena in discourses.

2. Toward a Cognitive Interpretation of Contexts

The term *context* has been used to refer to a variety of different notions that all shed some light on the complex cognitive mechanisms related to contextualization. In linguistics and cognitive psychology, the importance of the notion of context has

---

[1] This research is supported by the Natural Sciences and Engineering Research Council of Canada (grant OGP 05518) and by FCAR.

been acknowledged for the study of language comprehension and generation. Langacker [15] suggests that all linguistic units are context-dependent. "They occur in particular settings, from which they derive much of their import, and are recognized by speakers as distinct entities only through a process of abstraction" ...

Fauconnier [10] suggested that understanding a discourse is based on the creation or evocation of knowledge domains, called mental spaces, that we set up as we talk or listen and that we structure with elements, roles, strategies and relations. "These domains are not part of the language itself, or of its grammar; they are not hidden levels of linguistic representation, but language does not come without them... Language builds up mental spaces, relations between them, and relations between elements within them" [10]. Within mental spaces we can represent objects and relations holding between them, regardless of the status of those objects and relations in the real world. Mental spaces are evoked and accumulate information during the processing of discourses, primarily in response to the occurrence of linguistic structures called *space builders* such as 'John believes that ----', 'In 1995, ---', "When Peter will ---, ----", and so forth. For instance, processing the sentence "John regrets that Peter will leave school" involves either creating or locating a mental space containing John's regret and representing in it the information that Peter will leave school.

Objects in different spaces can be related by a *connector*, a particular "psychologically relevant relation" (called a *pragmatic function*) that allows reference to one object in terms of another appropriately linked to it. For example a locutor can say "Plato is on the top shelf" to mean "The books by Plato are on the top shelf". Considering the pragmatic function F1 linking authors with books containing their works, we have the relation between the object a="Plato" and F1(a) = b = "Books by Plato". Using such an approach, Fauconnier provided a "simple and uniform account of a wide variety of problems: referential opacity and transparency, specificity of reference, definite reference in discourse, the interaction of quantifiers with modalities, the projection problem for presuppositions, the semantic processing of counterfactuals and the use of comparatives in modal contexts [5].

Palacas [23] advocated the importance for text analysis to consider the notion of "cognitive point of view [14] where all meanings in text are attributed to their personal or purported personal sources". These sources are the agents that have performed (in a direct style) or are reported to have performed (in an indirect style) the speech acts that compose the text. Palacas introduced a *default context* in which utterances represent "real world truths" and are expressed by the speaker at the speaking time. Then, he defines a *linguistic world* as a structure that captures a cognitive point of view: a set of utterances associated with "a unique mentality", a "unique sentient source" and a "unique reference time" [23]. However, Palacas did not propose a precise characterization of temporal structures (such as temporal localizations and temporal relations) and did not provide a way of processing references (anaphoras, indexicals) within the context of linguistic worlds.

The importance of interpreting discourse contents with respect to locutors' perspectives has been recognized quite early. In a classical paper, Bar-Hillel [2] argued that indexicality is an inherent and unavoidable property of natural language, and speculated that more than 90% of the declarative sentences people utter are indexical in that they involve implicit references to the speaker, addressee, time and/or place of utterance. Kamp [13] developed the Discourse Representation Theory in an attempt to provide an explicit account of discourse connectedness using a truth-conditional semantics. In order to represent the knowledge extracted from a discourse by his so-called 'construction algorithm', he used *discourse representation structures* (a notion equivalent to a context) in which appear variables representing discourse referents as well as predicates denoting propositional knowledge extracted from a discourse.

The usefulness of context mechanisms to represent knowledge in discourses should be sought for in human cognitive abilities to generate and understand natural language. The so-called abstraction mechanisms found in knowledge representation approaches reflect similar capabilities of human cognition. For instance, the notion of *situation*, as it is used by Barwise and Perry [3] or by Sowa [25], reflects the intuition that people observe, think about or speak about such things that are spatially and temporally situated and that can be called "events", "processes" or "states". The intuitive notion of concept expansion into a context, proposed by [26] and clarified by Esch [6, 7,8, 9], reflects the intuition that a speaker can use a word in order to evoke in the addressee's mind a set of concepts (in the form of different situations for example). Hence, the mechanism of concept expansion is justified by its cognitive counterpart experienced by human agents.

Linguists have noticed that various expressions in discourses, called *cue phrases*, such as 'by the way' or 'in the first place' play an important role in signaling topic changes and are useful to recognize the structure of a discourse [12]. Cue phrases play a similar role as Fauconnier's space builders [10]. In their analysis of discourse, Grosz and Sidner [11] emphasized the importance of the *linguistic structure* which is composed of *discourse segments* of "naturally aggregating utterances". Discourse segments may be nested and associated with cue phrases. Allen [1] proposes an approach for segmenting a discourse. He also shows how segments are used to facilitate understanding of the content of a discourse as well as the interpretation of verb tenses and the identification of temporal and causal connections between situations described in a discourse.

Knowledge representation works should benefit from the studies of those cognitive mechanisms done by researchers in cognitive psychology and linguistics. Since the early nineties [17], we have been advocating such an approach for the representation of temporal knowledge in discourse, emphasizing the importance of explicitly introducing pragmatic constructs in our representations [18, 19]. Our approach which provides a practical adaptation of the notion of mental spaces [10] within the CG theory, has been extended to encompass a variety of different context types interpreted as discourse spaces [21]. An important assumption of our approach is that a human agent naturally uses contextual information because she cannot manipulate or communicate knowledge without positioning herself relatively to that knowledge. Indexicality is an inherent property of natural language [2] and any technique used to manipulate contexts or mental spaces should provide some mechanism to deal with it. In Moulin [21] we present in detail the main categories of discourse spaces that are useful to represent knowledge contained in discourses: *base DSs* (situations, definitional DSs, focal DSs, temporal localizations and temporal objects); *Agent-based DSs* (Agent's perspectives, agent's attitudes); *Inference-related DSs* (conditional, counterfactual, hypothetical, alternative and generalized DSs). In this paper we discuss the question of trans-space interactions of concepts: anaphoric references and indexicals, pragmatic functions, de dicto, de re, attributive and referential interpretations.

3. Anaphoric References and Indexicals

Discourse spaces partition the knowledge contained in a discourse according to various dimensions: temporal, spatial, hypothetical, etc. However, DSs are not independant of each other since the same concept can appear in different DSs associated with a given discourse. We say that there are trans-space interactions between concepts. In discourses these interactions correspond to various kinds of references which are used by locutors in order to minimize repetitions of the same words. Usually a concept is named in a given sentence, and we find in other sentences special words used to refer to that concept. The main referential categories are: personal pronouns (I, you, he, etc.); possessive adjectives (my, your, his, etc.); possessive pronouns (mine, yours, ours, etc.); reflexive pronouns (myself, ourselves, etc.); relative pronouns (who, whom, which, that, of which, etc.); demonstrative pronouns (this, that, those, etc.); temporal indexicals (now, tomorrow, yesterday, etc.); spatial indexicals (here, there); the definite article (the). These references are used to establish a correspondence between concepts appearing in different sentences. In [17, 18] we discussed how to deal with different reference categories (anaphoras and indexicals) with respect to agent's perspectives. These solutions still hold in the present framework. We will only give in this section some complementary information.

Anaphoric references (like he, she, it, them) are used to establish a correspondence between two concepts in whatever discourse space they appear. They correspond to the third person of singular or plural at any gender. To specify them, we use Sowa's convention [25] and attach to the coreferring concepts the same anaphoric variable marked by a star.

For example in figure 3 we have an anaphora [PERSON: *O] in EVENT: ev1 referring to [PERSON: Oedipus *O] in OEDIPUS-ATTITUDE: oa1.

Indexicals referring to the narrator and to her locutor (first and second persons of singular or plural in a text) must be resolved with respect to agent's perspectives or attitudes. This is the case for personal pronouns such as I, you, we, me, us, mine, yours, ours, etc. In terms of discourse spaces the following rules apply.

- Let DS_1 be an agent-related discourse space (agent's perspective or attitude) contained in the interior of another agent-related discourse space DS_0;

- Let $C_{0\text{-agnt}}$ and $C_{0\text{-pat}}$ be the concepts that respectively represent the agent and the patient in the situation contained in the envelope of DS_0;

- Let $C_{1\text{-agnt}}$ and $C_{1\text{-pat}}$ the concepts that respectively represent the agent and the patient in the situation contained in the envelope of DS_1;

- Any indexical appearing in a situation contained in DS_1's interior is resolved with respect to the situation appearing in DS_1's envelope: #I refers to $C_{1\text{-agnt}}$, #you refers to $C_{1\text{-pat}}$, #we refers to both $C_{1\text{-agnt}}$ and $C_{1\text{-agnt}}$.

- Any indexical appearing in DS_1's envelope is resolved with respect to the situation appearing in DS_0's envelope: #I refers to $C_{0\text{-agnt}}$ and #you refers to $C_{0\text{-pat}}$.

These rules recursively apply to any agent-related discourse space included in DS_1.

As an example consider the following short story:

Mary says to Peter: "I brush my teeth and I will come with you".

This is represented in figure 1. We have examples of the indexicals "I" and "you" in processes pr1 and pr2. [PERSON: #I] and [PERSON: #you] respectively refer to [PERSON: Mary] and [PERSON: Peter] in the envelope of Mary's perspective mp1. In figure 2 we have the indexical [PERSON: #we] in situation pr2 which refers to [PERSON: Paul] and [PERSON: Helen] in Paulr's perspective envelope.

There are similar rules for spatial and temporal indexicals such as #here, #now which refer to the place and time associated with the agent's perspective. For more details see [18, 19], especially to consider the impact of using direct or indirect styles in sentences.

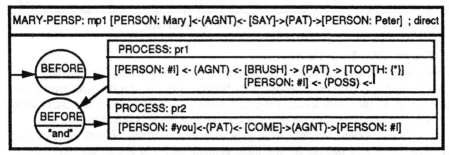

Figure 1: Personal pronouns and possessive adjectives

Possessive pronouns and adjectives and reflexive pronouns corresponding to the first and second persons (of singular or plural) are also resolved relatively to the proper agent's perspective or attitude. In the example of figure 1, notice that the possessive adjective "my" is represented by ->(POSS)-> [PERSON: #I] where #I is resolved with respect to Mary's perspective mp1.

In our approach, we do not make a semantic difference between a main clause and a relative clause, representing both of them as temporal situations. A *relative pronoun* is considered as a special case of an anaphoric reference that relates the corresponding concepts in the two temporal situations associated with the main and relative clauses (see [18] for a discussion on that subject). We denote this special anaphoric reference using a variable introduced by the symbol λ.

For example in figure 2 we have a representation of the sentence: *Paul says to Helen: "John who was visiting our friend Peter, met Mary at the hospital".*

The concept [PERSON: John λJ] corresponds to the antecedent ("John") of the relative pronoun ("who") in the text, while [T: λJ ; "who"] denotes the relative pronoun. T is the universal concept in the concept type lattice, and "who" is a linguistic annotation indicating which relative pronoun is used in the text.

Verb tenses can also be considered as indexical elements. The choice of a tense depends on how are situated the moment when an agent (or the narrator) utters a sentence and the time when a situation takes (or took) place, using a time axis (calendar dates for example) as a reference. For instance, if a situation took place before the moment of speech, the tense used to describe that situation is a past tense. In [20] we have showed how narrator's and agent's perspectives can be used to determine the verb tenses describing the situations included in the envelop of the corresponding discourse spaces.

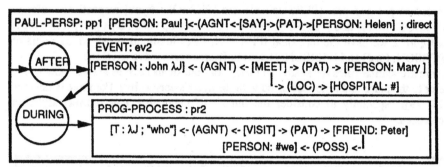

Figure 2: Relative pronouns

4 Pragmatic Functions

Fauconnier [10] used the notion of *pragmatic function* proposed by Nunberg [22] who showed that locutors establish links between objects of a different nature for psychological, cultural or locally pragmatic reasons and that the links thus established allow reference to one object in terms of another appropriately linked to it.

Fauconnier proposed the following general principle, called *Identification (ID) Principle*: "If two objects (in the most general sense), *a* and *b*, are linked by a pragmatic function F (*b* = F(*a*)), a description of *a*, d_a, may be used to identify its counterpart *b*... Call *a* the reference *trigger*, *b* the reference *target* and F the *connector*"... "For instance, one function F_1 links authors with the books containing their works. Taking as an example *a* = "Plato" and *b* = $F_1(a)$ = "books of Plato", the ID Principle allows (2) to mean (3):

(2) Plato is on the top shelf;   (3) The books by Plato are on the top shelf"...

... "The ID Principle states that in a connected situation a description of the trigger may be used to identify the target".

This type of pragmatic reference has not been introduced in the CG theory [25] which only allows coreference between identical concepts appearing in different graphs. We extend it to allow the use of pragmatic functions. Given a pragmatic function F linking a concept of type C1 (the trigger) and a concept of type C2 (the target) we use the following conventions:

- the trigger is marked by a variable with a star [Concept-C2: *x]
- the target is marked by a reference function π [Concept-C1: π(*x, F, Agent)]
    where Agent represents the agent or group of agents that established the pragmatic function F and *x is a variable coreferring to the target concept.

Applying the ID Principle, if a concept C1 is marked by function π, it refers through a pragmatic function F to another concept C2, and concept C2 is used to name concept C1 in the text. For example a waiter could say in a restaurant to the cashier *"The mushroom omelet* asks for the bill". This is an example of metonymy where the 'mushroom omelet' refers to the customer who ordered it. The waiter's sentence is represented by the following graph:

[CUSTOMER: *b π(*x, F1, Waiter)] <- (AGNT) <- [ASK] -> (OBJ) -> [BILL: #]
considering that the waiter knows that
[CUSTOMER: *b] <- (AGNT) <- [ORDER] -> (OBJ) -> [MUSHROOM-OMELET: *x].

5. De Dicto, De Re, Attributive, Referential Interpretations

Several authors have studied different problems encountered when handling anaphoric expressions in so-called referentially opaque contexts. Bernth [4] gives a good overview of these problems. "... Referential opacity covers the phenomenon that expressions denoting the same object are not mutually substitutable *salva veritate* in certain contexts. Furthermore, it may be the case that the expression does not have a referent in the real world. Referential opacity may occur in cases where somebody, whom we choose to refer as the *agent*, has a certain mental relation to a proposition, hence the common name for verbs denoting such relations: *propositional attitude verbs*. Examples are "think", "doubt", etc. Also some verbs expressing relations to individuals may introduce referential opacity, e.g. "seek"... For this reason we shall distinguish between the so-called *de dicto* and *de re* interpretations.
De dicto : The way the agent refers is important and is described by the embedded noun phrase which cannot be substituted for another - this gives opacity;
De re : The embedded noun phrase is used solely as identification of some object without claiming that this is the way the agent refers to it - this gives transparency"...
Let's take an example. Suppose that Lucas says: "Oedipus will marry *his mother*".
The *de re* interpretation of this sentence states that Oedipus will marry a person, and that Lucas knows that this person is Oedipus's mother, when Oedipus does not know it. The sentence "Oedipus will marry Jocasta" leads to a *de dicto* interpretation since we can consider that Oedipus knows that Jocasta is the name of the person he will marry.
Let's quote again [4] on the theme of attributive and referential interpretations:
"Definite descriptions appearing in transparent contexts have been explained as having two different uses: the attributive and referential uses. These can be combined with the de re and de dicto interpretations in interesting ways. The attributive use states something about whoever is the so-and-so:
    ex. Smith's murderer [whoever it is] is insane
whereas the referential use picks out a certain referent:
    ex. Smith's murderer [i.e. Mr. Jones] is insane.
The important characteristic of the referential use is that it picks out an individual; the attributive use gives a description, which some individual might satisfy. Barwise and Perry [3] explain this in terms of *value-loaded* and *value-free* interpretations. The value-free interpretation of a definite description is a relation, a partial function from situations to individuals; the value-loaded interpretation is the value of that function when applied to some particular context. The value-free interpretation hence gives a description, and the value-loaded an individual"... "Whereas the distinction between the referential and attributive uses is a matter of value-loaded vs. value-free interpretations, the *de re / de dicto* distinction is a matter of *scope*, i.e. whether the quantification introduced by the noun phrase in the embedded sentence scopes over the attitude verb or vice versa. The referential / attributive uses and the *de re / de dicto* interpretations are thus different kinds of distinctions, and combinations of them are possible. They can be combined in the following ways.

1. Referential-de re. A *de re* interpretation will always combine with a referential interpretation, since there has to be some specific object for the *de re* interpretation. Ex. *John believes that (a particular) president is powerful*

2. Referential-de dicto. There is a specific object which the agent refers to in a certain way. Ex. *Charlie Brown believes that the little red-haired girl is beautiful.*
There is a specific person Charlie Brown is thinking of, and he does not know her name, but refers to her as the little red-haired girl.

3. Attributive-de dicto. There is a non-specific object the agent refers to in a certain way. Ex. *John believes that the person who stole his bicycle (whoever it is) is red-haired.* John does not know who stole his bicycle, but he believes that whoever it is, he must be red-haired".

These distinctions are important if we want to deal with reference in a proper way. Our approach offers an appropriate framework to deal with referential problems in a similar way as Fauconnier [10] used pragmatic functions and mental spaces.

Let an agent X 's discourse space DS_x (either an agent's attitude or perspective) contain an agent A's attitude discourse space ADS_a. Let C be a concept used in a situation S contained in the interior of an agent A's attitude discourse space ADS_a.

- Referential - de dicto interpretation

This is the default option we have considered in our examples up to now. Attitude DSs can be considered as opaque contexts: concept C exactly illustrates the way agent A thinks about the corresponding entity.

- De re interpretation

We should be able to indicate that the concept C which appears in situation S embedded in agent A's attitude discourse space, illustrates the way agent X thinks about the corresponding entity, which is not necessarily shared by agent A. Hence, there is a pragmatic function F that associates concept C (the target) in agent A's discourse space to a concept C1 (the trigger) contained in Agent X's discourse space. Following the conventions of section 4 we have:
   - Concept C1 (the trigger) is marked by a variable with a star [Concept-C1: *x]
   - Concept C (the target) is marked by a reference function π [Concept-C: π(*x,
      F, Agent X)]   where *x is a variable referring to the target concept.
As an example consider the following short story. Lucas says to Helen:
   "Jocasta is Oedipus's mother. Oedipus believes that he will marry *his mother*".
Figure 3 presents the discourse spaces illustrating the de re interpretation of this story. Lucas establishes a pragmatic function F2 between the person whom Oedipus will marry (event ev1 in Oedipus' attitude oa1)

[PERSON: Oedipus *O] <- (AGNT) <- [MARRY] -> (PAT) -> [PERSON: π(*x, F2, Lucas)]
and the fact that he (Lucas) knows that this person is Oedipus's mother (state st1). Note that "Oedipus's mother" appears in a definitional discourse space.
As in the case of metonymy (section 4), the surface form of the discourse is obtained by replacing [PERSON: π(*x, F2, Lucas)] by the trigger concept [MOTHER] <- (POSS) <- [PERSON: Oedipus].

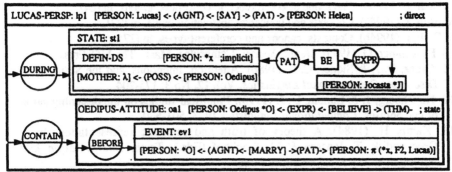

Figure 3: Opacity / transparency of contexts

- Attributive - de dicto interpretation

The attributive de dicto interpretation is supported by Sowa's approach. Let us consider the following example:

Lucas says: "Oedipus is persuaded that he will meet the queen of Thebes".

If Lucas does not know if Oedipus knows who is the queen of Thebes, we will use the # mark with the concept [QUEEN-OF-THEBES: #]. This concept may be interpreted as a role especially if the queen of Thebes does not exist. If Lucas knows that Oedipus knows who is the queen of Thebes, the #mark can be replaced by a variable coreferring to the proper concept in Oedipus' belief discourse space.

6. Conclusion

In discourse generation and comprehension we find several mechanisms which are used by locutors to establish references between concepts situated in different portions of the text: anaphoras, indexicals, metonymy, verb tenses. In this paper we have showed that introducing a pragmatic form of contexts (called discourse spaces) in the conceptual graph framework enables us to give an account of those referential mechanisms. This work shows that the introduction of contexts in semantic networks is not only useful for factoring knowledge, but that it provides a bridge between semantic and pragmatic representations.

References

1. Allen, J. F. (1995). *Natural Language Understanding*. Second Edition. Benjamin.
2. Bar-Hillel, Y. (1954). Indexical Expressions. *Mind*, 63, 359-379.
3. Barwise, J. & Perry, J. (1983). *Situations and Attitudes*. The MIT Press.
4. Bernth, A. (1989). Treatment of anaphoric problems in referentially opaque contexts. In *Natural Language and Logic*. Lecture Notes in A.I., n. 459. Springer Verlag.
5. Dinsmore, J. (1987). Mental spaces from a functional perspective, *Cognitive Science*, vol 11, 1-21.
6. Esch, J. (1993a). Contexts as white box concepts. In proceedings of the First International Conference on Conceptual Structures. Mineau, G. & Moulin, B. (Eds.). Quebec, Canada, August 1993.17-29.

7. Esch, J. (1993b). The scope of coreference in conceptual graphs. In [24].

8. Esch, J. (1994a). Contexts and concepts: abstraction duals. In [27].

9. Esch, J. (1994b). Contexts, canons and coreferent types. In [27].

10. Fauconnier, G. (1985). *Mental Spaces*. Cambridge, Mass. The MIT Press.

11. Grosz, B. J. & Sidner, C. L. (1986). Attention, Intentions, and the Structure of Discourse. *Computational Linguistics*, Vol. 12, No 3, 175-204.

12. Hirschberg, J. & Litman, D. J. (1993). Empirical studies on the desambiguation of cue phrases. *Computational Linguistics* 19, 3, 501-530.

13. Kamp, H. (1984). A theory of truth and semantic representation. In J. Groenendijk, T. Janssen & M. Stockhof (Eds.), *Truth, Interpretation and Information*. 277-322. Dordrecht: Foris.

14. Kuroda, S. Y. (1973). Where epistemology, style and grammar meet. In S. Anderson & P. Kiparsky (Eds.). *A Festschrift for Morris Halle*. Holt, Rinehart & Winston.

15. Langacker R. W. (1987), *Foundations of Cognitive Grammar*, Stanford U. Press.

16. Mineau, G. W., Moulin, B. & Sowa, J. F. (Eds.).*Conceptual Graphs for Knowledge representation*. Lecture Notes in A. I. 699. Berlin: Springer Verlag.

17. Moulin, B., Rousseau, D. & Vanderveken, D. (1992). Speech acts in a connected discourse: a computational representation based on conceptual graph theory. *Journal of Experimental and Theoretical Artificial Intelligence* 4 , 149-165.

18. Moulin, B. (1993a). Representing temporal knowledge in discourse: an approach extending the conceptual graph theory. In [24].

19. Moulin, B. (1993b). The representation of linguistic information in an approach used for modelling temporal knowledge in discourses. In [16].

20. Moulin, B. & Dumas, S. (1994).The temporal structure of a discourse and verb tense determination. In [27].

21. Moulin, B. (1995). Discourse spaces, a pragmatic interpretation of contexts. In Proceedings of the International Conference on Conceptual Structures ICCS'95, G. Ellis, R.A. Levinson, W. Rich, J. Sowa (Edts.), Lecture Notes in Artificial Intelligence, Berlin: Springer Verlag.

22. Nunberg, G. (1978). *The Pragmatics of Reference*. Bloomington: Indiana Univ. Linguistics Club.

23. Palacas, A. L. (1993). Attribution semantics: linguistic worlds and point of view. *Discourse Processes* 16, 239-277.

24. Pfeiffer, H. & Nagle, T. (Eds.) (1993). *Conceptual Graph: Theory and Implementation*. Lecture Notes in Artificial Intelligence 754. Springer Verlag.

25. Sowa, J. F. (1984). *Conceptual Structures*. Reading Mas: Addison Wesley.

26. Sowa, J. F. (1993). Logical foundations for representing object-oriented systems. In *Journ.of Experimental and Theoretical Artificial intelligence*. Vol.5, n.2, 237-261.

27. Tepfenhart, W. M., Dick, J. P., & Sowa, J. F. (Edts.) (1994). *Conceptual Structures: Current Practices*. Lecture Notes in A. I. 835. Springer Verlag.

Using the Conceptual Graphs Operations for Natural Language Generation in Medicine

J. C. Wagner, R.H. Baud, J.-R. Scherrer

Medical Informatics Centre, Geneva University Hospital,
Geneva, Switzerland

Abstract. Natural language generation systems for the medical domain have to take into account the specific domain vocabulary as well as the particular language style of a highly conventionalised language. This is particularily relevant for multilingual generation.

We have developed a system for multilingual natural language generation in medicine, based on the Conceptual Graphs formalism and using a semantic model of medicine. The latter is being developed as part of the European project GALEN and it intends to be a language-independent, semantically valid model of clinical terminology.

Within the system we make intensive use of the operations defined in the Conceptual Graphs formalism in order to deal with the specific requirements of the medical language, multilinguality and the specific modelling style encountered within the GALEN-model. The approach has been applied to several languages, with a main focus on English and French. It has also been used within a clinical demonstrator application.

Keywords. Natural Language Generation, multilingual systems, lexical choice, medical information systems, domain terminology.

1. Introduction

The generation of natural language has several uses in the medical domain. Re-usable and sharable, and therefore language-independent, concept models are of increasing interest for building model-based patient record systems, for integrating health care information systems and for sharing knowledge between decision-support systems [1-4]. Yet, language-independent concept systems need to be transformed in natural language in order to be used within applications. Such a transformation can be achieved by natural language generation tools. Furthermore, in the European context, multilinguality opens up additional possibilities for reusing applications. Finally, the gathering and reporting of medical facts is a very frequent task in the medical domain. All kinds of tools supporting this process by generating summaries may be of great use [5-7].

There exist several attempts in the medical domain to build terminological systems (terminologically oriented ontologies), as e.g. the UMLS (Unified Medical Language System) [8] or SNOMED [9]. GALEN (Generalised Architecture for Languages, Encyclopaedias, and Nomenclatures in Medicine) is a project of the AIM (Advanced Informatics in Medicine) programme of the European Union. GALEN aims at integrating terminological systems, it is "developing a 'Terminology Server' to support the development and integration of clinical systems through a range of key terminological services, built around a language-independent, re-usable shared system of concepts, the CORE model" [10, 11]. The CORE model is a compositional, generative model in a specific formalism called GRAIL (GALEN Representation and Integration Language). This formalism uses so-called 'sanctioning mechanisms' for the creation of sensible composite concepts. Furthermore, it classifies composite concepts automatically on the basis of their definition, comparable to the Conceptual Graphs formalism (generalization hierarchy, [12], p. 105). One of GALEN's terminological services is the generation of noun phrases to describe composite concepts in several natural languages. This service is provided by the so-called 'Multilingual Module' (MM) [13], a result of the subsequently described work.

This work is also part of the Natural Language Processing (NLP) project of the University Hospital of Geneva [14, 15]. This project investigates specific semantic-oriented approaches of natural language processing in the medical domain, using the Conceptual Graphs formalism [16, 17].

2. Multilingual Generation of Natural Language

2.1. Natural Language Generation as a terminological service

Within GALEN, there were several objectives for building a multilingual natural language generation tool. They have arisen from the overall goal of GALEN to build a language-independent medical model:

- to have a view on the language-independent CORE-model in several languages;
- to use the CORE-model within language-dependent applications and thereby to allow applications to become multilingual;
- to generate equivalent representations of medical facts in several languages.

The main task of the MM is to generate natural language expressions for generated, composite GRAIL-concepts (see example in Figure 1).

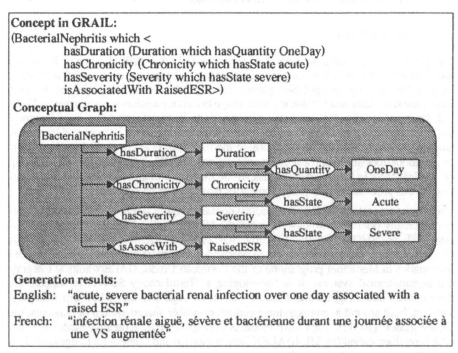

Concept in GRAIL:
(BacterialNephritis which <
        hasDuration (Duration which hasQuantity OneDay)
        hasChronicity (Chronicity which hasState acute)
        hasSeverity (Severity which hasState severe)
        isAssociatedWith RaisedESR>)

Conceptual Graph:

Generation results:

English: "acute, severe bacterial renal infection over one day associated with a raised ESR"
French: "infection rénale aiguë, sévère et bactérienne durant une journée associée à une VS augmentée"

Figure 1. Example of natural language generation for composite concepts.

2.2. The approach: annotation based generation

The specific natural language generation task at hand is different from standard language generation tasks. This is mainly due to the fact, that the system relies on a semantic medical model and not on a model which has been built under linguistic perspectives.

The main features to be met are:
- to deal with the complexity of the GRAIL expressions and corresponding grammatical structures;
- to deal with the difference between basic conceptual entities (concepts) to be found in the model and basic linguistic entities (words or lexemes) to be used for language: concepts are often expressed by 'multi-word' terms, and isolated words often correspond to complex conceptual entities;
- to close the gap between 'how things are defined' (in the model) and 'how things are expressed' (in language) (e.g. handling of implicit knowledge);
- to deal with the medical language usages (which strongly depend on the contexts);
- to deal with different usages in different languages;
- to deal with the fact that in different languages words are available at different levels.

A semantic-oriented approach to natural language generation has been implemented in order to deal with these requirements. The natural language expressions for concepts are generated from the words (or terms) for the basic concepts and rules about how to represent relationships by syntactic structures as well as some grammatical knowledge.

The approach uses the Conceptual Graphs formalism, as this formalism provides a well-known formal base with well-defined operations, as it is suitable for natural language purposes and as experiences in this domain exist already (e.g. [18,19]).

We follow Sowa's general guidelines ([12], p. 20) and try to map concepts into 'content words' (nouns, verbs, adjectives, adverbs) and relationships into syntactic elements (e.g. inflection) or 'function words' (e.g. prepositions, conjunctions). These guidelines reflect in the so-called 'annotations', which are basic to our approach: large parts of the linguistic knowledge are linked to the CORE-model by linguistic annotation. There are two types of annotations (see section 2.3):
- concept-annotations, i.e. annotation of concepts by corresponding words (or terms) in several languages including their syntactic properties.
- statement-annotations, i.e. annotation of relationships by syntactic structures qualified to express these relationships. The use of these syntactic structures depends on the context in which the relationship is found.

Different transformations are necessary to obtain a form of a graph on which the above guidelines can be applied. For example, the level of concept decomposition has to be adapted to the different languages with respect to the availability of words; relationships may not be directly expressible by function words or inflections. These transformations are described later (see section 3.)

2.3.   Knowledge for Natural Language Generation

Different kinds of knowledge are necessary for the generation process. One important part is the domain model itself. Here the CORE-model acts as the domain model. One other important part is linguistic knowledge with semantic contents. It was a prime issue of this approach to link this linguistic knowledge to the CORE-model and therefore to base the semantic extensions of the linguistic knowledge on the semantic model wherever possible. Hence, most of this kind of knowledge is integrated in the CORE-model or based on information which came out of it. However, beyond all integration efforts the linguistic knowledge is clearly distinguished from the purely semantic model. This distinction is intentional, as the CORE-model is not intended to be a linguistic model. Moreover, there is also some merely linguistic and language specific knowledge necessary for the generation which is not linked to the CORE-model.

The domain model

The CORE-model is used as domain model for the natural language generation process. It defines the entities on which the GRAIL-expressions are constructed. Therefore, it also defines the basic entities for the generation process. Three parts of the model are directly used by the generation system:

- the concept hierarchy
  defining the basic concepts, i.e. the basic model entities. The concept hierarchy also includes the composite concepts through formal subsumption.
- the hierarchy of relationships
  defining the relationships which may be used and their inverses.
- the concept definitions
  defining composite concepts by basic or other composite concepts and relationships, the composition of concepts being sanctioned by statements. The concept definitions are used for the transformation operations (see section 3.).

Other parts of the model might be used in future for several decisions, e.g. the decisions of using definite or indefinite articles or the relative positioning of adjectives when several adjectives are used for specifying a noun etc.

Concept Annotation

Every concept of the model, elementary or composite, may be annotated (Figure 2). Basic concepts need to be annotated, whereas composed concepts may be annotated following the availability of words in the respective language. A concept is annotated with the different synonyms and word type variants which serve to express this concept in the several languages. Each annotation includes the syntactic properties of the word (or expressions consisting of more than one word). The lexical semantics is defined by the conceptual model itself, i.e. the position of the concept in the type hierarchy or the definition of the concept. Additional information is necessary when the linguistic usage differs from the conceptual view (see section 3.). The concept annotations form the lexicons for the different languages, the so-called 'multilingual dictionaries'.

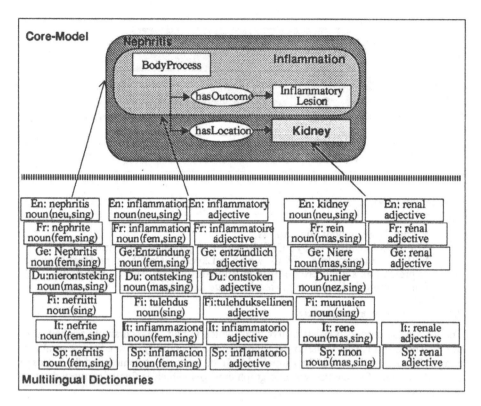

Figure 2. Annotation of basic and composite concepts.

Statement Annotation

The statement-annotations consist of rules, which have two constituents. The first part is formed by the so-called statement. A statement defines a concept-relationship-concept combination which is semantically sensible. It describes the context, where a rule applies, i.e the context for the annotation. The second part is the annotation itself consisting of syntactic structures qualified to express the relationship in the given context. For example, the localisation of an infection can be expressed by an adjectival structure (*renal infection*) or by a noun complement structure (*infection of the kidney*). However, as the medical language is a highly conventionalised language, there exist clear preferences for certain structures, even if all other structures are correct. Therefore, these rules define specifically for each language the preferences for syntactic structures by the order of arguments (Figure 3). This technique of statement annotations also permits us to respect specific properties of different languages, e.g. that the same relationship is expressed in English by a noun premodifier, in French by a noun complement (and in German by a compound noun). The concept hierarchy allows the rules defined by the statement-annotations to be inherited.

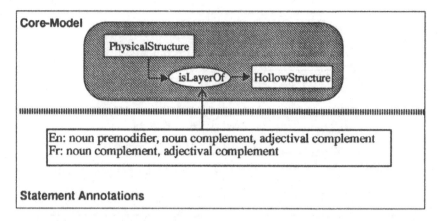

Figure 3. Statement annotations.
The relationship "isLayerOf" between a physical structure and a
hollow structure (e.g. "Wall which isLayerOf Cell") is preferably
expressed as a noun premodifier in English (e.g. *cell wall*) and a noun
complement in French (e.g. *membrane de la cellule*).

2.4.   The generation process

The generation process consists of five steps which in part are recursively applied (Figure 4). The graph is first transformed following the availability of words in several languages. The adapted graph is then supplied with the syntactic structures and the words to
be used. A kind of semantic-syntactic graph emerges, which is finally converted into a
noun phrase.

- Graph decomposition:
  The GRAIL-expression is transformed to a Conceptual Graph, the basic representation formalism of the whole generation process. The first concept is considered as
  the central concept of the expression (entry-concept, genus). Consequently, it will
  form the head of the generated noun phrase. The rest of the graph is decomposed in
  subgraphs by a top-down decomposition. The subsequent steps are applied to the
  resulting subgraphs. The generator works 'description-driven' resp. 'message-
  driven'.

- Graph transformations:
  Several operations transform the graph to a form where concepts can be expressed
  by (content) words in the respective language, and relationships can be phrased by
  grammatical structures as adjective complements, noun complements or function
  words etc. (see section 3.)

- Selection of syntactic structures and words:
  The above transformations allow the concepts to be expressed by words and the
  relationships to be expressed by grammatical/syntactic structures. Now, syntactic
  structures are selected in order to phrase the relationships, whereas, words are chosen in order to phrase the concepts. The syntactic structures are selected by rules
  (the statement annotations). Grammatical constraints limit the combination of syntactic structures to syntactically correct phrases. The words are selected under the
  annotations for the concept. Their lexical category has to correspond to the selected
  syntactic structure (e.g. an adjective is needed for an adjectival complement, a

noun for a prepositional phrase etc.). If no word exists conforming to the syntactic structure, another structure has to be chosen. Both the selection of syntactic structures and the selection of words, are intimately interlinked, as a syntactic structure can only be achieved if the corresponding words exist. Therefore, a backtrack mechanism applies to these selections.

- Modular generation functions:
Several modular generation functions realise the above syntactic structures with the respective selected words. These modules realise specifically for each language e.g. the declension of nouns, agreement of adjectives (essentially non-formative features). The modularisation at this level constitutes another important issue for the approach being open to several languages.

- Surface structure generation:
In this phase the final phrase is formed. The deep structure is transformed to the surface structure by ordering and positioning of the different elements following language specific (formative) rules. This phase includes also a post-treatment for elisions and contractions of words.

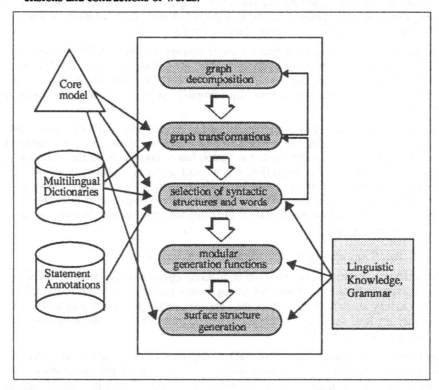

Figure 4. The phases of the generation process.

3. Transformation Operations

Several operations exist which allow us to transform a graph into a form where concepts can be phrased by words or terms and relationships by grammatical structures or function words. These operations close the gap between the way things are defined or described in the model, and the different ways of expressing the same medical fact in the different languages. We use the four operations, type contraction, type expansion, relational contraction and relational expansion being introduced by the Conceptual Graphs formalism [12].

3.1. Type expansion and type contraction

The basis of the type expansion and the type contraction operations are the concept definitions as found within the CORE-model. These definitions are transformed to Conceptual Graphs. The two operations are the basis for the right use of the vocabulary: they enable different levels of detail of the model to be shifted until a level is found where all concepts may be expressed by content words (Figure 5). They allow the detail of description in the model expression to be changed following the availability of words in the different languages and the desired language style.

The type expansion operation replaces a single concept (concept type) by its definition. This operation permits composite concepts which are not annotated, i.e. which have no direct term in natural language, to be handled. We use the minimal type expansion for preserving truth ([12], p. 109), followed by a simplify-operation for eliminating eventual redundancy.

The inverse operation, type contraction, is used to replace a subgraph by a single concept. It allows single words which represent a (complex) composite concept to be used. Two conditions have to be fulfilled: a subgraph has to correspond to the definition of such a composite concept (projection, but without admitting specialisation), and the composite concept has to be annotated by words or terms of the required syntactic category in the respective language. In this case the definition is replaced by the name of the composite concept and the annotations to the composite concept may be used for phrasing the whole subgraph.

3.2. Relational expansion and relational contraction

The relational expansion and relational contraction operations are based on definitions of relationships by graphs. These definitions cannot presently be found in the CORE-model, as they are only of interest for the natural language generation purposes. Therefore, they are held separately within the generation module. They might by integrated into the semantic model in future.

The relational contraction operation enables simplification of graphs: a subgraph containing several relationships may be replaced by a single relationship between two concepts, if the specific relationships of the subgraph are not relevant in natural language (Figure 5). These simplified relationships may be created specifically for the linguistic purposes.

The relational expansion operation is applied, whenever a relationship 'has too much content' to be expressed by a function word or inflections; parts of the semantics of the relationship have to be expressed by content words. In this case the concept included within the relationship is dissociated by the relational expansion operation.

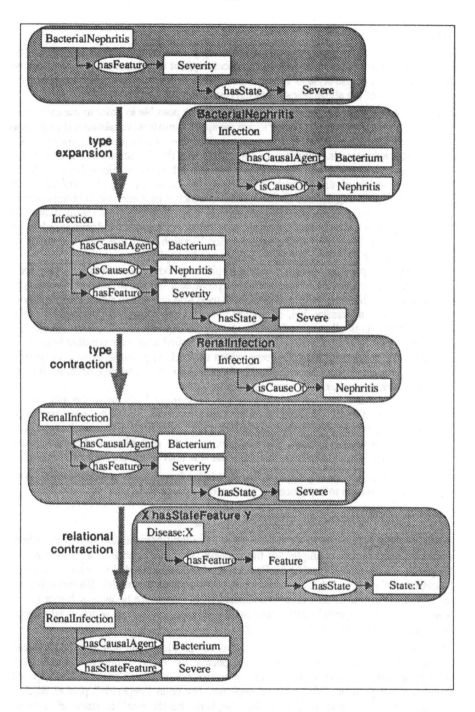

Figure 5. Example of the use of the graph operations.

3.3. Using the graph operations

The application of the graph operations is mainly guided by the desired language style. In our case the guideline is 'as detailed as necessary, but as concise as possible', following the general concise style of medical texts. This is reflected in several heuristics for the application of the above transformation operations:

- The type contraction operation is used whenever possible in order to use more concise words. Therefore, it serves to use the specific medical vocabulary. If more than one contraction is possible, the most specific definition is preferred, for using the most precise words. Figuratively, this means that the definition covering the 'largest graph' is preferred. Formally, we use the subsumption hierarchy in order to decide which of the possible concepts is the most specific one. For example, 'Nephritis' (being defined as 'BodyProcess which < hasOutcome InflammatoryLesion hasLocation Kidney>') is more specific than 'Inflammation' (being defined as 'BodyProcess which has Outcome InflammatoryLesion'). 'Inflammation' is an ancestor of 'Nephritis' in the subsumption hierarchy and therefore 'Nephritis' is preferred for the type contraction, if both can be applied.
- The type expansion operation is only applied whenever no annotation exists for a concept which would correspond to the possible syntactic structures. For the type expansion, the most general definition is used first in order to refine the graph only as much as necessary. Yet, composite concepts are recursively replaced by their definitions until all constituting concepts can be expressed in natural language.
- The relational contraction operation again is applied whenever possible in order to simplify the natural language expressions as much as possible. Also here, principally, the more specific relationship is preferred, when having several choices. However, this case rarely occurs at present.
- The relational expansion operation is also only applied whenever the relationship can not be directly expressed by a syntactic structure. As for the type expansion operation, the simplest definitions are tried first. However, until now the relational expansions are not nested. The complexity of the definitions and the necessary expansions is much lower on the relational side than on the concept side.

All operations are recursively applied. The heuristics for using the graph operations constitute a central means to influence the generated language style.

3.4. Differences between conceptual and linguistic view

Sometimes the definitions in the model differ from the linguistic decomposition which would be needed. For example, for the concept 'BacterialNephritis' a definition like 'Nephritis which < hasCausalAgent Bacterium >' would be needed in order to generate 'bacterial nephritis' or 'néphrite bactérienne'. However, in the model 'BacterialNephritis' is defined as 'Infection which < hasCausalAgent Bacterium isCauseOf Nephritis >', which results in 'bacterial renal infection' for English ('Infection which < isCauseOf Nephritis>' being contracted to 'RenalInfection').

There are three principal possibilities to solve this problem. The first, not the preferred one, is to annotate the concept 'BacterialNephritis' by 'bacterial nephritis' for English and by 'néphrite bactérienne' for French. In this case both terms have to be decomposed linguistically (syntactically): 'bacterial nephritis' is a noun ('nephritis') plus an adjectival complement ('bacterial'), and also 'néphrite bactérienne' consists of a noun ('néphrite') plus an adjectival complement ('bactérien'). Only with this information can the terms be used correctly in more complex phrases. This solution is generally valid.

However, the linguistic decompositions are needed for every language where a composed term is used, i.e. the annotation effort is multiplied by the languages in use.

The second one would be to transform the graph by transformation operations to the form 'Nephritis which < hasCausalAgent Bacterium >'. However, this would presume some different concept definitions in the core model, e.g. 'Infection which < isCauseOf Nephritis >' defining a 'Nephritis' and not a 'RenalInfection'. This would be the most convenient solution for the generation module, but would limit the modellers too much by language restrictions.

The third possibility is to have in addition to the conceptual definition a 'linguistic definition' (Figure 6), different from the conceptual definition. It has to represent in a language-independent way how the concept is phrased in language. 'In language' means here 'in medical language', often common to the several natural languages: 'Bacterial-Nephritis' may be phrased as 'bacterial nephritis' in English, 'néphrite bactérienne' in French, 'bakterielle Nephritis' in German etc. In the medical language many parallels of this type may be found between different languages, often due to the extended use of latin vocabulary. It is then up to the generator to bridge the differences between conceptual view and linguistic view by the transformation operations (recognizing the semantic definition and replacing it by the linguistic one): a type contraction using the conceptual definition is followed by a type expansion using the linguistic definition.

At the moment, we have implemented a combination of the third possibility and parts of the first possibility, which is formally more proper, in order to economise annotation effort.

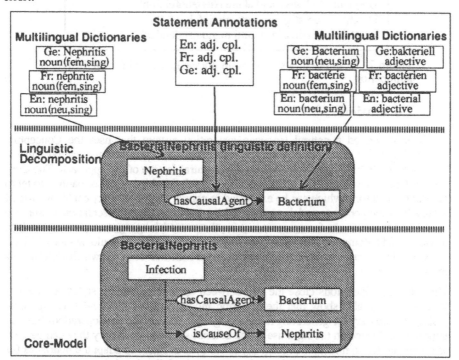

Figure 6. Difference between conceptual definition and linguistic decomposition for a composed concept.

4. Applications

Within the GALEN project, a demonstrator application has been implemented. This application SCUI (Structured Clinical User Interface) is based on the CORE-model and uses the language generation services [20]. It collects information about urinary infections using a structured input interface and shows summaries of the entered data which are generated by the MM. Due to the use of the language-independent model and the generation services, the application is both available in English and in French. As an experiment, the MM has also been extended to other languages (German, Dutch, Spanish, Italian, Finnish) for simple phrases and parts of the model necessary for the application have been annotated in these languages. It thereby was possible to have the input interfaces available in all these languages and to have an output in a language other than the input language (Figure 7).

```
(RenalInfection which
<          hasChronicity (Chronicity which hasState acute)
           hasSeverity (Severity which hasState severe)>)


English:   'severe and acute renal infection'
French:    'infection rénale sévère et aiguë'
Italian:   'infezione renale grave e acuta'
Spanish:   'infección renal grave y agúda'
German:    'schwere und akute Niereninfektion'
Dutch:     'ernstige en acute nierinfektie'
Finnish:   'vakava ja akuutti munuaistulehdus'
```

Figure 7. Example of a simple noun phrase generated for the SCUI application

5. Conclusion

We have described a natural language generation system that uses the Conceptual Graphs formalism and especially the graph operations in a practical application. Indeed, the use of the transformation operations are a central feature of the approach. The Conceptual Graph operations allow the concepts and relationships of the semantic model to be adapted to a form which can be expressed in natural language. They enable the use of a language-independent model, as they bridge between the representation within the model and the different forms necessary for the several languages. A purely linguistic model would automatically be more language-dependent - e.g. it would be more difficult to switch between word levels. The current approach has been shown flexible in this respect.

The relational operations close the gap between specific modelling styles and natural language. They also show that the semantic model contains more relationships than would be necessary for generating natural language. Though the type operations have an important influence on the right use of the domain specific vocabulary they can be seen as an instrument to tune the language style. We plan in future to adapt the heuristics for the use of these operations to different language styles following the users of an application.

Experiments with respect to the adaptation to new languages have been successful. They have shown the approach to be multilingual. The overall principles of the approach can be easily transferred to other language and grammatically correct phrases can be generated. The main problem remains to achieve the specific medical language style, in every language. The subtilised utilisation of the graph operations and the refinement of the statement annotations seem the right means to solve this problem.

An important perspective of this work is to share the experience of closing the gap between linguistic and conceptual view with other natural processing tools, especially for natural language analysis. Yet, also the linguistic knowledge sources shall be shared, in form of a so-called 'Medical Linguistic Knowledge Base' [21].

Furthermore, the experience with this system has showed that the features of the Conceptual Graphs formalism are well-suited to develop such an application.

ACKNOWLEDGEMENTS

The GALEN project (A2012) is funded by the European Union within the AIM (Advanced Informatics in Medicine) programme. The Swiss part of the project is fully supported by the Swiss government (CERS - Commission d'encouragement à la recherche scientifique).

We thank the GALEN members of the Medical Informatics Department of the University of Manchester for their support and cooperation.

References

[1]   Musen MA. Dimensions of Knowledge Sharing and Reuse. Comput Biomed Research 1992; 25:435-467.

[2]   Rector AL, Nowlan WA, Kay S. Conceptual Knowledge: The core of medical information systems. In: Lun KC et al. (eds). MEDINFO 92. Proceedings. Amsterdam: North-Holland, 1992: 1420-1426.

[3]   Evans DA, Cimono JJ, Hersh WR et al. Toward a Medical-concept Representation Language. J Am Med Informatics Assoc 1994; 1:207-217.

[4]   Ceusters W, Deville G, Buekens F. The Chimera of Purpose- and Language Independent Concept Systems in Health Care. In: Barahona P, Veloso M, Bryant J (eds). MIE 94. Proceedings. 1994: 208-212.

[5]   Campbell KE, Wieckert K, Fagan LM, Musen MA. A Computer-based Tool for Generation of Progress Notes. In: Safran C (ed). SCAMC 93. Proceedings. New York: McGraw-Hill, 1993: 284-288.

[6]   Bernauer J, Gumrich K, Kutz S et al. An Interactive Report Generator for Bone Scan Studies. In: Clayton PD (ed). SCAMC 91. Proceedings. New York: McGraw-Hill, 1991: 858-860.

[7]   Kuhn K, Zemmler T, Reichert M, Heinlein C, Roesner D. Structured Data Collection and Knowledge-Based User Guidance for Abdominal Ultrasound Reporting. In: Safran C (ed). SCAMC 93. Proceedings. New York: McGraw-Hill, 1993: 311-315.

[8]   Lindberg DAB, Humphreys BL, McCray AT. The Unified Medical Language System. Meth Inform Med 1993; 32: 281-291.

[9]   Rothwell DJ, Cote RA, Cordeau JP, Boisvert MA. Developing a Standard Data Structure for Medical Language - The SNOMED Proposal. In: Safran C (ed). SCAMC 93. Proceedings. New York: McGraw-Hill, 1993: 695-699.

[10]   Rector AL, Solomon WD, Nowlan WA, Rush TW, Zanstra PE, Claassen WMA. A Terminology Server for Medical Language and Medical Information Systems. Meth Inform Med 1995; 34: 147-57.

[11]   Rector AL, Nowlan WA, Glowinsky A. Goals for Concept Representation in the GALEN project. In: Safran C (ed). SCAMC 93. New York: McGraw-Hill, 1993: 414-418.

[12]   Sowa JF. Conceptual Structures: Information Processing in Mind and Machine. Reading, MA: Addison-Wesley Publishing Company, 1984.

[13]   Wagner JC, Solomon WD, Michel P-A, Juge C, Baud RH, Rector AL, Scherrer J-R. Multilingual Natural Language Generation as Part of a Medical Terminology Server. To be published in: Greenes RA, Peterson H, Protti D (eds). MEDINFO 95. Proceedings of the 8th World Congress on Medical Informatics, Vancouver, Canada, July 23-27, 1995.

[14]   Baud RH, Rassinoux A-M, Scherrer J-R. Natural Language Processing and Semantical Representation of Medical Texts. Meth Inform Med 1992; 31: 117-125.

[15]   Baud RH, Rassinoux A-M, Wagner JC et al. Representing Clinical Narratives using Conceptual Graphs. Meth Inform Med 1995; 34:176-86.

[16]   Rassinoux A-M, Baud RH, Scherrer J-R. A Multilingual Analyser of Medical Texts. In: Tepfenhart WM, Dick JP, Sowa JF (eds). Conceptual Structures: Current Practices. Second International Conference on Conceptual Structures, ICCS'94. Berlin: Springer-Verlag, 1994: 84-96.

[17]   Wagner JC, Rassinoux A-M, Baud RH, Scherrer J-R. Generating Noun Phrases from a Medical Knowledge Representation. In: Barahona P, Veloso M, Bryant J (eds). MIE 94. Proceedings. 1994: 218- 223.

[18]   Nogier J-F, Zock M. Lexical Choice as Pattern Matching. In: Nagle T, Nagle J, Gerholz L, Eklund P (eds). Current Directions in Conceptual Structures Research. Reading, MA: Addison-Wesley, 1991.

[19]   Nogier J-F. Génération automatique de langage et graphes conceptuels. Paris: Hermès, 1991.

[20]   Alpay LL, Lovis C, Nowlan WA et al. Constructing Clinical Applications: The Galen Approach. To be published in: Greenes RA, Peterson H, Protti D (eds). MEDINFO 95. Proceedings of the 8th World Congress on Medical Informatics, Vancouver, Canada, July 23-27, 1995.

[21]   Baud R, Lovis C, Rassinoux A-M et al. Towards a Medical Linguistic Knowledge Base. To be published in: Greenes RA, Peterson H, Protti D (eds). MEDINFO 95. Proceedings of the 8th World Congress on Medical Informatics, Vancouver, Canada, July 23-27, 1995.

Table 10 left without paying the bill ! a good reason to treat metonymy with conceptual graphs

Tassadit Amghar, Françoise Gayral, Bernard Levrat

LIPN, URA 1507 CNRS
Université de Paris-Nord
Institut Galilée, Dpt informatique
Av. J.-B. Clément
93430 Villetaneuse (France)
e-mail: {ta,fg,bl}@ura1507.univ-paris13.fr

Abstract. Metonymy is a valuable test bed to appreciate the adequacy of a model of referential mechanisms occurring in natural language surface forms. Our concern in this paper is to propose an alternative treatment to deal with metonymy. We benefit considerably from the ability of conceptual graphs to express semantic norms for implementing a kind of preferential semantics and make special use of its normative apparatus including the type lattice used in the application, canonical graphs, conformity relation...The originality of our approach concerns two main points:
First, we treat not only nominal metonymies, as traditional works do, but also verbal ones. Whereas verb is always, in these approaches, considered as steady, we consider that the verbal operator itself can be a metonymical reference. This not only enlarges the field covered but also allows us to reach more realistic solutions.
The second point concerns the use of classical metonymy typologies. Canned metonymies may intervene in two ways, as guides and as filters; in the first case, they render the combinatorial problem manageable by focusing the search during the resolution; in the second case, they are a means to rank the relevance of the provided solutions.

1 Introduction

It is now commonly accepted that the conceptual graph formalism (henceforth cg) introduced by John Sowa [10] is well suited for natural language (henceforth NL) applications. As a result a lot of NL treatment systems use it as knowledge representation language ([16], [1], [11], for example). Nevertheless many difficult problems still remain to be resolved such as the treatment of some language figures, metonymy or metaphor for example, or the resolution of ellipsis, etc.
In this paper, we adress the problem of metonymy and propose a treatment to deal with it. This treatment relies on a kind of preferential semantics and so benefits considerably from the ability of cg to express semantic norms through their normative apparatus including the type lattice used in the application, canonical graphs, conformity relation, ...

The first part of the paper is devoted to the characterization of metonymy. We see it primarily as a referential phenomenon and emphasize this view by contrasting it with another trope, metaphor. Both of them imply a shift, but we show that the shift concerns meaning in the case of metaphor, and reference in the case of metonymy.

The second part is the heart of the paper. We describe our propositions then go on to detail the treatment with some illustrative examples. We dissociate ourselves from traditional approaches on two major points: first, on the nature of the element which could be metonymical, second in the way we use classical typologies of metonymies. Canned metonymies intervene as guides and as filters; respectively, they help render the problem tractable by focusing the search during the resolution and they permit us to embody a preferential principle to rank the relevance of the provided solutions.

In the third part, we present some related works and then we conclude briefly with some promising perspectives.

2 A characterization of metonymy

Metonymy can be defined as *a NL figure in which a concept is expressed with a term generally associated with another one, on the basis of the existence of a necessary relation between them.*

So, for example, the term *glass* used in *He drinks his glass* is metonymical with respect to this definition as it stands for *the contents of his glass*.

The following are examples of these relations which give rise to classical metonymies:

- cause/effect: *This **fever** carried him off at the age of 30* for *This **illness** carried him off at the age of 30*

- container/contents: *The **theater** applauds* for ***People in the theater** applaud.*

- a sign/the thing it symbolizes: *The English **crown** is ill-treated by the press* for *The English **royalty** is ill-treated by the press*

- abstract/concrete: ***weak sex** is strong* for ***women** are strong*

- producer/product, and more specifically artist-art form: *He bought a **Van Gogh*** for *He bought a **picture by Van Gogh***

- material/object made of this material: *What an interesting **paper** !* for *What an interesting **article** !*

These examples show how the existence of a form of conventional relation between two terms allows the speaker to use one term for the other, and the recipient of the communication to understand it, since these relations enable him to restore the intended term.

Such relations are often qualified as necessary since they may be observed in daily life, and exist independently of the terms used to express them. We say that they describe the world organization, the structure of world-objects, the internal structure of processes or states of facts in which these objects occur. From these considerations, three conclusions result:

• First, since metonymy appears to be the substitution of a term for another one, it is a surface phenomenon and so concerns reference and not semantics. In as much, we differ from those who treat it as a shift of meaning. For example, it seems to us more realistic to present the metonymical term *glass* in *to drink a glass* as an economic way to express *the contents of a glass* rather than as a shift of meaning of *glass* with its sense becoming *the contents of a glass*. So if a shift occurs - and that is a good image of the occurrence of a change-, it has to be qualified as referential as opposed to semantic.

• Second, extreme variability of the conventional aspect of metonymies has to be considered.
Some of them are entirely lexicalized and figure as ordinary entries in the dictionaries, occasionally mentioned as metonymical[1]. In some other cases, although conventional, they are not at all explicitly mentioned in dictionaries. They may also be purely contingent and dependant on the context of the speech. For example, in the context of a restaurant it is possible to refer to a patron by using the number of the table he is sitting at. It is more probable that you will say *Table 10 left without paying the bill !* if you are a waiter in a restaurant than *The patron sitting at the table number 10 left without paying the bill !* to express the same state of the world. If you have no doubt of it, you will perhaps have some chance to catch **him** up ! Here is a good reason to accept and treat such metonymies ! More seriously, such contingent metonymies are so commonly used that it seems to us very important to treat them and so, we differ from other approaches which restrict the phenomenon to the conventional [5].

• Third, metonymy rests on a kind of contiguity relation between concepts corresponding to metonymical terms. As such, it can be opposed to the metaphor phenomenon, since in that case, implied terms entertain some sort of similarity relation. From that relation results a modification of the meaning of one term which is overloaded with some semantic features of the other. In the classical example, *The car drinks gasoline* [15], using *drinks* in place of *uses* creates an analogy between the animate type, the preferred agent type of *to drink*, and the car type and similarly between *gasoline* and *beverage*. Some features of animates thus impregnate the car while at the same time gasoline becomes thing drinkable. It is then possible to draw the metaphor out and the car may revive, have a bit of a cough, and so on.
So, in metonymy and metaphor, the recipient of the communication has to operate a shift. But if he can be helped in the case of metonymy by a set of registered ones and by conventional relations, it is not the case when he is faced with a metaphor. Analogy relations are completely created by metaphoric sentences: the analogy between a car and a person does not exist outside the metaphor itself. This is why it seems easier to treat metonymy with better chance of success.

---

[1] Note that dictionaries do not agree with each other: the same metonymical term which has an ordinary entry in one dictionary, may be tagged as metonymical in a different one, and not even be mentioned in yet another one.

3 Our proposal

This is not the place to present conceptual graphs as a knowledge representation language well suited to NL applications. The great number of studies, models and NL systems built around this formalism attests to this now well-known fact. So we will just describe our proposal by first exposing the general principles underlying our approach, then by providing a more detailed description, followed by the treatment of some illustrative examples.

3.1 The methodology

Normative apparatus of cg offers a good framework to different kinds of norms associated with NL. The type requirement of relations and casual structures associated with verbal operators allows us to detect conceptual conflicts. Hierarchy and canonical graphs which indicate prototypical states of affairs, help us to retrieve the necessary relation which permitted metonymy to occur.

The choice of a preferred interpretation.

It is often the case that surface forms in NL may be associated with multiple meanings. The choice of an internal representation free of ambiguity can be obtained by applying an *isotopic principle* which, briefly, states that elements of a sentence tend to put overlapping constraints on the linguistic environment they constitute for each other. The emergent interpretation is the one which fits best these interweaving constraints.

For example, in *Paul plays Bach*, Fass [5] takes the twelfth registered meaning for the verb *to play* in his lexicon without justifying that choice. It is patent that such a choice rests on the knowledge of J-S.Bach as a composer. If the complement had been *baseball*, another sense for *to play* would have been chosen. We encounter here what is called *co-compositionality* by Pustejovsky [9]: it is an extension of the *selectional restrictions* [7] that an operator puts on its argument types, to the symmetric case where arguments themselves put similar constraints on the predicate they are governed by[2].

We use such an isotopic principle to guide ourselves towards a preferential interpretation among the possible ones in case of ambiguous expressions.

So, the first step of our treatment consists in a preliminary semantic analysis using a methodology inspired by [Veronis Ide 90]. It mainly consists in the propagation of activation markers. Each possible representation of lexical items occurring in the sentence is activated and then activates the adjoining concepts, either in canonical graphs or in the types lattice. If no collision occurs, the corresponding interpretation of the different items in the sentence is excluded. In case of multiplicity of possible interpretations, a preferential choice has to be made. This choice has yet to be grounded on a criterion which takes into account the number of hierarchical and relation links crossed. It would be some sort of semantic weighing of a configuration,

---

[2] Note that this is the basis of embodiment of a preferential semantics, see [15] for an early approach in this direction.

some weakening factor rendering the semantic distance for joining the chosen interpretations for each term during the marking process.

Treatment of metonymy.

A preferred meaning having been chosen for each item by this preliminary treatment, some breach of the semantic norm may persist, in particular when the sentence includes a metonymy. There begins the principal phase of treatment of possible metonymies. It occurs when the cg representation which has been obtained after the first phase conflicts with the semantic requirements of the casual structure of the verbal concept[3]. In the latter case, we apply traditional treatments, but we propose two ways to restore the coherence.

• First, the verb (the operator) is considered as steady and its arguments as possibly metonymical. In concrete terms, we seek a concept which fits the casual constraint of the verb and look for in the canonical cg[4], a path between the faulty concept (the target), which is the one resulting from the parse and the source which is the one predicted by the verbal requirement.

• Second, we consider the dual method: the nominal expressions are the "good" ones and the verb is metonymical. This way is original since, to our knowledge, no work proposes a search in that direction. Indeed, traditional treatments always interpret semantic abnormalities as possible discrepancies between given and intended concepts occurring in argument positions of the verbal structure. Nevertheless, we think that it is possible to find conventional relations between two verbs: looking at the relation between *to blush* and *to be ashamed* or between *to tremble* and *to be afraid* gives an idea of what underlying relations are at play ; generally they are of type antecedent/consequent or effect/cause. Consider for example the sentence *The radio was playing a tune.* The preferential interpretation *to play* as *to play music* requires a human filling the agent case, and a tune filling the object one. A traditional treatment of metonymy would search for human beings linked to radio by conventional relations in order to satisfy the requirement of the agent case and would find radio technicians, disk jockeys, speakers.... But these possible solutions, although apparently correct, disagree with the intuitive interpretation.
On the other hand, if instead of searching for a substitute of the agent concept *radio*, we search for one of the verb *to play*, something like *The radio was **giving out** a tune* becomes reachable. Accessing to typical knowledge associated with this meaning of *to play*, permits us to link the action of playing to its intended effect of giving out music. This gives rise to this last interpretation and it is more satisfactory than the previous one.

3.2 The detail of the treatment

As we have explained before, our treatment is twofold solving both nominal and verbal metonymies. In the two cases we use a set of classical registered metonymies. It permits us to gain in two ways: first they help control the complexity of the search

---

[3]We say verbal concept to mean concept corresponding to the verbal operator.

[4] Notice that an indexing mechanism has to be used to retrieve relevant canonical graphs for the process to be tractable.

for solutions by limiting the length of chains taken into consideration in the case of successive metonymies, second they establish a preferential criterion to choose among competing solutions.

They are stored as a set of *metonymical schemata* which link two metonymical concepts by a relation labeled *meto*.

Following are the metonymical schemata corresponding to a relation between:

- a container and its contents:
    relation meto1(x,y) **is**
    [entity:*x] --> (part) --> [interior:*z] <--- (loc)<---[entity:*y]

- a cause and its effect:
    relation meto2(x,y) **is**
    [state[5]:*x]--->(cause)--->[state:*y]

- a producer and its product:
    relation meto3(x,y) **is**
    [productor:*x]<---(agt)<---[create]--->(obj)--->[product:*y]

- an instrument and its user:
    relation meto4(x,y) **is**
    [person:*y]<----(agt)<----[act]--->(inst)--->[entity:*x]

- the part and the whole:
    relation meto5(x,y) **is**
    [entity:*y]<----(part)<----[entity:*x]

- an artist and his work:
    relation meto6(x,y) **is**
    [artist:*x]<---(agt)<---[compose]---(obj)--->[work:*y]

We take advantage of the homogeneous representations given by cg to verbal operators and to their arguments, both being concepts. So cg used to represent registered metonymies (concepts linked by a relation labeled *meto*) are identical in both cases and they can be used indifferently for first or second order concepts such as those which correspond to events, states, or action, for example.

Resolving nominal metonymies.

Linking a faulty concept (the target) and a source concept may require two kinds of processes depending on the knowledge source wich is used. Such a connection between the target and the source can be found either in the type definition of the target concept, or in the canonical graphs which contain a type compatible with it.

A• If the target concept corresponds to a defined type, the expansion of this type is achieved.
B• If not, a directed join on the target concept is done between the initial graph and the canonical graphs which contain the target concept, the compatible concept having been marked.

---

[5] The type 'state' gathers the types event, process, activity...

C• In the case where the source concept doesn't occur in the "added part" resulting from the directed join, the following operations are processed on the "added" concepts:

    C1--> their expansion, if possible.

    C2--> a search for the graphs containing them.

    But there may be a huge number of added concepts, and we have to limit the combinatorial explosion. For this purpose, we use registered metonymies for guiding the choice of the relevant concepts. We retain only the concepts which are in a relation meto with the target concept. So, when a canned metonymy of the form meto(target_concept, x) is detected, x becomes the new target and the treatment resumes the search for the source concept. So we may explain a metonymy as a string of successive elementary ones.

D• If the source concept has been found

    D1--> The source concept becomes the new value of the initially faulty relation. In the graphic representation that corresponds to have the outgoing edge of the relation node pointing to the source concept.

    D2--> Then we try to match a registered metonymy of the form

        meto(target_concept, source_concept).

Note that in order to detect a registered metonymy meto(x,y) we try to abstract the part of the graph linking concepts x and y and to match it with the definition of meto(x,y).

Succeeding cases are thus the following:

- The metonymical reference was immediately solved: the source concept was found directly in the typical knowledge associated with the target concept. In this case, registered metonymies are used first as a way to sort solutions when they are multiple[6], second as a kind of justification which both underlies and serves as a foundation of this path.

- The metonymical reference was obtained as the result of a sequence of registered metonymies those which guide the search. It may seem paradoxical to exclude some metonymies, the un-registered ones, as it can be the case here, since we proposed initially to solve all of them. In fact, it allows us to control the process of weighing the distance between initial information and information used to access to the result.

Treating verbal metonymies.

Here the approach consists in taking the casual structure associated with the predicate (i.e. the verbal operator) as being the source of the semantic norm definition. This relational environment defines the steady points, some kind of anchor which doesn't have to be reconsidered in the process of restoring the norm by having some part of the representation changed. So instead of looking for argument replacements, we look for verbal operator replacements.

The treatment follows three main steps:

    A'• Each concept which doesn't fit the casual types requirements of the verbal operator is marked in order to process the convenient directed join in the following step.

---

[6] A way to prefer the "normal" is to prefer the "registered".

B'• The concepts linked with the initial verbal operator and the relations which connect them compose the *relational environment* of the verbal operator. This environment determines the pattern we look for among the canonical graphs. In this environment the location of the verbal operator has to be considered as a gap. It will eventually be filled by a verbal operator during the match between the environment and a canonical graph.

C'• Finally, each canonical graph selected is joined on the marked concept with the initial graph. We then try to establish a correspondence between the two verbal concepts in the resulting graph of the directed join.

3.3 Examples

The proposed treatment is illustrated through the following examples:

(1) *the theater applauds*
(2) *Paul plays Bach*
(3) *The radio plays a musical piece of Bach*
(4) *The radio was playing Bach*

In (1) and (2) metonymies concern concepts having an argument position in the casual structure of the sentence (they are cases of "nominal metonymies"). Both correspond to registered metonymies, the first one known as the *container-content* one, the second known as the *artist-art form one*.

(3) is a case of a simple metonymy bearing on a concept corresponding to a verb (verbal metonymy) and is an example of the *cause-effect* metonymy.

In (4) a double metonymy occurs, on the verb (*cause-effect*) and on argument (*artist-art form*).

The theater applauds* for *People in the theater applaud.

The sentence corresponding to:

(1.1) [applaud]-(agt)→[theater: *x]

doesn't fit with the norm defined by the canonical graph (1.2)

(1.2) [applaud]-(agt)→[person:{*}]

Among the canonical graphs where [theater] occurs, let us mention:

(1.3) [contain]-
    (loc)→[theater: *x]→(part)→[interior: *z]
    (obj)→[person:{*}]→(loc)→[interior: *z]

(1.4) [contain]-
    (loc)→[theater: *x]→(part)→[interior: *z]
    (obj)→[furniture]→(loc)→[interior: *z]

A directed join of (1.1) on [theater] with graphs (1.3) leads to (1.5):

```
(1.5) [contain]-
        (obj)→[person:{*}]→(loc)→[interior: *z]
        (loc)→[theater: *x]-
                →(part)→[interior: *z]
                ←(agt)←[applaud]
```

The source concept occurs in (1.5). The graph (1.6) results from a move of the outgoing edge of the (agt) relation towards the concept ([person: {*}]) of the relevant type in the graph (1.5).

```
(1.6) [contain]-
        (loc)→[theater:*x]→(part)→[interior: *z]
        (obj)→[person:{*}]-
                →(loc)→[interior: *z]
                ←(agt)←[applaud]
```

A particularization of the meto1 relation gives

```
(1.7) [theater:*x]→(part)→[interior: *z]←(loc)←[person:{*}]
```

and abstracting the relevant part of (1.6) gives rise to (1.8) where the meto1 relation appears.

```
(1.8)    [person:{*}]-
                ←(agt)←[applaud]
                ←(obj)←[contain]→(loc)→[theater:*x]
                ←(meto1)←[theater:*x]
```

Thus it becomes possible to explain the substitution of *people* (more exactly *set of persons*) for *People in the theater*.

Paul plays Bach for Paul plays a musical piece composed by Bach.

(2.1) is the representation of the initial sentence in terms of cg

```
(2.1) [person: 'paul']←(agt)←[play12⁷]→(obj)→[composer: 'bach']
```

The type composer clashes with the type of the complement required by play12 as given by (2.2)

```
(2.2) [person: *]←(agt)←[play12]→(obj)→[musical_piece]
```

So we take into account the type definition (2.3) for *composer*:

---

⁷ *play12* corresponds to the meaning *play music* used by [Fass 91] for the verb *to play*. This interpretation was chosen in a previous step which had applied the isotopic principle.

```
        type composer(x) is
(2.3)   [artist: *x]←(agt)←[compose]→(obj)→[musical_piece]
```

The expansion of this type leads to (2.4):

```
(2.4) [artist: 'bach']-
          ←(obj)←[play12]→(agt)→[person: 'paul']
          ←(agt)←[compose]→(obj)→[musical_piece]
```

With the following knowledge:

• The definition of the meto6 relation

```
        relation meto6(x,y) is
(2.5)   [artist: *x]←(agt)←[compose]→(obj)→[work: *y]
```

•the position of the concept *musical_piece* in the type lattice

musical_piece < work

• the particularization of the meto6 relation which links an artist to one of his works, and specifically an artist to one of his musical_pieces. We reach (2.6)

```
(2.6) [musical_piece]←(meto6)←[artist]
```

Since the meto6 states a relation between an artist and his musical_piece, a possible solution goes from the substitution of the link between play12 and artist in (2.4) to a link between play12 and musical_piece, giving (2.7):

```
(2.7) [musical_piece]-
              -(obj)←[play12]→(agt)→[person: 'paul']
              -(obj)←[compose]→(agt)→[ artist: 'bach']
```

(2.7) which roughly corresponds to *Paul plays a musical piece composed by Bach* agreeing with the semantic norm and explicates the originally implied relation in (2.1).

The radio plays a piece composed by Bach **for** *The radio broadcasts a musical piece composed by Bach.*

(3.1) is the initial representation

```
(3.1) [musical_piece]-
              -(obj)←[play12]→(agt)→[radio:*]
              -(obj)←[compose]→(agt)→[artist: 'bach']
```

The concept [radio] breaches the agent case of the concept [play12] and gives access to the canonical graph (3.2).

(3.2) [radio]←(loc)←[work]→(agt)→[sound_engineer]

The positions of musical_piece and sound_engineer in the type lattice are given by

musical_piece<music & sound_engineer<person

(3.3) is the result of the directed join on [radio] between (3.2) and (3.1):

(3.3) [radio:*]-
    ←(loc)←[work]→(agt)→[sound_engineer]
    ←(agt)←[play12]→(obj)→[musical_piece]-
                            -(obj)←[compose]→(agt)→ [artist: 'bach']

"sound_engineer<person" allows to redirect the outgoing (agt) link of [play12] towards [sound_engineer]:

(3.4) [play12]-
    -(agt)→[sound_engineer]←(agt)←[work]→(loc)→[radio:*]
    -(obj)→[musical_piece]←(obj)←[compose]→(agt)→[artist: 'bach']

Once the norm is restored, an abstraction is attempted on the (3.5) graph contained in (3.4):

(3.5) [radio]←(loc)←[work]←(agt)←[sound_engineer]

We found no justification for the implied relation in the metonym use of *radio*; such a justification would be the existence of a registered metonymy (location of work for the employees). It is a contingent but nevertheless accepted metonymy.

The search focuses now on [play12] to try to find a concept which would have a similar relational environment (i.e. the same relations between the search concept and compatible types in their corresponding concept).

[radio] indexes (3.6) in the canonical basis.

(3.6) [radio]←(agt)←[broadcast]→(obj)→[sound]

The directed join on [radio] between (3.6) and (3.1) gives (3.7)

(3.7) [radio]-
    (agt)←[broadcast]→(obj)→[sound]
    (agt)←[play12]→(obj)→[musical_piece]←(obj)←[compose]-
                                   -(agt)→[artist: 'bach']

The type lattice indicates musical_work<sound, so we look for a relation linking [broadcast] and [play12]

```
(3.8) [State: [play12]→(obj)→[musical_piece]]-
                -(caus)→[State: [broadcast]→(obj)→[sound]]
```

An abstraction operation leads to meto2 (Cause-effect) and, since it is a registered metonymy, it is preferred to the previous one.

```
(3.9) [musical_piece]-
        -(obj)←[broadcast]→(agt)→[radio]
        -(obj)←[compose]→(agt)→[artist: 'bach']
```

The radio plays Bach for ***The radio plays a piece composed by Bach.***

Here the treatment is only outlined.

```
[Radio]←(agt)←[play12]→(obj)→[composer: *'bach']
```

Here there are two breaches of the type requirement on the casual structure associated with [play12]. Solving the "nominal metonymies" requires the following steps:

```
(4.1) [sound_engineer]-
        -(agt)←[work]→(loc)→[radio:*]
        -(agt)←[play12]→(obj)→[composer: *'bach]
```

Then:

```
(4.2) [musical_piece]-
        -(obj)←[compose]→(agt)→[ artist: 'bach']
        -(obj)←[play12]→(agt)→[sound_engineer]←(agt)←[work]→(loc)→[radio:*]
```

And finally through the search of a verbal metonymy:

```
(4.3) [musical_piece]-
        -(obj)←[broadcast]→(agt)→[radio]
        -(obj)← [compose]→(agt)→[artist: 'bach']
```

4 Some related works

4.1 Pustejovsky's generative lexicon

[9] treats metonymy as a particular case of type coercion, a semantic operation which converts an argument to the type expected by a function in case of a type error. In case of conflict between a complement and the expected type as in (1) *John began a book* (*begin* requires an argument of type transition while *book* is a physical object), the complement is coerced to the correct type ; this coercion is obtained by using the information associated with the 'incorrect' item in the lexicon (in the example, *book*),

in a structure called the *qualia*. A minimal description of the item *book*, for example, includes in its *qualia* structure, the verb *write* in the agentive role (since writing is the way to create a book) and the verb *read* in the telic role (since to be read is the primary function of a book). In sentence (1), the verb *begin* coerces the word *book* in one of these two verbs, both of the required *activity* type.

This approach, though interesting, will have to tackle the problem of chains of metonymies. This will occur when the qualia structure does not include the "good" information, the one which rightly fits with the sought type.

4.2 Two outlines of solutions

[12] places the treatment of the metonymy in an intermediary stage of the analysis. After the parsing stage, the presence of metonymical references in a representation is revealed by the existence of a relation having a syntactic label (*subject, complement*, for example) and not a conceptual one. So, these "bad" relations trigger treatments related to the restoration of the "good" concepts. The suggested treatment consists in introducing a graph between the two concepts linked with the faulty relation ; this graph is, in fact, the expansion of a lambda-expression. Although the crucial point, i.e. the building of this lambda-expression, is not described, this work gives some promising tracks to explore.

Another way to treat the metonymy [4] is to group all the potential metonymical meanings of an item in a single structure indexed by it. This structure is a lambda-abstraction with one variable for each metonymical use of the item. Searching for one of these particular uses consists in projecting the function described by the lambda abstraction onto the adequate variable.

Factoring the knowledge necessary to deal with metonymical shifts seems a good trick, but destroys natural groupings. What is gained on one side is lost on another. And what about "new" senses that could arise generatively by composition with other words ?

4.3 *Collative semantics*: a preferential semantic theory to deal with met*aphors* and met*onymies*

The program of Dan Fass [Fass 91] meta5 which recognizes both metonymy and metaphor[8] relies on a preference semantics where the verb, the pivot of the analysis, determines the type requirements on its arguments.

Faced with a type breach, meta5 tries to apply metonymical inference rules as 'part for whole', 'artist for art form'... These rules, when triggered off, build a path between the target (the type of the complement present in the sentence) and the source (the object preference). They involve lexical semantics, hierarchy of types, basic knowledge related to the source and the target and are directed either from the source to the target, or vice-versa, depending on the nature of the metonymy.

So meta5 tests iteratively every metonymical rule and their potential chains until it finds a compatibility ; if this is not the case, it tries a metaphorical interpretation. The list of classical, pre-defined metonymies used to choose a solution among several,

---

[8] This explains the name of the method: met*.

embodies the principle 'prefer the normal, even in the abnormality'. On an other side it includes a constructive use in which this list leads by itself to the resolution of the metonymy.

A bias to this method is that it prohibits the resolution of contingent metonymies which are, by definition, not classical and so, not present on the list.

Moreover, we consider that this treatment is partial since metonymy is always considered as occurring in the presence of a breached selection restriction enforced by the verb and so is treated solely as a nominal phenomenon.

5 Conclusion

Metonymy is such a common language phenomenon that it occurs even in simple NL applications. So it can be considered as a good subject to test the adequacy of methodologies, tools or knowledge representation languages used to treat realistic NL applications.

The cg formalism is well suited for our treatment of metonymy since it enables us to integrate preferential principles in our model ; it also offers a set of operations and homogeneous representations for knowledge of different kinds.

The present paper is intended to outline the main principles for the resolution of metonymy and implementation has now begun in the Cogito platform environment[9] [2].

Some works have already tackled metonymy but our approach differs from them in the following points:
- first, in our use of registered metonymies which serve both as guides to generate possible paths during the search and as filters to rank eventually multiple solutions.
- second, in our treatment of both nominal and verbal metonymies. Verbs are usually considered as steady. They determine the casual structure requirements of the sentence, and only concepts filling argument roles may vary in order to re-establish the conformity with that norm in case of a breach. By considering the concept corresponding to a verb as a possible metonymical reference, we not only enlarge the field covered but also possibly reach more realistic solutions.

Metonymy responds to a kind of economical principle governing expressions in NL, and can be seen as a stylistic device, but its use goes far beyond these two functions. A third one, which is often neglected, is the accomplishment of a communication goal through the choice of a specific form of expression to refer to a state of the world. This last function may be the source of some promising extensions to this work.

6 References and bibliography

1. Anne Bérard-Dugourd, Jean Fargues, Marie-Claude Landau]: "*Natural Language Analysis using Conceptual Graphs*", Proceedings of the International Computer Science Conference, Hong Kong, 1988.
2. Boris Carbonneill, Ollivier Haemmerlé: *Implementing a CG Platform for Question/Answer and DataBase capabilities*. Proceedings of the 2nd International Workshop on PEIRCE, Québec 1993, pp29-32.

---

[9] The platform Cogito is a conceptual graph managing system.

3. Jaime Carbonnell.: *Metaphor as a key to extensible semantic analysis*, in Lehnert & Ringle eds, *Strategies for natural langage processing*, Lawrence Erlbaum Associates, publishers, pp.415-434, Hillsdale, New Jersey, 1982.

4. Jean Fargues: *Sur le rôle de la pragmatique pour la résolution de problèmes difficiles d'analyse du langage naturel.* 4° Ecole d'été sur le traitement des langues naturelles, organisée par le CNET, pp VII-1, VII-9. Lannion, 5-9 juillet 1993.

5. Dan Fass: *Met*: a method for discriminating metonymy and metaphor by computer*, computationnal linguistics, vol 17, n°1, pp49-90.

6. Gilles Fauconnier: *"Espaces mentaux, aspects de la construction du sens dans les langues naturelles"*, 216 pages, Les éditions de Minuit, Paris, 1984.

7. Jerrold Katz, Jerry Fodor: *The structure of the semantic theory*, Langage n°39, pp 170-210, 1963.

8. Daniel Kayser: *What kind of thing is a concept ?*, Computational intelligence, Vol.4, num. 8, 1988.

9. James Pustejovsky: *The generative lexicon*, computationnal linguistics, vol 17, n°4, pp409-441, 1991.

10. John Sowa: *Conceptual Structures - Information in Mind and Machine*, 481 pages, Addison Wesley, 1984.

11. J., E.C. Way: *", IBM Journal of Research and Development "*, Vol.30, n°1, pp 57-69, 1986.

12. John Sowa: *Logical structures in the lexicon*, Knowledge-Based Sytems, vol 5, n°3, pp 173-182, sepembre 1992.

13. J. Veronis, N.M. Ide: *Word sense disambiguation with very large neural networks extracted from machine readable dictionaries*, COLING 90, Helsinki, pp389-394, 1990.

14. E. C. Way: *Metaphor as a mechanism for reorganizing the type hierarchy*, Knowledge-Based Sytems, vol 5, n°3, pp 223-232, septembre 1992.

15. Yorick Wilks: *"A preferential pattern-seeking semantics for natural language inference"*, Artificial intelligence, n°6, pp 53-74, 1975.

16. Pierre Zweigenbaum: *"MENELAS: an access system for medical records using natural language"*, in *Computer Methods and Programs in Biomedicine,*à paraître,1994.

Object-Oriented Conceptual Graphs

Gerard Ellis

Department of Computer Science, Royal Melbourne Institute of Technology,
GPO Box 2476V, Melbourne, Victoria, 3001, AUSTRALIA Tel: 61-3-660-5090
FAX: 61-3-662-1617 Email: ged@cs.rmit.edu.au,
WWW://http.cs.rmit.edu.au/~ged/.

Abstract. In this paper a state based view of conceptual graphs borrowed from the Object-Z program specification language is introduced. This new view is contrasted with the object-oriented model developed by Sowa [5]. The new model is demonstrated by reducing Sowa's example proof from 18 steps to 4 steps. The new model uses relations connecting object pre-states and post-states to represent object methods, rather than using messages as concepts as in Sowa's version. We argue that the new model is clearer because it is based on a simple state transition and that this leads to more efficient theorem proving and programming.

1 Introduction

Object-oriented (OO) logics, programming languages and modelling languages are popular today because they offer useful characteristics such as encapsulation and inheritance which encourage information hiding and reuse of designs, proofs or programs. Logic is the foundation of most formal approaches to object-oriented methods, so it is not surprising that Sowa [5] took the object-oriented wand to conceptual graphs [4]. In this paper, an alternative state based view of conceptual graphs is presented which is proposed as an object-oriented model for conceptual graphs. Below it is argued that this model is more illustrative than Sowa's approach and is more efficient in theorem proving and programming.

2 Sowa's Proof of Turning an Engine On

Sowa [5] gives an 18 step proof that a car's engine will be running after receiving an *ignition turn on* message. Using the new OO model a 4 step clearer proof is given.

Fig. 1 shows the definition of a car given by Sowa [5].[1] A car is an automobile which has a kind which is a model, and has parts an engine, a set of 4 wheels,

---

[1] This new notation was suggested by Sowa at ICCS'94. The concept definition consists of a graphical equality sign (double line) partitioning a concept box. The top section of the box contains the concept/relation being defined. The lower section of the box contains the *differentia* or definition of the concept/relation. In this example a line of concept identity connects the concept being defined and the genus concept in the body.

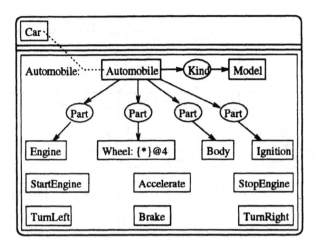

Fig. 1. Sowa's definition of a car

a body and an ignition (we added the ignition to the definition since it is an essential part of the definition for the problem which was omitted in [5]). A car object also has methods for starting an engine, turning left, accelerating, stopping the engine, braking, and turning right. The major difference between the form of this definition and previous notation in conceptual graphs [4] is the encapsulation of the definition in a (Automobile) context.

In the Sowa's proof, the car instance PCX999 in Fig. 2 was used. The car is shown at the point in time (PTim) of 9:51:02 Greenwich Mean Time. PCXX999 is a kind of Mustang, which has an Engine with number S6F901T which is in the state of not running, 4 wheels not rotating, body RJ88107 which is not moving, and an ignition #999999.

The main point at which our approach diverges from Sowa's is in the definition of the method for starting an engine. Whereas Sowa defines a method to be a concept, we define a method as a relation on object state (pre-state and post-state). As we will show this leads to simple but powerful proof techniques.

Fig. 3 defines a StartEngine process involving a car *c with a part an ignition *i, and a time #now *t1. The process describes a state transition, from the time *t1 to the time *t2 5 seconds later. The rule reads as follows: if the engine *e (which should be a part of car *c, but is not modelled in Sowa's example) which is in the state (Stat) of not running (¬ Running) and the ignition *i is the patient (Ptnt) of a TurnOn message (turning the ignition), then there is an interval of 5 seconds from time *t1 to the point in time *t2 when the engine *e is running at 750 revolutions per minute.

Assertion: A message is sent to a car by drawing the message in the car's context. Fig. 4 shows a StartEngine message whose context contains a graph describing a situation where the car's ignition is turned on.

We now look at the inference rules Sowa used in the proof that turning the ignition on results in the engine of PCXX999 running at 750 r.p.m. at 9:51:07 G.M.T.

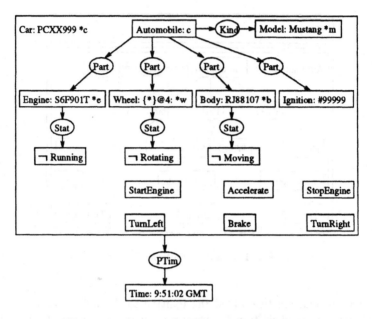

Fig. 2. Car instance PCXX999 at 9:51:02 GMT

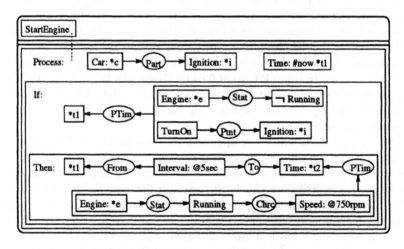

Fig. 3. Sowa's [5] definition of a StartEngine process

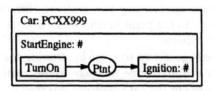

Fig. 4. A TurnOn ignition message for car PCXX999

2.1 OO Rules of Inference

Sowa [5] defined the following rules of inference in addition to those already defined in his book [4].

Definition 2.1 *Additional object-oriented rules of inference for conceptual graphs [5].*

Export rule:
If a proposition p describes an object x,
and if the description is true,
then p may be assumed to be true.

Import rule:
If a proposition p describes an object x,
and if the proposition q is true of x,
then the conjunction (p ∧ q) describes x.

Derived rule of inference: *If a message m is sent to an object described by a graph g, then m may be inserted into the same context as g and be combined with g by the standard rules of inference for conceptual graphs*

Time Iteration:
If a context c has a time specification s
and a concept x nested in c has no explicit time specification,
then a copy of s may be linked to the concept x.

Time Deiteration:
If a context c has a time specification s_1
and a concept x nested in c has a time specification s_2
where s_1 is contained in s_2,
then s_2 may be erased from the nested concept x.

Persistence:
Let Δc be a change to the description of a context c.
Any aspects of c that are disjoint from Δc may be
assumed to be left unchanged by the events that caused Δc.

In the proof of the engine running in Fig. 8, Sowa took 18 steps: export StartEngine message; import StartEngine message; resolve StartEngine message with method; expand definition of StartEngine method; iterate time in StartEngine precondition; resolve ignition; join on ignition; iterate time in StartEngine rule consequent; deiterate time and PTim relation in precondition; deiterate engine not running from precondition; deiterate turning ignition from precondition; erase double negation [If: [Then: 'graph']] to get 'graph'; compute time *t2 and join duplicate concepts *t1; *5 seconds elapse* and so update time concept attached to [Car: PCXX999]; deiterate time specification in StartEngine context; export running engine from StartEngine context into the outer Car context; join engine *e; generalise StartEngine context to general method to produce the car in Fig. 8.

This proof seems to be long and clumsy, and the logical rules like *"5 seconds elapse and so update time concept attached to [Car: PCXX999]"* seem problematic. Further the example shows this approach does not provide a simple proof mechanism for inference on method application. The main objection to the approach is that it encourages destructive updating of the original axiom states by trying to produce a single car state at each step. In a logical view of object-orientation, operations on objects must produce a history of object state changes. Earlier states can be "garbage collected" using the logical operation of erasure [4] if required later.

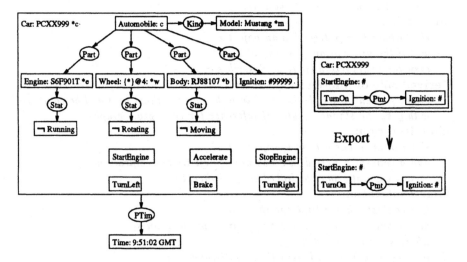

Fig. 5. Step 1: Exporting a message

Sowa Proof Step 1: Fig. 5 shows the first step in the proof where the StartMessage is exported. The StartEngine message is a proposition asserting that a StartEngine message was received.

Sowa Proof Step 2: In Fig. 6 the StartEngine message is then imported into the car context. Since the graph inside the car context describes the car PCXX999 and the proposition StartEngine is true of car PCXX999, then the conjunction of the original graph describing PCXX999 with the StartEngine graph describes car PCXX999.

... **Sowa Proof Step 14:** In Fig. 7 we show the state of the graph after 14 steps. The 14th step *"5 seconds elapse and so update time concept attached to [Car: PCXX999]"* causes the referent of the Time concept attached to the car context to be changed to the value 09:51:07 G.M.T. This operation should be a series of operations, such as "copy the car context", "join a (PTim) relation" and "restrict the time to 09:51:07 G.M.T."

... **Sowa Proof Step 18:** After 4 more steps the graph in Fig. 8 is obtained.
□

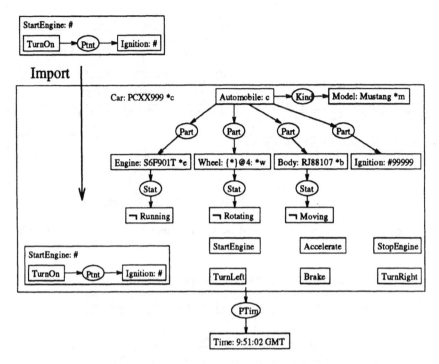

Fig. 6. Step 2: Importing a message from the ether

3   Quick Ignition

Sowa's proof [5] described above took at least 18 "steps". In the following section
we offer a proof of 4 "steps": *method/relation expansion, modus ponens, computation of time constraint,* and *erasing previous state information and links.*

In the following discussion an informal mapping between the Object-Z [1]
language and Conceptual Graphs is used. This methodology has been demonstrated on reactive objects in a vending machine case study [2]. Fig. 9 shows
the mapping of a generic object schema into a concept definition in conceptual
graphs. The subobjects $1 \ldots m$ in the Object-Z schema map to concepts attached
via part relations to the genus concept. A method with n arguments maps to
a $n + 2$ relation where the first and $(n + 2)$th arguments are concepts of type
Object. The first argument refers to the pre-state of the object when the message
was received and the $(n + 2)$th argument refers to the post-state of the object
after the message is acted upon. The body of the definition is a state transition
similar to that suggested by Sowa for state-transition diagrams [6]. The pre-state
is succeeded by an event described by some proposition involving n arguments,
which is then succeeded by the post-state of the object.

In Fig. 9 we use dotted lines and dashed lines to show the diffence between
concept identity and object identity. Fig. 10(a) and (b) show the meanings of
dotted and dashed lines, respectively. A dotted line (or line of concept identity)

Car: PCXX999 *c
Automobile: c
Kind
Model: Mustang *m
Part
Engine: S6F901T *e
Wheel: {*}@4: *w
Body: RJ88107 *b
Ignition: #99999
Stat
¬ Running
¬ Rotating
¬ Moving
Accelerate
StopEngine
TurnLeft
Brake
TurnRight
StartEngine: #
Car: *c
TurnOn
Ptnt
Ignition: # *i
From
Time: 9:51:02 GMT *t1
Interval: @5sec
To
Time: 9:51:07 GMT: *t2
PTim
Engine: *e
Running
Chrc
Speed: @750rpm
5 seconds elapse
Time: 9:51:07 GMT

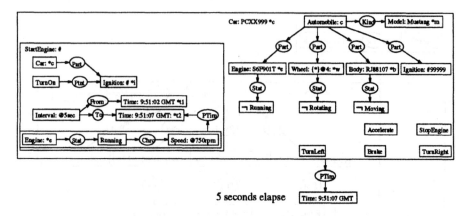

Fig. 7. Step 14: *5 seconds elapse* and so update time concept attached to [Car: PCXX999]

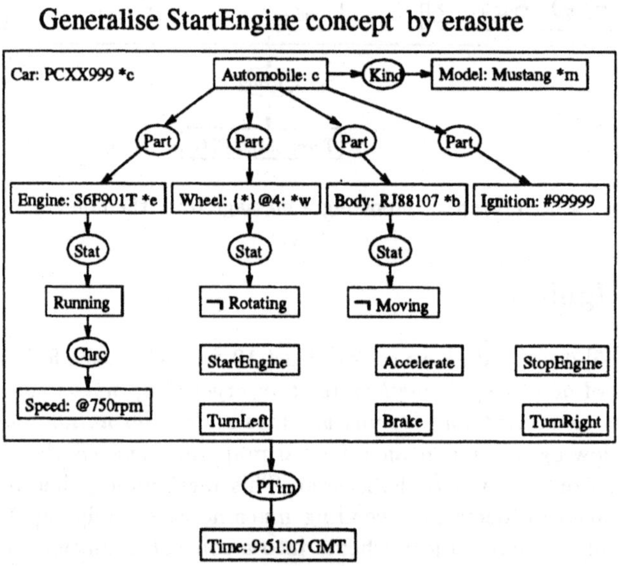

Generalise StartEngine concept by erasure
Car: PCXX999 *c
Automobile: c
Kind
Model: Mustang *m
Part
Engine: S6F901T *e
Wheel: {*} @4: *w
Body: RJ88107 *b
Ignition: #99999
Stat
Running
¬ Rotating
¬ Moving
Chrc
StartEngine
Accelerate
StopEngine
Speed: @750rpm
TurnLeft
Brake
TurnRight
PTim
Time: 9:51:07 GMT

Fig. 8. Step 18: Generalise StartEngine concept to generic method

connecting two concepts and indicates they are the same concept and hence can be joined. In the example in Fig. 10(a), the left Engine is running and right Engine is not running and the concept states are the same, representing the impossible situation where an Engine is running and not running simultaneously. Fig. 10(b) shows a dashed line (line of object identity) connecting Engines ***o** and ***p** which are running and not running, respectively. This means that the concepts ***o** and ***p** are (possibly) different states of the same object: state ***o** where the engine is running and state ***p** where the engine is not running. Each

Object-Z   Conceptual Graphs

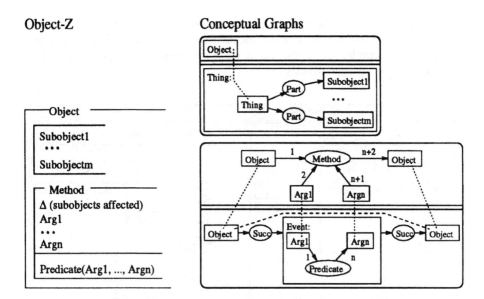

Fig. 9. Mapping Object-Z specifications to Conceptual Graphs

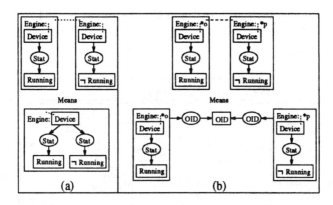

Fig. 10. Line of concept identity versus line of object identity

object state concept can have at most one OID relation determining the concept's object identifier, that is, OID must be a function. Line of object identity (dashed) is a weaker line than line of concept identity, since line of concept identity means same concept and hence same object identifier, and line of object identity only means same object identifier. Concepts can only refer to one object, but an object identifier can have many conceptualisations.

Our definition of the concept car is similar to that given by Sowa [5] in Fig. 1, except in our model methods are relations rather than concepts and attributes of subobjects are contained in the subobject context. In Fig. 1 methods are contained in the Automobile context. These contexts are unlike general conceptual graph contexts in that the concepts do not mean the existence of such a concept, since this would mean there is some instance of each method in an instance of a car: for example, this would mean there are StartEngine and StopEngine messages in the car concurrently.

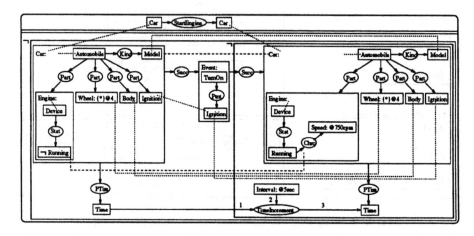

Fig. 11. Full Definition of TurnOn ignition method

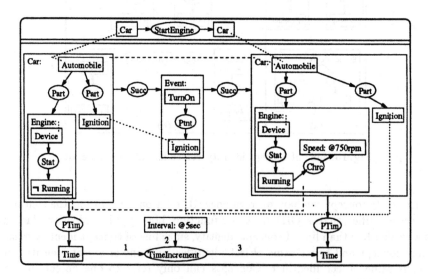

Fig. 12. Definition of TurnOn ignition method

In Fig. 11 a definition of the StartEngine method is given using a state based view. It takes as an input a car and outputs a car. There is a line of object identity (dashed line) connecting the two car objects indicating the concepts refer to the same car (object identifier), but (possibly) different states. The pre-state of the car is that the engine is not running. The succeeding event is described by the proposition that an ignition which is part of the car is turned on. The post-state of the car asserts that the engine is running at 750 r.p.m and that the state of all other parts and kind are preserved. The post-state occurs 5 seconds from the time of receiving the message. The whole state transition is overlaid on a graphical implication ($p \Rightarrow q$ is the same as ⌐[p ⌐[q]]). The pre-state and the event are the antecedant of the implication and the post-state is the consequent of the implication. Which reads *if the pre-state and event are true, then the post-state is true.*

Fig. 12 shows an alternative definition where the principle of language design of omitting any subobjects in the object state schema which were not affected by the method's operation. For example, the car subobjects Wheels and Body are not affected by the StartEngine message and are hence omitted. The post-state of these subobjects is assumed to be identical to the pre-state: that is, every subobject of the Car state has a line of object identity between the occurrence in the pre-state and the post-state. The rule that *post-state and event implie the post-state* could be assumed for modelling purposes. This means the semantics of the short graph definition Fig. 12 would be that of Fig. 11.

In Z the point in time of the state is usually modelled as a variable in the object state. Here point in time (PTim) is an attribute of the whole object state. This is in accordance with Sowa's rule of time iteration in definition 2.1, where each subobject such as Engine and Ignition have the same PTim as the whole Car object context.

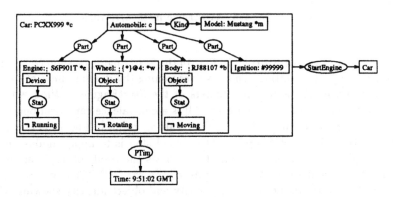

Fig. 13. Send StartEngine message to car PCXX9999

In Fig. 13 the message is sent to (asserted on) the car PCXX999 by joining a StartEngine relation to the car concept. That is, joining of a basis graph [Car]->(StartEngine)->[Car] to the [Car: PCXX999] concept asserts that a StartEngine message is received by PCXX999 and there is some as yet undetailed post-state of the car.

The following proof of "car PCXX999 with engine running" uses only the original rules from [4] and takes 4 steps: relation expansion, modus ponens, relation expansion, erasure.

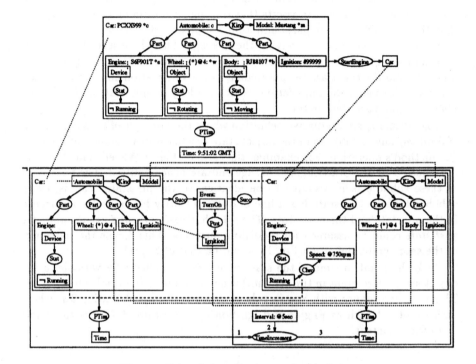

Fig. 14. Expand StartEngine method/relation

State Based Proof Step 1: In Fig. 14 the StartEngine method (relation) is expanded using its definition. This involves joining (by drawing lines of concept identity between) the actual parameters of the StartEngine method with the formal parameters. Relation expansion [4] is a equivalence canonical operation, hence if the graph in Fig. 13 is true, the resulting graph in Fig. 14 is also true.

State Based Proof Step 2: The second step is to apply modus ponens p and ⁻[p ⁻[q]] implies q. Fig. 15(a) shows the result of applying modus ponens. This graph shows the new state of the car PCXX999 succeeding the previous state with lines of concept identity and object identity showing which parts have changed and which have stayed the same.

State Based Proof Step 3: In Fig. 12 we have assumed that TimeIncrement relation relates a time *t1 (argument 1) and a time interval *i (argument 2) to the third argument a time *t2 which is the result of adding the time interval *i to time *t1. Here we use a relation expansion operation on TimeIncrement to compute the time 9:51:07 G.M.T. In [3] we show how it is possible to write

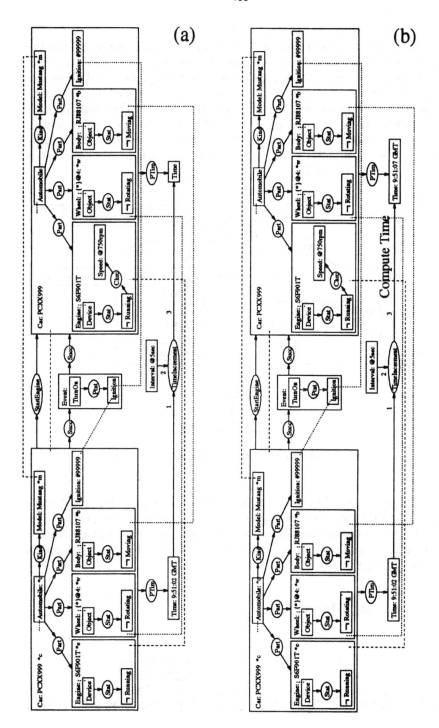

Fig. 15. Modus Ponens and Compute Increment Time

conceptual graph programs based on relation expansion.

At this point Fig. 15(b) gives a "movie" of the action resulting from the StartEngine message. This is a logical view, it shows the state of the car object at points in time, whereas the approach taken by Sowa [5] discussed above was trying to update the car context by doing insertions, erasures etc. This is reasonable for logical operations such as insertion and erasure, but the proof approach falls apart when the time operation is executed "*5 seconds elapse* and so update time concept attached to [Car: PCXX999]". What is intended is that a copy of the graph is taken and a PTime relation is joined and the Time concept is restricted to 09:51:07 G.M.T.

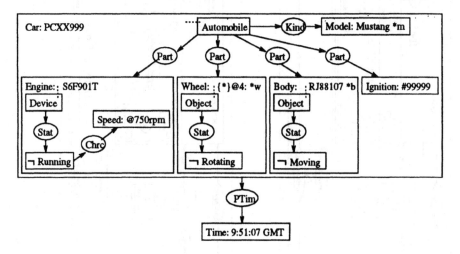

Fig. 16. Erase previous state information and links

State Based Proof Step 4: To derive the state graph in Fig. 16(b) which describes car PCXX999 with an engine running at 750 r.p.m., the previous prestate information and event information are erased. Erasure of a graph or parts of a graph is a logical operation, hence the graph in Fig. 16(b) is a consequent of the previous graph. □.

4 Composing Messages

Fig. 17. Composing messages

It is possible to compose messages by joining the post-states of messages to the pre-states of other messages, thus "pipelining" the state information in the transition. Fig. 17 shows a composition of StartEngine, PutInGear, and AccelerateTo (100 k.p.h) messages. The resulting rightmost car state would be expected to be a car with its engine on, in gear and travelling at 100 k.p.h. This could then be used as the body of a definition for a high-level message such as StartCruisingAt100KPH.

5 Summary

We have introduced a state based view of conceptual graphs borrowed from the Object-Z program specification language. We have contrasted this with the object-oriented approach developed by Sowa [5] and shown the advantages of a state-based methodology by reducing Sowa's example proof from 18 steps to 3 steps.

References

1. Roger Duke, Paul King, Gordon Rose, and Graeme Smith. The Object-Z specification language. In *Technology of Object-Oriented Languages and Systems: TOOLS 5*, pages 465–483. Prentice-Hall, 1991.
2. Gerard Ellis, Stephen R. Callaghan, and James Ricketts. Modelling reactive objects in conceptual graphs. In Christine Mingins and Bertrand Meyer, editors, *Proceedings of TOOLS PACIFIC 94 Conference*, pages 217–225. Prentice-Hall, November 1994.
3. Gerard Ellis, Robert A. Levinson, and Peter J. Robinson. Managing complex objects in Peirce. *International Journal on Human-Computer Studies*, 41:109–148, 1994. Special Issue on Object-Oriented Approaches in Artificial Intelligence and Human-Computer Interaction.
4. John F. Sowa. *Conceptual Structures: Information Processing in Mind and Machine*. Addison-Wesley, Reading, MA, 1984.
5. John F. Sowa. Logical foundations for representing object-oriented systems. *Journal of Experimental & Theoretical Artificial Intelligence*, 5, 1993.
6. John F. Sowa. Relating diagrams to logic. In Guy W. Mineau, Bernard Moulin, and John F. Sowa, editors, *Conceptual Graphs for Knowledge Representation*, number 699 in Lecture Notes in Artificial Intelligence, pages 1–35, Berlin, 1993. Springer-Verlag. Proceedings of the 1st International Conference on Conceptual Structures, Quebec City, Canada, August 4-7.

A Direct Proof Procedure for Definite Conceptual Graph Programs

Bikash Chandra Ghosh* and Vilas Wuwongse**

Computer Science, School of Advanced Technologies,
Asian Institute of Technology, P.O. Box 2754, Bangkok 10501, Thailand

Abstract. Conceptual graphs form the basis of a graph-based existential-conjunctive logic. In this paper, first, we illustrate the problems associated with a proof procedure for conceptual graph programs and then specify the definitions of a normal form representation for a conceptual graph program, an anti-normal form representation for a goal, and a conceptual graph unification procedure called CG unification. Next, we develop a direct proof procedure for definite conceptual graph programs, called CGF-derivation. The proof procedure takes advantage of the normal form of a definite conceptual graph program and the anti-normal form of a goal, and utilizes the CG unification procedure for matching conceptual graphs. Finally, we prove that the proposed CGF-derivation procedure is sound and complete.

1 Introduction

The notion of a *conceptual graph program* was introduced in [1] based on the *conceptual graph* theory of J. F. Sowa ([9]). A conceptual graph program can be viewed as a graph-based order-sorted logic program that consists of a finite set of *clause conceptual graphs* from a graph-based logic programming language. The declarative semantics of conceptual graph programs was presented in [1, 2].

There have been several attempts in devising an inference procedure for conceptual graph based systems, in general, in [5, 8, 4]. Each of them reports some interesting results about implementation of some Prolog-like deduction procedure, but lacks any formal account of the procedure as well as its soundness.

Outlines of two deduction procedures for conceptual graph programs were presented in [3]. The first of the procedures is based on the *refutation-based resolution* ([7]) proof, while the second one is based on *direct derivation* ([10]) or *direct proofs* ([6]) that is used to derive the formula to be proved instead of deriving a refutation. Although both the proof procedures are sound, they are not complete. In fact, they are very much limited because of the use of *projection* for matching conceptual graphs.

The purpose of the present work is to develop a more concrete and formal proof procedure based on the proposal for a direct proof in [3]. The new procedure

---

* Internet:*bikash@cs.ait.ac.th*
** Internet:*vw@cs.ait.ac.th*

should be more powerful, and complete with respect to the declarative semantics of conceptual graph programs.

A brief review of the syntax of the conceptual graph programs is presented in Section 2. Two of the major issues arisen from the original proposal of the proof procedures for conceptual graph programs in [3] are discussed in Section 3. Section 4 is devoted to the development of the new direct proof procedure for *definite* conceptual graph programs. The soundness and completeness properties of the proposed proof procedure are presented in Sections 5. Finally, Section 6 summarizes the results and concludes the paper.

2 Definite Conceptual Graph Programs

In this section we present the syntax of conceptual graph language, and the notions of conceptual graph programs and definite conceptual graph programs.

2.1 Conceptual Graph Language

A conceptual graph program (CGP) consists of a set of "statements" or *conceptual graphs* (CGs) from a conceptual graph language or CG *language*. A CG is a bipartite, directed, and connected graph with two types of nodes: *concepts* and *conceptual relations*. Every concept node consists of a *concept type*, and an optional *referent*.

Concept types, referents, and conceptual relations (or simply relations) for a CG language are provided by the *concept universe* ([3]) of that language. The CGs that follow the selectional constraints on the permissible combinations of concepts and relations are known as *canonical* CGs ([9]), and all "well-formed" CGs in a CG language must be canonical. We, therefore, use simply CG to refer to a canonical CG unless explicitly specified otherwise. There is a special concept, called *context* that is extensively used in the CG language. The referent field of a context contains a set of conceptual graphs. A context connected to a monadic relation NEG is known as a *negative context*. Now the definition of concept universe follows.

Definition 1. The *concept universe* is defined as, $C_u = \langle (T_c, \leq_c), T_r, S, M, :: \rangle$ where,

- C_u is the concept universe;
- $(T_c, \leq_c)$ is the concept type hierarchy - the *type lattice*, where T_c is a set of concept types, and $\leq_c$ is the subsumption relation on T_c;
- T_r is a set of conceptual relation types;
- S is a set of *signature expressions* for concept types in T_c and relation labels in T_r;
- $M = I_m \cup \{*, ?, \chi, 0\}$ is a set of markers, where I_m is a non-empty set of *individual markers*, * and ? are the *generic markers*, χ is the *context marker*, and 0 is the *absurd marker*; and
- :: is the *conformity relation* between concept types and markers in C_u. □

Conceptual graphs constitute a kind of existential-conjunctive logic. That means, the only basic quantifier in a CG language is the existential quantifier, and the basic connective is conjunction. Every CG occurs in some context. All the graphs in a CG language occur in the *outermost* context which has a *depth* 0. If there is more than one CG in a context, then they are in logical conjunction. Now, we define the well-formed CGs with respect to (wrt) a given concept universe, and the CG language.

Definition 2. Let C_u be a concept universe. The *well-formed* conceptual graphs wrt C_u are classified into the following three categories: ACG, NCG and NNCG, which are defined as follows, provided all the concepts and relations are in C_u.

ACG. A CG g is called an *atomic conceptual graph* or ACG if it does not contain any context as its concept node.

NCG. A CG g is called a *negative conceptual graph* or NCG if it consists of a single negative context whose referents are only ACGs.

NNCG. A *nested negative conceptual graph* or NNCG is defined inductively as follows: (1) a NCG is a NNCG; and (2) a negative context whose referent contains a set of ACGs and NNCGs, is also an NNCG. □

Definition 3. A *conceptual graph language* or CG *language* consists of:

1. a concept universe, C_u;
2. a *canonical base*, $\mathcal{B}_c$, which is a finite set of conceptual graphs that are declared to be canonical with all types and referents in C_u;
3. a set of rules and operations on conceptual graphs that includes *canonical formation rules*, *maximal join*, and *constrained join*[3]; and
4. the set of all well-formed CGs that can be formed from the concept universe and canonical base using the rules. □

2.2 Conceptual Graph Programs

Intuitively, a CGP consists of a subset of the well-formed CGs. Now, we identify two specific sub-classes of well-formed CG in the following definitions.

Definition 4. A *clause conceptual graph* or CCG is a nested negative conceptual graph whose maximum depth of nesting is 2. The general form of a CCG is, $\neg[[u_1 \ldots u_l \neg[v_1 \ldots v_m] \neg[w_1 \ldots w_n] \ldots]$, where all of u_i, v_j and w_k are ACGs. □

Definition 5. A *definite clause conceptual graph* or *definite* CCG is a CCG which has exactly one negative context at depth 1 and exactly one ACG in the referent of that negative context. For a definite CCG $g = \neg[[u_1 \ldots u_n] \neg[w]]$, $head(g)$ gives the CG w at the inner negative context of depth 2, $body(g)$ gives the set of CGs $\{u_1 \ldots u_n\}$, and $order(g)$ gives the number of CGs in the outer negative context of g. □

Note that an ACG g can be represented as a definite CCG $\neg[\neg[g]]$. We now define a CGP and a definite CGP as follows.

---

[3] the set of canonical formation rules (that includes *copy*, *restrict*, *join*, and *simplify*) and *maximal join* operation are defined in [9], and the *constrained join* operation is defined in [1]

Definition 6. A *conceptual graph program* or CGP consists of a (finite) set of clause conceptual graphs from a conceptual graph language.   □

Definition 7. A *definite conceptual graph program* or *definite* CGP consists of a (finite) set of definite clause conceptual graphs from a CG language.   □

In the sequel, we use the term CCG to refer to a definite CCG, and only CGP to refer to a definite CGP, as the rest of the paper is concerned only with the definite CGPs.

3 Problems and Issues with the Proof Procedures

In this section, two of the major issues arisen from the proposal of inference procedures in [3] are discussed, and solutions proposed for them.

3.1 Representation of a Conceptual Graph Program

Due to the topological nature of the CGs, the set of graphs provable from a CGP using one of the proof procedures in [5, 8, 3] may vary depending on the representation of the CGs in the CGP.

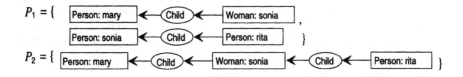

Fig. 1. Two different representation of CGP

Let us consider two representations, P_1 and P_2, of a CGP in Fig. 1 and two *query graphs* q_1 and q_2 in Fig. 2. It can be shown that both q_1 and q_2 are logically implied by (both the representations of) the CGP in Fig. 1. However, q_2 is not provable from P_1 using either of the procedures in [5, 8, 3]. On the other hand, q_2 is provable from P_2 using the same procedures. Moreover, q_2 is provable from P_1 if we break down q_2 into two CGs: q_{21} and q_{22} as in Fig. 2.

Thus, the provability of a CG wrt a CGP may be affected by the representation used by the proof procedure. Based on this observation, we propose two forms of representation for a set of CGs: the *normal* form, and the *anti-normal* form. The transformation of a CGP P from the user representation into the normal form, P_N, may be considered as a *pre-compilation* process. Similarly, the transformation of a goal G from the user representation into the anti-normal form, G_A, may be considered as a *query pre-processing* step. Now, the definitions follow.

Definition 8. Let C_u be the concept universe of a CG language L and S be a set of CGs of L that belong to the same context. The *normal form* of S, represented as $S_N = normal(S)$, is obtained by the following procedure:

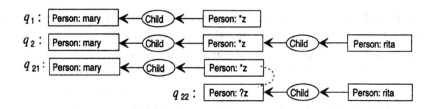

Fig. 2. Example query graphs

1. For every individual concept c in S with $referent(c) = m$[4], where m is an individual marker in C_u, let $D = \{d_1, \ldots, d_n\}$ denote the set of all individual concepts in S (excluding c) that contain the same individual referent m. None of $\{c\} \cup D$ may be nested inside a context of S. The concepts in $\{c\} \cup D$ are combined together using the *join* rule followed by the *simplify* rule, provided the following conditions hold: there is a concept type[5] t in C_u such that,
   (a) $t = (type(c) \cap type(d_1) \cap \ldots \cap type(d_n))$[6],
   (b) Absurd $\leq_c t$[7], and
   (c) either $t = type(c) = type(d_1) = \ldots = type(d_n)$, or all immediate super-types of t are in $\{type(c), type(d_1), \ldots, type(d_n)\}$.
2. For every concept c in S, let $E = \{e_1, \ldots, e_n\}$ denote the set of all concepts in S (excluding c) that are connected to c by coreference links. None of $\{c\} \cup E$ may be nested inside a context of S. The concepts in $\{c\} \cup E$ are combined together using the join rule followed by the simplify rule, provided the following conditions hold: there is a type t in C_u such that,
   (a) $t = (type(c) \cap type(e_1) \cap \ldots \cap type(e_n))$,
   (b) Absurd $\leq_c t$,
   (c) either $t = type(c) = type(e_1) = \ldots = type(e_n)$, or all immediate super-types of t are in $\{type(c), type(e_1), \ldots, type(e_n)\}$, and
   (d) if there is any individual concept in $\{c\} \cup E$ that contains an individual marker m, then every other individual concept in $\{c\} \cup E$ must contain the same individual referent. $\square$

Definition 9. Let S be a set of ACGs of a CG language L that belong to the same context. The *anti-normal form* of S, denoted by $S_A = anti\text{-}normal(S)$, is obtained as follows.

1. If there is any CG u in S that contains more than one conceptual relation, then the following steps are repeated:
   (a) let c be a concept in u that is connected to m relational arcs of m distinct relations, where $m > 1$;

---

[4] for any concept c, $referent(c)$ gives the marker in its referent field
[5] for any concept c, $type(c)$ gives its type label
[6] $\cap$ stands for the *maximal common sub-type*, which is the *greatest lower bound* (glb) operator for the type lattice
[7] Absurd is the bottom concept type of the concept type lattice

(b) m concepts $c_1, \ldots, c_m$ are inserted in S, each of which is a copy of c;

(c) all the relational arcs connected to c are detached and each one of them is reattached to one of the inserted concepts $c_1, \ldots, c_m$ such that the ith arc previously attached to c is now reattached to c_i;

(d) c is erased;

(e) a coreference link is drawn connecting all the concepts in $\{c_1, \ldots, c_m\}$;

(f) if c was a generic concept with the generic marker *, then the referent field of all but one of the concepts in $\{c_1, \ldots, c_m\}$ are changed to the ? marker; and

(g) all coreference links previously attached to c, are reattached to one of $c_1, \ldots, c_m$.

2. S_A is set to S. $\qquad\qquad\qquad\qquad\qquad\qquad\qquad\qquad\qquad\qquad$ □

The two procedures presented above for deriving the normal form and anti-normal form, respectively, of a set of CGs can be considered as equivalence rules. This is formally stated in the following theorems.

Theorem 10. *Let S_N be the normal form of a set S of CGs of a CG language L that belong to the same context. Then S and S_N are logically equivalent.* □

Theorem 11. *Let S_A be the anti-normal form of a set S of CGs of a CG language L that belong to the same context. Then S and S_A are logically equivalent.* □

The proofs are simple from the basic semantic properties of CGs, and are left out here. An example of normal form representation follows.

Example 1. The CGs in the set S in Fig. 3 are from the same context. The set S_N is the normal form of S. Note that [Person : *ram*], [Staff : *ram*], and [Student : *ram*] are joined together and the resulting combined concept is [EmpStudent : *ram*], since all of them contain the same individual referent *ram*, EmpStudent = Person ∩ Staff ∩ Student, and all immediate supertypes of Emp-Student are in {Person, Staff, Student}.

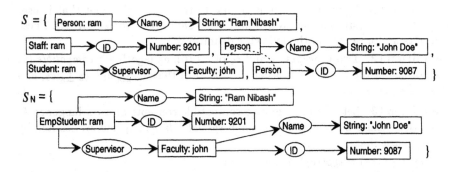

Fig. 3. A set of CGs in non-normal form and normal form

Similarly, the three coreferent concepts are also joined together, and the resulting concept is [Faculty : *john*]. $\qquad\qquad\qquad\qquad\qquad\qquad\qquad$ □

3.2 Conceptual Graph Unification

In the proof procedures in [3], projection is the only "mechanism" used for matching two CGs in the derivation process. This limits the set of provable CGs from a given CGP. In order to illustrate the problem, let us consider the CGP and query graphs in Fig. 4.

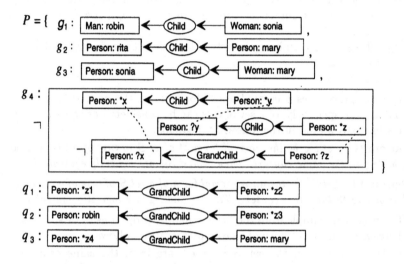

Fig. 4. An example CGP and three query graphs

Only q_1 has a projection onto the head of g_4. Thus q_2 or q_3 are not provable from P using projection, although both of them are logical consequence of P. This shows that a generalized technique is necessary to handle such cases. The proposed CG matching procedure is similar to that in [5, 8] except that it uses the notion of a *restricted sub-CG* in addition to that of *compatible concepts*. These concepts are presented in the following definitions.

Definition 12. Two concepts c_1 and c_2 of a CG language L are *compatible* (to each other) iff there exists a concept c_3 in L, such that, $c_3 \leq_g c_1$[8], $c_3 \leq_g c_2$, and $Absurd <_c type(c_3)$. □

Definition 13. Let c_1 and c_2 be two compatible concepts of a CG language L. Then there exists an *immediate common specialization* of c_1 and c_2, denoted by $c_1 \sqcap c_2$, such that,

- $type(c_1 \sqcap c_2) = type(c_1) \sqcap type(c_2)$, and
- either $referent(c_1 \sqcap c_2) = referent(c_1) = referent(c_2)$, $referent(c_1 \sqcap c_2) = m$ (m is an individual marker in L), where referent field of one of the concepts, c_1 or c_2 contains m, and the referent field of the other concept contains a generic marker. □

---

[8] $\leq_g$ is the *generalization-specialization* relation among CGs

Definition 14. Suppose g_1 and g_2 are two ACGs in a CG language L. The CG g_1 is a *restricted sub-CG* of g_2, if g_1 is a sub-graph modulo restriction of g_2. That means, there exists a sub-graph (in usual sense) g_2' of g_2 such that, for every relation r in g_1, there is a relation r' in g_2' where,

- $type(r) = type(r')$ and $degree(r) = degree(r')$[9], and
- $\forall i \in degree(r)$, if $neighbor(r, i) = c_i$[10] in g_1 and $neighbor(r', i) = d_i$ in g_2, then c_i and d_i are compatible concepts. ☐

Definition 15. Suppose g_1 and g_2 are two ACGs in a CG language L. Then g_1 has a CG *matching* with g_2, if g_1 is a restricted sub-CG of g_2 and there is an ACG w that can be obtained by join of g_1 and g_2 such that there are two *projections* [11] π_1 and π_2 where, (1) $w \leq_g g_1$, $\pi_1 : g_1 \to w$; and (2) $w \leq_g g_2$, $\pi_2 : g_2 \to w$. ☐

Note that a CG matches with itself with a pair of *identity projections*. Besides, when a CG g_1 has a projection onto another CG g_2, then g_1 and g_2 have a CG matching with the pair of projections $\theta = (\pi, \varepsilon)$, where ε denotes the identity projection.

In order to ensure the soundness, the CG matching operation has to be restricted with some additional constraints. We call the resulting operation as the *conceptual graph unification* or CG *unification*. The definition of the CG unification procedure follows.

Definition 16. Suppose g_1 and g_2 are two CGs of a CG language L, and g_1 has a CG matching with g_2 with a pair of projections $\theta = (\pi, \pi')$. The CG *unification* of g_1 and g_2, denoted by $cg\text{-}unify(g_1, g_2)$, produces a pair $\langle \theta, S_a \rangle$, called a CG *unifier* of g_1 and g_2, where θ is a pair of projections and S_a is a set of ACGs, which are determined as follows.

1. If g_2 is a *specialization* of g_1, i.e., $g_2 \leq_g g_1$ with $\pi : g_1 \to g_2$, then $\theta = (\pi, \varepsilon)$ and $S_a = \emptyset$, where ε is the identity projection.
2. If g_2 is not a specialization of g_1 but g_1 is a restricted sub-CG of g_2, then $\theta = (\pi, \pi')$, and $S_a = \{v_1, \ldots, v_m\}$, with $0 \leq m$, which are calculated as follows.
   (a) A CG g is obtained by joining g_1 and g_2, such that $\pi : g_1 \to g$ and $\pi' : g_2 \to g$. The procedure for joining g_1 and g_2 to obtain g follows.
      i. For each pair of compatible concepts (c_1, c_2) that are joined together, let c be the joined concept in g, then $c = c_1 \cap c_2$.
      ii. If c is a generic concept, then the (optional) variable followed by the generic marker in c can be either the variable in c_2 or a unique variable not occurring in any other concept of L.
      iii. All coreference links attached to c_1 and c_2 are reattached to c in g.
      iv. Any coreference link, whose both ends are attached to the same concept in g, is erased.

---

[9] for any relation r, $degree(r)$ gives the arity of r ([1])

[10] for any relation r in a CG g and a positive integer i, $neighbor(r, i)$ is the concept in g that is connected to the ith arc of r

[11] a projection is a mapping from the set of nodes of one CG into that of another CG ([9, 1])

(b) The set S_a is obtained as follows, starting with $S_a = \emptyset$.

   i. Let S_c be the set of all concepts c in g such that: c comes from the join of c_1 and c_2 in g_1 and g_2, respectively; $c <_c c_2$; and there is no coreference link to c from any of the concepts dominating c.

   ii. For every c in S_c, let g_c be the sub-graph of g that consists of the branches containing the node c and the following nodes: every relation r_i in g that is attached to c, and every concept c_j (different from c) that is connected to an arc of r_i. If there is no CG w in S_a, such that $w \leq_g g_c$, then g_c is added to S_a.

   iii. $\forall u, v \in S_a$ such that $v \leq_g u$, u is erased from S_a.

   iv. If there are more than one instances of any generic concept in S_a, then a coreference link is drawn connecting all instances of that concept. □

The informal reading of $cg\text{-}unify(g_1, g_2) = \langle (\pi, \pi'), S_a \rangle$ is that, g_1 is unifiable with g_2 provided the CGs in S_a are provable. Clearly, this is similar to, but more general than the notion of unification in first-order logic. Now, an example of CG unification follows.

Example 2. Let us consider two ACGs q_1 and q_2, and a CCG g as shown in Fig. 5. The CG w in the head of g is not a proper specialization of q_1 or q_2, but each of q_1 and q_2 has a CG matching with w. However, the CG unification of q_2 and w is different from that of q_1 and w because of the individual marker *john* in the concept c_4 in q_2, and the absence of any dominating coreference concept to c_7 in w. The CG unification of q_1 and w returns $\langle (\pi_1, \pi_1'), \{\} \rangle$ where, $\pi_1 =$

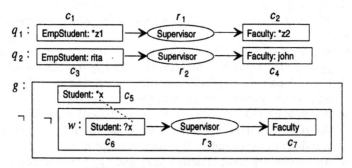

Fig. 5. Example conceptual graphs

$\{(c_1, d_1), (r_1, r_4), (c_2, d_2)\}$ and $\pi_1' = \{(c_6, d_1), (r_3, r_4), (c_7, d_2)\}$; and the CG unification of q_2 and w returns $\langle (\pi_2, \pi_2'), \{v\} \rangle$ where, $\pi_2 = \{(c_3, d_3), (r_2, r_5), (c_4, d_4)\}$ and $\pi_2' = \{(c_6, d_3), (r_3, r_5), (c_7, d_4)\}$; which are obtained as shown in Fig. 6, where v is the CG: $[EmpStudent : rita] \rightarrow (Supervisor) \rightarrow [Faculty : john]$. □

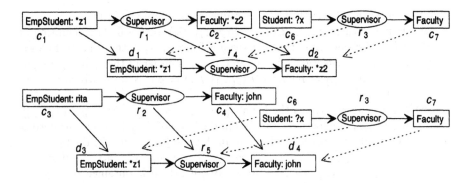

Fig. 6. Examples of CG matching

4 A Direct Proof Procedure for Definite CGPs

The main idea of a *direct proof procedure* is that it derives the CG to be proved instead of deriving a refutation as in resolution based pocedures ([5, 8, 3]). However, the derivation process is directed by the *goal graph* to be proved. For this reason, we call it a *goal-directed forward derivation* procedure, or simply a *forward derivation* procedure.

An outline of such a procedure was presented in [3]. In this section we develop a more general form of the forward derivation procedure for CGPs. As in [3], we call the top level forward derivation procedure as CGF-*derivation*. A goal for a CGF-derivation is the set of queries to be proved, which are essentially in logical conjunction.

Definition 17. Let P be a CGP of a CG language L. A CGF-*goal* consists of a set of CGs $\{q_1, \ldots, q_m\}$, where each of the graphs q_i is an ACG in L, called a CGF-*goal graph*. □

4.1 The CG-derivation Procedure

The selected goal graph in the CGF-derivation procedure is derived by another procedure called CG-*derivation*. We now define the procedure and other related concepts. The CG-derivation procedure is defined inductively.

Definition 18. Let P be a CGP of a CG language L, $P_{\rm N} = normal(P)$, and q be a CGF-goal graph. The CG-*derivation* procedure is defined as follows.

1. Let $S_{\rm sp}$ denote the set of CGs, given by an operation, called *PROJECT*:
   $S_{\rm sp} = PROJECT(q, P_{\rm N}) = \{g | g \in P_{\rm N}, order(g) = 0, g \leq_g q\} \cup$
   $\{g' | g \in P_{\rm N}, g = \neg[u_1 \ldots u_k \neg[w]], k > 0,$
   $\langle(\pi, \pi'), \{v_1, \ldots, v_m\}\rangle = cg\text{-}unify(q, w), g' = \neg[\pi(u_1 \ldots u_k) v_1 \ldots v_m \neg[\pi'w]]\}^{12}$.
2. If $S_{\rm sp} \neq \emptyset$, then a CG $g_{(0)} \in S_{\rm sp}$ is selected, otherwise the CG-derivation procedure terminates.

---

[12] there may be coreference links among concepts in $v_1 \ldots v_m$ and $\pi'w$ of g'

3. Starting with the graph $g_{(0)}$, the sequence of graphs $g_{(0)}, g_{(1)}, \ldots, g_{(n)}, \cdots$ are derived, where $g_{(i)} = REDUCE(g_{(i-1)}, P_N)$ for $i > 0$. The definition of the $REDUCE$ operation that derives $g_{(i)}$ from $g_{(i-1)}$ in the ith step follows,

   (a) $k = 0$, i.e., $order(g_{(i-1)}) = 0$, and $g_{(i-1)}$ is either of the form $\neg[\neg[w]]$ or just w and $g_{(i)} = REDUCE(g_{(i-1)}, P_N) = g_{(i-1)} = w$, where w is an ACG.

   (b) $k > 0$, and $g_{(i-1)}$ has the form $\neg[u_{(i-1)1} \cdots u_{(i-1)j} \cdots u_{(i-1)k} \ \neg[w_{(i-1)}]]$, where $u_{(l)m}$ refers to the mth CG in the body of the lth CCG in the derivation sequence, whose head is $w_{(l)}$. In this case, $g_{(i)} = REDUCE(g_{(i-1)}, P_N)$ $= \neg[\pi_i u_{(i-1)1} \cdots \pi_i u_{(i-1)j-1} \pi_i u_{(i-1)j+1} \cdots \pi_i u_{(i-1)k} \ \neg[\pi_i w_{(i-1)}]] =$ $\neg[u_{(i)1} \cdots u_{(i)j-1} \ u_{(i)j+1} \cdots u_{(i)k} \ \neg[w_{(i)}]]$, which is obtained from $g_{(i-1)}$ in the following way,

      i. u_j is an ACG in the body of $g_{(i-1)}$, called a selected graph,

      ii. u_j has a *successful* CG-derivation with a projection π_i,

      iii. $g_{(i)}$ is obtained by erasing $\pi_i u_j$ from $\pi_i g_{(i-1)}$.

   (c) $k > 0$, and there are two CGs $u_{(i-1)j}$ and $u_{(i-1)l}$ in the body of $g_{(i-1)}$ such that $u_{(i-1)l} \leq_g u_{(i-1)j}$ with a projection $\pi_i : u_{(i-1)j} \to u_{(i-1)l}$, then $g_{(i)} = REDUCE(g_{(i-1)}, P_N) =$ $\neg[\pi_i u_{(i-1)1} \cdots \pi_i u_{(i-1)j-1} \pi_i u_{(i-1)j+1} \cdots \pi_i u_{(i-1)k} \neg[\pi_i w_{(i-1)}] =$ $\neg[u_{(i)1} \cdots u_{(i)j-1} \ u_{(i)j+1} \cdots u_{(i)k} \neg[w_{(i)}]]$, which is obtained by erasing $\pi_i u_{(i-1)j}$ from $\pi_i g_{(i-1)}$.

   (d) $k > 0$, but there is no CG in the body of $g_{(i-1)}$ that has a successful CG-derivation. In this case, $g_{(i)} = g_{(i-1)}$.

4. A finite CG-derivation terminates under one of the following conditions:

   (a) in step (2) when S_{sp} is empty, in which case the *length* of the CG-derivation is 0; or

   (b) at a minimum value of n such that $g_{(n)} = REDUCE(g_{(n-1)}, P_N)$ and $order(g_{(n)}) = order(g_{(n-1)})$ where the graph $g_{(n)}$ is called a *terminal graph*, and n is called the *length* of the CG-derivation. □

Definition 19. Let P be a CGP and q be a CGF-goal graph. A *successful* CG-derivation of q from P is a CG-derivation of q from P that terminates with a terminal graph $g_{(n)}$ such that $order(g_{(n)}) = 0$ and $g_{(n)} \leq_g q$. If $\pi : q \to g_{(n)}$, then the CG-derivation is said to be successful with the projection π. □

Definition 20. Let P be a CGP and q be a CGF-goal graph. A *(finitely) failed* CG-derivation of q from P is a CG-derivation of q from P that terminates under one of the following conditions: (1) the length of the CG-derivation is 0, or (2) the terminal graph is $g_{(n)}$ such that n is finite and either $order(g_{(n)}) > 0$ or $order(g_{(n)}) = 0$ but $g_{(n)} \leq_g q$ does not hold. □

4.2 The CGF-derivation Procedure

Now we define the forward derivation proof procedure for a CGF-goal and other associated concepts. The CGF-derivation procedure works as follows. At each step, it selects one of the goal graphs from the CGF-goal and tries to derive it using the CG-derivation. Upon a successful CG-derivation of the selected CGF-goal graph, it is erased from the goal that is specialized by applying the projection obtained from the derivation to produce the CGF-goal for the next step.

Definition 21. Let P be a CGP and $G = \{q_1, \ldots q_m\}$ be a CGF-goal of a CG language L, and A_0 be an empty set of CGs, . A CGF-*derivation* of G from P consists of a finite sequence $\langle G_0, A_0 \rangle, \langle G_1, A_1 \rangle, \ldots, \langle G_n, A_n \rangle$ where the following conditions hold:

1. $G_0 = anti\text{-}normal(G)$.
2. Each pair $\langle G_i, A_i \rangle$ is obtained from $\langle G_{i-1}, A_{i-1} \rangle$ in the following way:
   (a) a CGF-goal graph $q_{(i-1)k}$ is selected from G_{i-1} called the selected goal graph;
   (b) the selected goal graph $q_{(i-1)k}$ has a successful CG-derivation with a projection π_i, and ρ_i is the *referent specialization operator* (rso)[13] wrt π_i;
   (c) G_i is the CGF-goal obtained by erasing $\pi_i q_{(i-1)k}$ from the $\pi_i G_{i-1}$; and
   (d) $A_i = A_{i-1} \cup \{\rho_i q_{(i-1)k}\}$.
3. The last pair $\langle G_n, A_n \rangle$ is called the *terminal pair*. The CGF-derivation terminates with one of the following conditions:
   Success pair. In this case, $n = m$, $G_n = \{\}$ and $A_n = \{q'_1, \ldots, q'_m\}$ where each q'_i is a *referent-specialization* of q_i in G.
   Failure pair. In this case, $n < m$ and G_n is non-empty, because one of the graphs in G_n does not have any successful CG-derivation. $\qquad\Box$

Definition 22. Let P be a CGP and G be a CGF-goal. A *successful* CGF-derivation of G from P is a CGF-derivation of G from P that terminates with a success pair $\langle\{\}, A_n\rangle$. $\qquad\Box$

Definition 23. Let P be a CGP and G be a CGF-goal. A *failed* CGF-derivation of G from P is a CGF-derivation of G from P that terminates with a failure pair $\langle G_n, A_n \rangle$. $\qquad\Box$

Definition 24. Let P be a CGP and G be CGF-goal. If there is a successful CGF-derivation of G from P then G is said to be *provable* from P using CGF-derivation, which is denoted by $P \vdash_f G$. $\qquad\Box$

Definition 25. Let P be a CGP and $G = \{q_1, \ldots, q_m\}$ be a CGF-goal. If G has a successful CGF-derivation from P with a terminal pair $\langle\{\}, A_n\rangle$ where $A_n = \{q'_1, \ldots, q'_m\}$, then A_n is called a CGF-*answer* for G from P, and each q'_i is called a CGF-*answer graph* for the CGF-goal graph q_i. $\qquad\Box$

5 Soundness and Completeness of the Proof Procedure

In this section we present the soundness and completeness of the proposed proof procedure. At first, we prove the soundness of the CG-derivation procedure.

---

[13] the rso is defined in [1], as a special case of projection mapping, that only specializes referent fields of some of the concepts

5.1 Soundness of CG-derivation

We present the soundness property of the CG-derivation procedure in the following theorem.

Theorem 26. *Let P be a CGP and q be a CGF-goal graph of a CG language L. Suppose q has a successful CG-derivation from P with a projection π. Then πq is a logical consequence of P.* □

Proof. Suppose P_N denotes the normal form of P, that means, for any CG g in L, if $P_N \models_g g$[14] then $P \models_g g$ (by Theorem 10). Now, every CG-derivation of q starts with a graph $g_{(0)} \in PROJECT(q, P_N)$. The proof is done by induction on the length n of the CG-derivation. Suppose $n = 1$, which means $g_{(0)}$ is an ACG in P_N, and $g_{(0)} \leq_g q$ with $\pi : q \to g_{(0)}$. Thus $P_N \models_g g_{(0)}$, and hence $P \models_g \pi q$.

Now, suppose the result holds for CG-derivations of length $n-1$. That means, the derivation of $g_{(n)}$ from $g_{(1)}$ is sound. Thus, if $P_N \models_g g_{(1)}$ then $P_N \models_g g_{(n)}$ and $P_N \models_g q$. Since $g \in P_N$, $P_N \models_g g$. Therefore, it is sufficient to prove that the derivation of $g_{(1)}$ from g is sound.

In getting $g_{(0)}$ from g, for each concept in $g_{(0)}$ that is a proper specialization of the corresponding concept in g, either it has a coreference link from a dominating concept or it is in some CG in $\{v_1, \ldots, v_m\}$[15]. Thus, if all subsequent steps in the derivation of $g_{(n)}$ are proved to be sound, then $P_N \models_g g_{(0)}$.

In deriving $g_{(1)}$ from $g_{(0)}$, let $u_{(0)j}$ has a successful CG-derivation of length i where $1 \leq i \leq n-1$ with a projection π_1. Then, by induction hypothesis, $u_{(0)j}$ as well as $\pi_1 u_{(0)j}$ is a logical consequence of P_N. The graph $g_{(1)}$ is derived by erasing $\pi_1 u_{(0)j}$ from $\pi_1 g_{(0)}$, which is equivalent to using a collection of sound inference rules of CGs ([9, 3]).

Thus deriving $g_{(1)}$ from $g_{(0)}$ is a sound operation. Therefore, the derivation of g_n from g is sound, and since $g_{(n)} \leq_g q$ with a projection $\pi : q \to g_{(0)}$. Thus $P_N \models_g g_{(n)}$, and hence $P \models_g \pi q$. □

5.2 Soundness of CGF-derivation

Now we prove the soundness of the CGF-derivation procedure.

Theorem 27. *Let P be a CGP and $G = \{q_1, \ldots, q_m\}$ be a CGF-goal. Every CGF-answer of G from P is a logical consequence of P.* □

Proof. Suppose G_0 denotes the anti-normal form of G, that means, $P \models_g G_0$ implies $P \models_g G$ (by Theorem 11). The proof is done by induction on the length n of CGF-derivation. Suppose $n = 1$, that means, the initial pair has the form $\langle G_0 = \{q_1\}, A_0 = \{\}\rangle$, the terminal pair is of the form $\langle G_1 = , A_1 = \{q_1'\}\rangle$ and $A_1 = \{q_1'\}$ is the CGF-answer. By definition, $q_1' = \rho q_1$ where ρ is the rso wrt π, and π is the projection obtained by a successful CG-derivation of q_1 from P. By Theorem 26, $P \models_g \pi q_1$, and hence $P \models_g \rho q_1$. Thus, $P \models_g \{q_1'\}$.

Suppose the result holds for the CGF-answer that comes from a CGF-derivation of length $n-1$. Now, let us consider a CGF-derivation $\langle G_0, A_0 = \{\}\rangle, \langle G_1, A_1\rangle, \ldots,$

---

[14] $\models_g$ stands for the logical consequence relation for CGs ([1, 3])
[15] in $PROJECT$ operation in Definition 18

$\langle G_n, A_n \rangle$ of length n. If q_i is the selected goal graph in the first step which has a successful CG-derivation with a projection π_1 and ρ_1 is the rso wrt π_1, then by theorem 26, $P \models_g \pi_1 q_i$, and hence, $P \models_g \rho_1 q_i$. Thus adding $\rho_1 q_i$ to A_0 and erasing q_i from G_0 do not change the truth values of A_0 and G_0, respectively. Since the rest of the CGF-derivation has a length of $n - 1$, thus by induction hypothesis, $P \models_g A_n$. □

5.3 Completeness of the Direct Proof Procedure

In investigating the completeness of the proposed proof procedure, at first, we set out the criteria to be considered for completeness.

Let P be a CGP of a CG language L. The declarative semantics of P is defined as the *least model* of P ([1, 2]), which is also the *least fixpoint* of the *immediate consequence operator*, F_P, defined in [1]. In fact, the least model of P is the normal form representation of the set of all CGs in L that are logical consequence of P. It is, therefore, sufficient to show that every CG in the least model of P has a successful CGF-derivation. We now present the following completeness theorem based on this criteria.

Theorem 28. *Let P be a CGP and $\mathcal{M}_P$ be the least model of P. Then $\mathcal{M}_P$ has a successful CGF-derivation from P.* □

Proof. Let $G_A = anti\text{-}normal(\mathcal{M}_P)$ and $P_N = normal(P)$. It is sufficient to prove that every CG in G_A has a successful CG-derivation. Suppose q is any CG in G_A. From the definition of the declarative semantics, q is contained in $F_P \uparrow n$ (wrt the CG *inclusion relation*)[16], for some $n \in \omega$, where ω is the first limit ordinal. That means, there is a CG u in $F_P \uparrow n$ such that $u \leq_g q$. We prove by induction on n that $F_P \uparrow n \sqsubseteq q$ implies q has a successful CG-derivation from P.

Suppose $n = 1$, then $F_P \uparrow 1 \sqsubseteq q$, which means that there is a CG v in P_N, such that $v \leq_g q$. Clearly q has a successful CG-derivation from P.

Now, suppose the result holds for $n - 1$. Let $F_P \uparrow n \sqsubseteq q$. By the definition of F_P, there is a more specific instance $g = \neg[u_1 \dots u_n \neg[w]]$ of a CCG in P, such that, $w \leq_g q$ with a projection $\pi : q \to w$ and $F_P \uparrow n - 1 \sqsubseteq \{\pi u_1, \dots, \pi u_n\}$. By induction hypothesis, each of πu_i has a successful CG-derivation from P. Thus w has a successful CG-derivation from P. Since q is a generalization of w, q has a successful CG-derivation. □

6 Conclusions

In this paper, a direct proof procedure, called the CGF-derivation is developed for definite conceptual graph programs based on the proposal of such a procedure in [3]. This procedure is a formal extension of the proof procedures in [5, 8, 3].

The major contributions of this work include: definition of a unique normal form representation for CGPs, development of the CG unification procedure, definition of an anti-normal form representation for the goal graphs, and finally using all these notions in the new direct proof procedure.

---

[16] the CG inclusion relation($\sqsubseteq$) is defined in [1]

Although the procedure developed in this paper is sound and complete, it is still limited in the sense that, it applies only to the definite CGPs which includes a limited subset of well-formed CGs in a CG language.

Further works related to this work would be to look into the implementability of an interpreter incorporating the procedure in order to provide some concrete guidelines for a practical implementation of a CG-based inference engine. These include developing tractable algorithms and associated data structures for implementing CG unification, converting a given CGP into its normal form, and converting a given CGF-goal into its anti-normal form. Moreover, extending the proposed deduction procedure for the CGPs that include negative CGs and other well-formed CGs, remains an open problem.

References

1. B. C. Ghosh and V. Wuwongse. Declarative Semantics of Conceptual Graphs Programs. In *Proceedings of the Second Workshop on Peirce held in conjunction with the First International Conference on Conceptual Structures*, August 1993.
2. B. C. Ghosh and V. Wuwongse. Conceptual Graph Programs and Their Declarative Semantics. Technical report, Asian Institute of Technology, Bangkok, Thailand, July 1994.
3. B. C. Ghosh and V. Wuwongse. Inference Systems for Conceptual Graph Programs. In W. M. Tepfenhart, J. P. Dick, and J. F. Sowa, editors, *Conceptual Structures: Current Practices*, Lecture Notes in Artificial Intelligence No. 835, pages 214–229. Springer-Verlag, 1994.
4. J. E. Heaton and P. Kocura. Presenting a Peirce Logic Based Inference Engine and Theorem Prover for Conceptual Graphs. In B. Moulin G. W. Mineau and J. F. Sowa, editors, *Conceptual Graphs for Knowledge Representation*, Lecture Notes in Artificial Intelligence No. 699, pages 381–400. Springer-Verlag, 1993.
5. A. Dugourd J. Faruges, M. Landau and L. Catach. Conceptual Graphs for Semantics and Knowledge Processing. *IBM Journal of Research and Development*, 30(1):70–79, January 1986.
6. S. Keronen. Natural Deduction Proof Theory for Logic Programming. In E. Lamma and P. Mello, editors, *Extensions of Logic Programming - Proc. ELP'92*, Lecture Notes in Artificial Intelligence No. 660, pages 265–289. Springer-Verlag, 1993.
7. J. W. Lloyd. *Foundations of Logic Programming*. Springer-Verlag, second, extended edition, 1987.
8. A. S. Rao and N. Y. Foo. CONGRES: Conceptual Graph Reasoning System. In *Proceedings of the 3rd Conference on Artificial Intelligece Applications*, pages 87–92. IEEE Computer Society Press, 1987.
9. J. F. Sowa. *Conceptual Structure: Information Processing in Mind and Machine*. Addison-Wesley Publishing Company, Inc., 1984.
10. V. Wuwongse and B. C. Ghosh. Towards Deductive Object-Oriented Databases Based on Conceptual Graphs. In H. D. Pfeiffer and T. F. Nagle, editors, *Conceptual Structures: Theory and Implementation*, Lecture Notes in Artificial Intelligence No. 754, pages 188–205. Springer-Verlag, 1993.

Conceptual Clustering of Complex Objects: A Generalization Space based Approach

Isabelle BOURNAUD and Jean-Gabriel GANASCIA

LAFORIA-IBP, Université Paris 6
4, place Jussieu - Boite 169
75252 Paris Cedex 05 - France
email: {bournaud, ganascia}@laforia.ibp.fr

Abstract. A key issue in learning from observations is to build a classification of given objects or situations. *Conceptual clustering* methods address this problem of recognizing regularities among a set of objects that have not been pre-classified, so as to organize them into a hierarchy of concepts. Early approaches have been limited to unstructured domains, in which objects are described by fixed sets of attribute-value pairs. Recent approaches in structured domains use a first order logic based representation to represent complex objects. The problem addressed in this paper is to provide a basis for the analysis of complex objects clustering represented using conceptual graphs formalism. We propose a new clustering method that extracts a hierarchical categorization of the provided objects from an explicit space of concepts hierarchies, called *Generalization Space*. We give a general algorithm and expose several complexity factors. This algorithm has been implemented in a system called COING. We provide some empirical results on its use to cluster a large database of Chinese characters.

1 Introduction

A key issue in learning from observations is to build a classification of given objects or situations. This kind of inductive learning, where the objects have not been pre-classified by a teacher, is often called *unsupervised learning*. It consists in searching for regularities in the training objects, so as to organize them into categories [6] [12] [24]. Conceptual clustering and traditional techniques developed in cluster analysis and numerical taxonomy differ in two major aspects. Firstly, the conceptual clustering problem is not only to build a partition of the set of objects into separated clusters, but also to associate a *characterization* to each cluster acquired. This characterization, called intensional description of cluster [30], has to be expressed in the terms used to describe the objects. Secondly, the clustering quality in the conceptual clustering paradigm is not solely a function of individual objects, but is dependent on concepts that describe the clusters, i.e. the intentional descriptions of the clusters [4].

More precisely, conceptual clustering has been defined by Michalski and Stepp [24] as:

Given:

- a set of objects and their associate descriptions

Find:

- clusterings that group these objects into *concepts*;
- an intentional definition for each concept,
- a hierarchical organization for these concepts.

The classification scheme resulting of a conceptual clustering process usually corresponds to a *concept hierarchy*. The task of *concept formation* is very similar to that of conceptual clustering, with the added constraint that learning be incremental [14].

Early systems in unsupervised concept learning have been limited to unstructured domains, in which objects are described by fixed sets of attribute-value pairs, like in CLUSTER/2 [24], COBWEB [9] or UNIMEM [20]. However, objects frequently have some natural structure, i.e. have components and relations among these components. Attribute-value based representations are not well suited to represent such complex objects [30]. Recent approaches in structured domains use a first order logic (FOL) based representation to represent complex objects. As examples, we may refer to systems such as CLUSTER/S [29] or LABYRINTH [30]. The major problem with such representations is that of the number of matching, i.e. the number of different ways two descriptions structurally match. In FOL based language, there is potentially an exponential number of matchings [27]. One way to limit the matching complexity is to restrict the expressiveness of the language used to characterize clusters [21]. This type of restriction is easier to express and to formalize in a graph-based formalism than in a FOL-based representation [32]. However, no work has analyzed in depth the use of conceptual graphs in conceptual clustering.

The problem addressed in this paper is thus to provide a basis for the analysis of complex objects clustering represented using conceptual graphs formalism [28]. Conceptual graphs are formally defined with a model theoretic semantics. They have all the expressive power of logic but are more intuitive and readable, with a smooth mapping to natural language [8]. Moreover, conceptual graphs are often used in data knowledge representation.

The main differences between our approaches and related works are first that the complex objects are represented using the conceptual graphs formalism, and secondly, the conceptual clustering process consists in searching through an *explicit* space of clusters of objects, called *Generalization Space*. A Generalization Space is an inheritance network, whose construction is based on a restricted graphs' generalization. We propose to extract from a particular Generalization Space different classifications corresponding to hierarchical organizations of the acquired concepts.

In section 2, we briefly review the classical clustering methods and present our approach based on the notion of Generalization Space. In the following section we describe a method called MSG, originally developed by Mineau [25], to build a Generalization Space. In section 4, we describe the conceptual clustering method proposed on an example and give a general algorithm used in an implemented system called COING. The last section presents some experimental results on conceptual clustering of Chinese characters.

2 Conceptual clustering

2.1 Related approaches

Different clustering methods have been developed in the area of research on unsupervised learning. On the one hand, the classic *agglomerative* clustering methods build a binary tree in a "bottom-up" manner. They begin with as many categories as the number of objects. They compute the similarity between all pairs of categories, and merge the two most similar categories into a single category. The merging process is repeated until all objects have been placed in a single category representing the root of the tree. Examples of systems based on this method are KBG [1] or MK10 [31].

On the other hand, the result of *divisive* methods is a tree built in a "top-down" manner. These methods begin with a single category containing all the objects, and repeatedly subdivide this category until all categories contain only a single object. The partitioning algorithm ("Nuées dynamiques") developed by Diday [5], and which has been used as a basis for developing the CLUSTER's family of systems, uses a top-down approach. Divisive methods are also used in concept formation approaches [14].

These two methods have been both used in Numerical Taxonomy and in Machine Learning. Although they differ in the building process, they both can be viewed as a search through a *non explicitly* built space of concept hierarchies [11] [14]. This is summarized on figure 1.

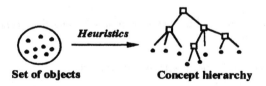

Set of objects   *Heuristics*   **Concept hierarchy**

Fig. 1. Classical approaches principle

2.2 A new approach

We propose a new approach to conceptual clustering, called COING, which consists in searching a concept hierarchy of a set of objects through a space of concept hierarchies *explicitly* built: the Generalization Space. This approach is based on a representation of complex objects using conceptual graphs [28]. We will first introduce some terminology and an example used along this article.

Knowledge representation Our work being related to the MSG developed by Mineau [25], we have considered the same restrictions on the conceptual graphs: existential graphs where only binary relations are used. Such graphs may be decomposed without ambiguity as a set of triplets $<$ *concept*$_1$, *relation*, *concept*$_2$ $>$. We call *conceptual arc* or *arc* such a triplet of a conceptual graph. Throughout this paper, we will use as an example a set of objects to be clustered (see figure 2). This set of objects, taken from [14], provides a good example of an application of our approach in spite of the fact that the objects do not require the full expression power of the conceptual graphs formalism. In this

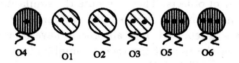

Fig. 2. An example of set of objects

example, we represent each object using conceptual graphs; figure 3 presents the representation of an object.

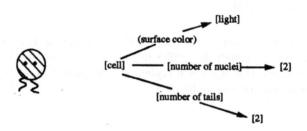

Fig. 3. Representation of the O3 domain object using three conceptual arcs

The term *concept* will be used within two contexts:

- an *acquired concept* in the conceptual clustering paradigm designs the entity "cluster + cluster description",
- a *domain concept* corresponds to a concept of the conceptual graphs formalism.

Generalization Space Structure A Generalization Space (GS) is an inheritance network. In this network, a node n is a pair (Cov(n), Desc(n)) where Cov(n), the *coverage* of n, is the set of objects covered by n; and Desc(n), the *description* of n, corresponds to the common characteristics of nodes of Cov(n) (i.e. the set of the common arcs).

A node p is a *father* of a node n iff: Desc(p) is more general than Desc(n) (i.e. arcs of Desc(p) are arcs that are more general than arcs of Desc(n)) and Cov(n) is strictly included in Cov(p). The set of m fathers of a given node n are noted $F(n)=\{f_1(n), f_2(n), ..., f_k(n), ..., f_m(n)\}$. For example, F(O3)={n2, n3} and F(O2)={n4} on figure 4. We call *elementary node* a node covering exactly one of the provided objects, and *generalizing node* any other node.

The Generalization Space is different from the "generalization hierarchy" introduces by Sowa [28]. Indeed, in the GS, nodes' descriptions have not to be a connected graph but are only a set of arcs [26], possibly disconnected. Figure 4 presents the generalization space obtained with the set of objects presented in figure 2.

A Generalization Space may have more than one root node. In order to guaran-

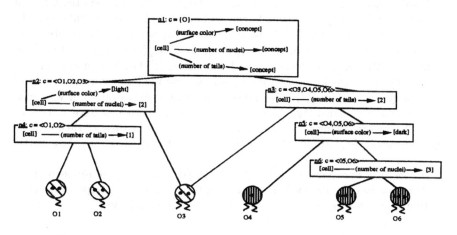

Fig. 4. The Generalization Space for the set of objects of fig.2: GS1

tee that the GS has a unique root, and to provide the GS with an upper lattice structure, we define, as in Mineau [25], a *minimal description* which generalizes all the objects' descriptions. The root of the GS corresponds to the root of the concept hierarchy that will be extracted from the GS.

2.3 Clustering objects using a GS

Any set of nodes of the GS, that cover all the provided objects, may be considered a priori as a classification of the set of objects. In that way, the GS may be viewed as a *space of classifications*. In fact, the GS contains a sub-set of all the potential classifications of the training objects, according to the restriction used in the building of GS (see section 3.1.). COING's approach consists in extracting one concept hierarchy from the GS. This approach is summarized in the figure 5.

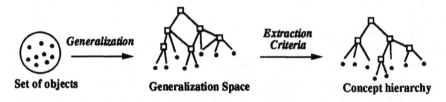

Fig. 5. Coing's principle

To build a GS from a set of objects represented using conceptual graphs, we use the MSG method developed by Mineau in [25]. This method is briefly sketched in the next section. In contrast to Mineau who considers the GS as a classification by itself, our goal is to extract from the GS one particular classification scheme. The main classification schemes that one meets in the literature are partition (mutually exclusive categories), clumping (overlapping categories), hierarchy with overlap, or hierarchy without overlap [4] [11]. As in major conceptual clustering methods, we are interested in finding the latter kind of structure, i.e. concept hierarchies [14] [23]. Indeed, hierarchical organization is very natural for many domains [19]. Moreover, hierarchical structures have proven to be useful for organizing and managing large databases because of their searching efficiency [7] [22]. Finally they are the easiest to be interpreted by the experts.

To the best of our knowledge, no work has yet dealt with the problem of extracting a concept hierarchy from an explicitly built space of concept hierarchies. Section 4 presents more precisely COING's approach to this problem.

3   Generalization Space

3.1   Building a GS

Building the GS requires to group into clusters all objects that have similarities. One way to "evaluate" similarity between objects is to generalize the objects by searching the common features subsuming the objects. Generalizing objects, in the representation used, requires to match graphs, which is a problem known to be NP complete [13].

Among the several approaches developed to find a partial solution to this problem, one with a low complexity is the one developed by Mineau in the MSG [25]. This approach consists in searching the largest sub-sets of common *arcs* and not the largest common *sub graphs*. In doing so, it ignores the fact that any two arcs may be connected. In other words, a graph is considered as a set of arcs; *matching graphs* is viewed as *matching arcs*. It consists in restricting the generalization language to a one-to-one matching [18]. As a result of this restriction, the complexity of the graph matching process is polynomial in the number of objects and arcs. Nevertheless, the GS does not contain all the existing generalizations of objects (only the ones that may be expressed in this restricted language).

3.2 Size of the GS

Let N be the number of objects and k the average number of arc in objects' descriptions.

The GS does not contain as many nodes as the number of all possible sub-sets of objects, but only nodes for sub-sets of objects having common features. The GS nodes are obtained in grouping in the same node all the arcs covering the same objects. The number of arcs in the GS, elementary and generalized arcs, is less or equal than $2^3 \times N \times k$ [25]. Since the GS nodes contain at least one arc, the number of nodes of the GS (GS size, noted $|GS|$) is linear with the number of objects:

$$|GS| \leq 8 \times N \times k$$

Our method to build the GS differs from the one proposed by Mineau since we introduce background knowledge in the GS's construction [15]. This background knowledge, represented as a hierarchy of the domain concepts, allows the descriptions of the generalizing nodes to contain arcs which do not explicitly appear in the objects' descriptions. For example, the domain concepts "black" and "grey" may be generalized by the concept "dark". The size of the GS obtained with this method is linear with the number of objects, given by the following formula: (where P is the maximum depth of the domain hierarchy).

$$|GS| \leq P^3 \times N \times k$$

In Mineau's work, all concepts get generalized in "?"; the maximum depth of the domain hierarchy, P, is thus equal to 2.

4 Coing's approach

In [11], the authors explain that the conceptual clustering process is "composed of three distinct sub processes: the process of deriving a hierarchical classification scheme, the process of aggregating objects into individual classes; and the process of assigning conceptual descriptions to object classes". In our approach, the aggregation and characterization processes are made during the building of the GS. The remaining problem is to derive a classification scheme from the GS. The goal of COING's approach is to extract from the GS a classification, corresponding to a *hierarchical organization* of the acquired concepts. We shall briefly present a two-stage process for such an extraction. We first give a simple extraction algorithm. We then describe the criteria used to choose one hierarchy over another, and give the algorithm complexity.

4.1 A two-stage process

The process of extracting a hierarchy from the GS is composed of two stages: *skeleton selection* and *skeleton reduction*. Let n be the given node, $f_i(n)$ be a father of n and $s_j(n)$ a son of n; let Cov(n) be the coverage of n and Desc(n) its description.

Skeleton selection As previously said, the GS has an inheritance network structure, with a unique root thanks to the *minimal description* used during the GS's construction. Thus, each node of the GS, excluding the most general and its immediate sons, may have more than one father, i.e. may be generalized by more than one node of the GS (see object O3 of figure 4). In a hierarchy, each node has only *one* father. Extracting such a structure from the GS may be viewed as the problem of *choosing a unique father* for each node having more than one father. We call *skeleton*, the structure obtained after a single father has been chosen for each node of the GS.

The selection of a skeleton consists in:

- choosing for each node n a unique father $f_c(n)$ from its set of fathers F(n) = $\{f_1(n), f_2(n), .., f_m(n)\}$. For example, the two potential skeletons for the GS1 of figure 4 are presented in figure 6. The skeleton sk1 is the result of the choice of the node n3 as a father for the node O3, and the skeleton sk2 the choice of the node n2 as a father for O3.

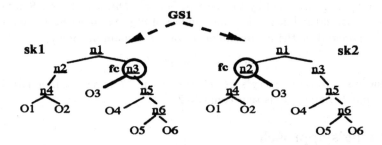

Fig. 6. Two potential skeletons, sk1 and sk2, for GS1 of fig.4

- updating the coverage of each father $f_i(n) \in$ F(n), $i \neq c$, in the skeleton where father $f_c(n)$ has been selected. If the resulting coverage of the node $f_i(n)$ is the same as the coverage of one of its sons $s_j(f_i(n))$ (for example, n2 and n4 of sk1, and n3 and n5 of sk2), then $f_i(n)$ and $s_j(f_i(n))$ are grouped in the same node, and the description of the resulting node is updated: arcs of the Desc($f_i(n)$) that generalize an arc of Desc($s_j(f_i(n))$) are removed, other arcs of Desc($f_i(n)$) are added to Desc($s_j(f_i(n))$) in the resulting node. Figure 7 presents the updated skeleton sk'1 of sk1 (figure 6): nodes n2 and

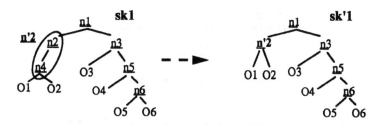

Fig. 7. The resulting skeleton for sk1 of fig.6

n4 have been grouped in a "new" node n'2. The updated content of the node n'2 is presented on figure 8.

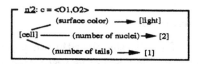

Fig. 8. Content of the node n'2 of fig.7 (different of node n2 of fig.4)

Skeleton reduction Given one skeleton sk_i, different *hierarchies* may be extracted. These hierarchies differ in the number of generalizing nodes they contain. In order to generate all the different hierarchies, we define an *absorption operation* over nodes. The process of restricting a skeleton to a hierarchy consists in choosing which nodes of the skeleton to absorb. Absorbing a node n into its father node $f_k(n)$ consists in propagating the description of n "down into" the descriptions of its sons that are not elementary nodes, and linking the sons of n to the absorbing node $f_k(n)$. The hierarchy h1 presented in figure 9 has been obtained from the skeleton sk'1 of figure 7 by absorbing the node n5 into its father node n3. The content of the node n"6 is presented on figure 10.

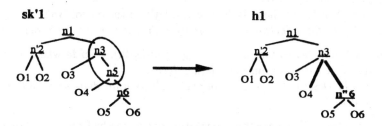

Fig. 9. One hierarchy extracted from sk'1: h1

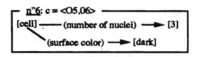

Fig. 10. Content of the node n"6

As a result of the absorbing operation, the resulting hierarchy does not contain as much information as the initial skeleton; when a node n is absorbed into its father, the common features between the sons of n are lost; consequently the resulting hierarchy does not contain a node for each sub-set of objects having common features. The objective of the reduction process is similar to the one exposed by Fisher in [10], i.e. aggregating nodes to simplify the results of the conceptual clustering process.

4.2 An extraction algorithm

Let us now present more formally the algorithm to perform such a two-stage extraction of hierarchies from the GS. Let us first introduce the notion of GS's *layer* that will be used in the algorithm. A GS's layer corresponds to the set of nodes of the GS that have the same depth, the depth of a node corresponding to the number of ancestors of this node. For example, the object O3 and the nodes n4 and n5 of GS1 belong to the GS1's layer of depth 3, noted L(3). Let $depth_{max}$ be the maximum depth of the GS .

Stage 1: Skeleton selection
For each depth, from depthmax to 2, analyze the GS's layer L(depth):

- *For each node n of layer L(depth) that has more than one father $\{f_1(n), f_2(n), ..., f_m(n)\}$:*
  1. *build the pair $(n, \{f_1(n), f_2(n), ..., f_m(n)\})$*
  2. *choose a unique father $f_k(n)$ in $\{f_1(n), f_2(n), ..., f_m(n)\}$ for the node n*
  3. *for each node $f_c(n)$, $c \neq k$, remove Cov(n) from Cov($f_i(n)$), if $f_i(n)$ has a unique son, agglomerate Desc($f_i(n)$) with its sons description.*

Stage 2: Skeleton reduction
The notion of skeleton's layer is defined similarly to the one of GS's layer.
For each depth, from depthmax to 2, analyze the skeleton's layer L(depth):

- *For each non elementary node n of layer L(depth), **decide** whether or not to absorb it with its immediate father $f_k(n)$.*
- *If so, add Desc(n) to the descriptions of all its immediate sons and update the links.*

As presented in this algorithm , extracting one hierarchy from the GS requires to make different choices: choosing one father from a set, choosing which are the nodes to be absorbed. We explain, in the next section, which criteria we used in order to make these choices.

4.3 Extraction criteria and computational complexity

The most classical criterion prefer one concept hierarchy over another one is the general criterion that may be expressed as: *choose categories that maximize similarity within categories and that concurrently minimize similarity between categories* [16] [17]. This criterion is in fact a good tradeoff between within-category similarity and inter-category dissimilarity of objects used in systems that do not build explicitly a space of concept hierarchies.

In our framework, we may define a set of criteria that will allow to define more explicitly the type of control desired on the building of the hierarchy. This approach may be particularly interesting in the context of conceptual clustering revision where potential users interactively request revisions of the provided hierarchy [2].

Choice for skeleton selection As previously explained, selecting a skeleton may be viewed as choosing a unique father for each node of the GS. Choosing a father for a node corresponds to decide which father is the "important" one, according to the clustering task. We have identified two types of criteria that may be used for choosing one father:

- a *preference function*, based on the coverage and the description of the nodes, that orders the father (for example the depth of the fathers in the GS). The preference function consists in choosing as father for a node the one with the minimum index value. Here are two different preference functions we have used: $f1(n) = \alpha \times Cov(n) + \frac{\beta}{Desc(n)}$ et $f2(n) = \frac{\alpha}{Cov(n)} + \beta \times Desc(n)$. The function $f1$ favors the nodes with small coverage and long descriptions, whereas the second function $f2$ favors the nodes with large coverage and short descriptions. The function $f2$ with $\alpha = \beta = 1$ has been used to extract sk'1 of GS1 of figure 4.
- a *preference bias*, based on user defined knowledge, that constrains the choice of father. This background knowledge may help to determine which features are relevant in a given situation [17]. For example, the relation "number of tails" is more important than "number of nuclei". It may also express some relations between objects and clusters; for example the fact that the user does not want the object O_i to be a son of the node p or objects O_i and O_j to have the same father.

In stage 1 of the algorithm, local criteria have been used to extract a skeleton. In stage 2, a global criterion is used to output one concept hierarchy.

Choice for skeleton reduction We have identified a set of parameters that describe the structure of a concept hierarchy: depth, width, maximum number of sons per node, ... These parameters have the great advantage to be easily interpretable by the user. We used these parameters to constrain the number of nodes to be absorbed. The choice of which nodes to absorb is based on an information measure criterion similar to that of *Category Utility* [16].

Computational complexity We will briefly give some time complexity results of the proposed method. Let N be the number of objects.

GS's construction

In the worst case, the time complexity for the construction of the GS in $\Theta(N^2)$.

Skeleton selection

Let SK be the set of potential skeletons for a GS. The size of SK is given by the following formula:

$$|SK| = \prod_{n_i \in GS} card(F(n_i))$$

In practice, this number may be small, e.g. the number of skeletons for the generalization space GS1 of the figure 4 is 2.

The proposed method does not explore all the possible skeletons. In fact, the time complexity of the skeleton extraction is, in the worst case, in $\Theta(N^2)$.

Skeleton reduction

The size of the skeleton (i.e. number of nodes) grows linearly with the number of objects, the time complexity of the reduction process is also growing linearly with the number of objects.

5   Experimentation

The algorithm presented above has been implemented in the system COING developed in LeLisp.

One of our domain of application of the COING's approach is related to Chinese characters. Chinese characters are complex structured objects that are naturally represented using conceptual graphs. Figure 11 presents a Chinese character represented using conceptual graphs. In an interdisciplinary research group, we explore with Chinese teachers the discovery of concept hierarchies of Chinese

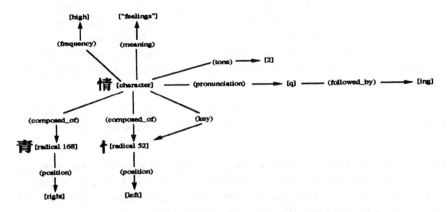

Fig. 11. Representation of a Chinese character

characters for two major reasons. Firstly, the skeleton nodes' descriptions correspond to similarities between characters that may be used in order to facilitate the learning of Chinese characters [3]. Secondly, the concept hierarchies offer a good mean to study the Chinese characters' evolution and structure.

To perform a large-scale experimentation of our system, we have built a database of 6768 Chinese characters (a standard set of characters) represented as conceptual graphs. The Chinese experts have found interesting not to build a general hierarchy for all characters but to first analyze the results on smaller sub-sets. The experimentation we report has been made on the basis of 416 characters randomly selected. COING has generated 473 generalizing nodes and 2370 links in its GS. The skeleton produced contains only 40 generalizing nodes and 39 links between generalizing nodes.

6 Conclusion

Conceptual graphs are well suited to represent complex objects but, to our knowledge, the use of conceptual graphs in conceptual clustering had not yet been analyzed. In this paper, we propose a basis for describing clustering of complex objects represented using the conceptual graph formalism. We present an algorithm to extract a concept hierarchy from a particular GS. Because the GS, under certain restrictions, has a tractable size, it may well be used as a basis for clustering complex objects described using conceptual graphs.

Even if some "classic" clustering methods consider complex objects, comparing our approach to these methods is not so easy principally for two reasons. Firstly, our method is original since it uses domain knowledge. Secondly, our approach has been developed in the perspective not only to build concept hierarchies but also to revise them [2]. Thus, the parameters used to extract concept hierarchies are meant to be both understandable and offer various ways to adapt a hierarchy to one's needs.

On an concrete domain, we have experimented the extraction of concept hierarchies from a sub-set of Chinese characters. Both the interest of the Chinese experts in the extracted hierarchies and the ability of our system to handle large datasets of conceptual graphs are encouraging results to continue our research.

References

1. G. Bisson. *KBG: Induction de Bases de Connaissances en Logique des Prédicats.* PhD thesis, Université Paris XI - Orsay, France, 1993.
2. I. Bournaud. Conceptual clustering revision. In *Annual Meeting of the Classification Society of North America*, Denver, June 1995.
3. I. Bournaud and J.D. Zucker. Découverte de similitudes entre objets structurés par exploration d'un espace de généralisations. In *Environnements Interactifs d'Apprentissage avec Ordinateur.* Eyrolles, Paris, 1995.
4. C. Decaestecker. Apprentissage et outils statistiques en classification incrémentale. In *Revue d'intelligence artificielle 7 - n1*, pages 33–71. 1993.

5. E. Diday, J. Lemaire, J. Pouget, and F. Testu. *Elements d'Analyse des données.* Dunod, 1982.

6. T. Dietterich. Unsupervised concept learning and discovery. In *Readings in Machine Learning*, pages 263–266. Morgan Kaufmann, 1990.

7. G. Ellis. Efficient retrieval from hierarchies of objects using lattice operations. In n-699 Lecture Notes in AI, editor, *Proc. First International Conference on Conceptual Structures*, pages 274–293. 1993.

8. G. Ellis. Managing complex objects. Technical report, University of Quennsland, Australia, 1993.

9. D. Fisher. Knowledge acquisition via incremental conceptual clustering. In *Machine Learning 2*, pages 139–172. 1987.

10. D. Fisher. Iterative optimization and simplification of hierarchical clusterings. Technical report, Vanderbilt University - Nashville, USA, 1995.

11. D. Fisher and P. Langley. Approaches to conceptual clustering. In *Proc. Ninth International Joint Conference on Artificial Intelligence*, pages 691–697, 1985.

12. D. Fisher and M. Pazzani. Computational models of concept learning. In *Concept Formation: Knowledge and Experience in Unsupervised Learning.* Morgan Kaufmann, San-Mateo, California, 1991.

13. M. Garey and D. Johnson. *Computers and intractability: A guide to the theory of NP- completeness.* CA: W. H. Freeman, 1979.

14. J.H. Gennari, P. Langley, and D. Fisher. Models of incremental concept formation. In *Artificial Intelligence 40*, pages 11–61. 1989.

15. O. Gey. Saturation et généralisation de graphes conceptuels. In *Actes des Neuvièmes Journées Francaise de l'Apprentissage*, pages C1–C14, Strasbourg, France, 1994.

16. M. Gluck and J. Corter. Information incertainty and the utility of categories. In *Proc. Seventh Annual Conference of the Cognitive Science Society*, pages 283–287, 1985.

17. S.J. Hanson and M. Bauer. Conceptual clustering, categorization, and polymorphy. In *Machine Learning 3*, pages 343–372. 1989.

18. F. Hayes-Roth and J. McDermott. An interference matching technique for inducing abstractions. In *Communications of ACM - 21(5)*, pages 401–410, 1978.

19. P. Langley. Editorial - machine learning and concept formation. In *Machine Learning 2*, pages 99–102. 1987.

20. M. Lebowitz. Experiments with incremental concept formation: Unimem. In *Machine Learning 2*, pages 103–138. 1987.

21. H. Levesque and R. Brachman. A fundamental tradeoff in knowledge representation and reasoning (final version). In *Readings in Artificial Intelligence*, pages 41–71. Morgan Kaufmann, 1985.

22. R. Levinson. A self organizing retrieval system for graph. In *Proc. AAAI'84,vol 1*, Austin, Texas, 1984.

23. R.S. Michalski and R.E. Stepp. An application of ai techniques to structuring objects into an optimal conceptual hierarchy. In *Proc. Seventh International Joint Conference on Artificial Intelligence*, pages 460–465, 1981.

24. R.S. Michalski and R.E. Stepp. Learning from observation: Conceptual clustering. In R.S. Michalski, J.G. Carbonell, and T.M. Mitchell, editors, *Machine Learning, An Artificial Intelligence Approach, Volume I*, pages 331–363. Springer-Verlag, 1983.

25. G. Mineau. *Structuration des Bases de Connaissances par Généralisation.* PhD thesis, Université Montréal, Canada, 1990.

26. G. Mineau, J. Gecsei, and R. Godin. Structuring knowledge bases using automatic learning. In *Proc. Sixth International Conference on Data Engineering*, pages 274–280, Los Angeles, USA, 1990.

27. F. Neri and L. Saitta. Knowledge representation in machine learning. In *Proc. ECML-94*, pages 20–27, Catania, Italy, April 1994.

28. J. Sowa. *Conceptual Structures: Information Processing in Mind and Machine.* Addison Wesley, Massachusetts, 1984.

29. R.E. Stepp and R.S. Michalski. Conceptual clustering: Inventing goal-oriented classifications of structured objects. In R.S. Michalski, J.G. Carbonell, and T.M. Mitchell, editors, *Machine Learning, An Artificial Intelligence Approach, Volume II*, pages 471–498. Morgan Kaufmann, San-Mateo, California, 1986.

30. K. Thompson and P. Langley. Concept formation in structured domains. In *Concept Formation: Knowledge and Experience in Unsupervised Learning.* Morgan Kaufmann, San-Mateo, California, 1991.

31. J. Wolff. Data compression, generalization, and overgeneralization in an evolving theory of language development. In *Proc. AISB-80 Conference on Artificial Intelligence*, pages 1–10, 1980.

32. J.D. Zucker and J.G. Ganascia. Selective reformulation of examples in concept learning. In *Proc. International Conference on Machine Learning*, pages 352–360, New-Brunswick, 1994.

Using Empirical Subsumption to Reduce the Search Space in Learning

Marc Champesme

Laboratoire d'Informatique de Paris-Nord (LIPN)
CNRS URA 1507
Institut Galilee
Universite Paris-Nord
Av. J.B. Clement
93430 Villetaneuse
e-mail : champ@ura1507.univ-paris13.fr
Tel : 49 40 35 78

Abstract. In the traditional learning framework, hypothesis that are not equivalent with respect to the standard subsumption relation can be equivalent from the learning's point of view. We define in this paper a new subsumption relation, called empirical subsumption, that allows to take into account this fact. This new subsumption relation is then used to define a particular kind of search space reduction that do not reduce the class of learnable concepts. Then, we show that theses theoretical results can be applied when the knowledge representation formalism is the conceptual graph formalism.

1 Introduction

In machine learning, the most commonly used model considers a learning problem as a search problem in a set of formulas ordered by a subsumption (or generalization) relation. More precisely, given a set of examples and counter examples of a target concept that are represented as logical formulas, the learning algorithm search for a hypothesis (i.e., a logical formula) which generalize all examples and none counter example. Then the search is performed by successive generalizations of too specific hypothesis (i.e. which are not generalizations of some examples) or by successive specializations of too general hypothesis (i.e. which are generalizations of some counter examples). Most often, this search is done using refinement operators. Such operators are therefore essential components of the learning algorithm.

Concerning the knowledge representation formalisms used in learning, many works have used and still use propositional logic. Precursory works of Plotkin [12], Shapiro [14] or Winston [16] which used more expressive formalisms have remained without any extensions for a long time. This fact was largely due to algorithmic inefficiency reasons that seemed to be unavoidable. However, more recent works (see for instance [13]) have shown that it was possible to design efficient learning systems using representation languages nearby first order logic. Hence, researches concerning this kind of approach have rapidly come along again. This new interest was materialized by the creation of a new research area called Inductive Logic Programming (or ILP), which groups together researchers that represent knowledge by clauses. At the same time other researchers have been interested in other languages of first order logic and some of them in the conceptual graph's formalism (see [9], [2], [10] or [8]).

It has appeared that many learning systems using languages of first order logic favor algorithmic efficiency to the detriment of learning effectiveness. Indeed, many learning systems use refinement operators that are not complete concerning the representation language they use. As a consequence, the class of learnable concepts is implicitly restricted without clear definition of the class of effectively learnable concepts. We propose here to adopt the opposite approach, namely to identify restrictions of first order logic as expressive as possible but in which we may learn efficiently. In this way, it is essential to bring to the fore languages for which there exists refinement operators that have "good" properties. Van der Laag and Nienhuys-Cheng's works [5][6][7] constitute a first approach to this problem. First they have proposed a minimal set of "good" properties. Then they have shown that it is not possible to design a learning system based on such an operator when knowledge is represented by unrestricted clauses. Furthermore, we have shown in [3] that these negative results could be extended to conceptual graphs. The aim of this paper is to study language restrictions that are necessary to allow the existence of refinement operators with such "good" properties.

We remark then, that in most learning algorithms the main search criterion used depends only on the number of examples and counter examples subsumed by the hypothesis. With such a criterion, hypothesis that are not equivalent concerning the subsumption relation can be equivalent from the learning algorithm's point of view. Hence some hypothesis could be discarded from the search space without any damage to the quality of the learning results. Therefore we define in this paper the notion of *empirically conservative reduction* that states properties that such reductions should fulfill and then the corresponding notion for ideal refinement operators, i.e. *empirically ideal refinement operators*. Indeed, when such "good" reductions are possible, the existence of refinement operators that can go all over the complete search space is not essential and we can content ourselves with operators that explore only the reduced search space provided that they have "good" properties concerning this search space. The main result of this paper is that, although ideal operators do not exist for conceptual graphs, such reductions are possible for non connected conceptual graphs and allow us to define empirically ideal refinement operators, i.e. operators that are not ideal but behave as ideal operators would do regarding the expected results of the learning.

Section 2 and 3 define respectively the conceptual graph framework and the learning framework of this paper. Then section 4 gives a precise definition of the empirical subsumption relation that serves as a basis for the results presented in the following sections. Section 5 presents the notion of empirically conservative reduction that is used in section 6 to remedy to the problem of nonexistence of ideal refinement operators.

2 Conceptual Graphs Framework

The conceptual graph's formalism offers mainly two advantages over more traditional logical formalisms. First, its "graphical" look improves human understandability of the represented knowledge as well as theoretical result's presentation or proof using the graph theory framework. And secondly, conceptual graphs offer a great variety of mechanisms for the representation of background knowledge. However, the drawback is that the conceptual graph's formalism suffers from a lake of logical foundations. As a consequence, results obtained for conceptual graphs can not be directly translated into the traditional logical formalisms (and reciprocally from logical formalisms into the conceptual graph's formalism).

In this paper, all results are about *simple* conceptual graphs. A simple conceptual graph is a conceptual graph whose referents are either generic or individual and where neither negations nor contexts may appear. We will talk about two kinds of simple graphs, ordinary simple graphs or S-graphs as defined by Chein and Mugnier [4] and S_{nc}-graphs that are defined as S-graphs except that the connectivity constraint is removed. Since the canonical formation rules of Sowa [15] are defined only for connected conceptual graphs, we define below the additional formation rule needed for S_{nc}-graphs. This new rule, called addition rule, allows derivation of non connected graphs from connected ones.

Definition 1. Let G and G' two S_{nc}-graphs. Doing the *addition* of G and G' consists in building the S_{nc}-graph union of G and G'.

The subsumption relation defined for conceptual graphs is the relation induced by the existence of a projection between two conceptual graphs. In this paper this relation will be called π-*subsumption* and will be denoted by $\leq_\pi$. So we will say that an S-graph (resp. S_{nc}-graph) G π-subsume an S-graph (resp. S_{nc}-graph) H, noted $H \leq_\pi G$, if and only if there exists a projection from G to H.

3 Learning Framework

Following the traditional Inductive Logic Programming settings, a learning problem can be stated as follows in the conceptual graph's framework :

Let $O = O^+ \cup O^-$ be a set of conceptual graphs, find a hypothesis H (i.e. a conceptual graph) such that :

- $\forall o \in O^+, o \leq_\pi H$
- $\forall o \in O^-, o \not\leq_\pi H$

In this definition, the sets O^+ and O^- refer respectively to the set of positive instances or examples and to the set of negative instances or counter-examples. The following figure shows a learning problem with two examples $(O^+ = \{o_1, o_2\})$ and one counter-example $(O^- = \{o_3\})$. In this case, the four hypothesis H_2, H_3, H_4 and H_5 are solutions of the learning problem as stated above. The hypothesis H_0 and H_1 are too general (they generalize both positive and negative instances) and the hypothesis H_6 and H_7 are too specific (they do not generalize the positive instance o_2).

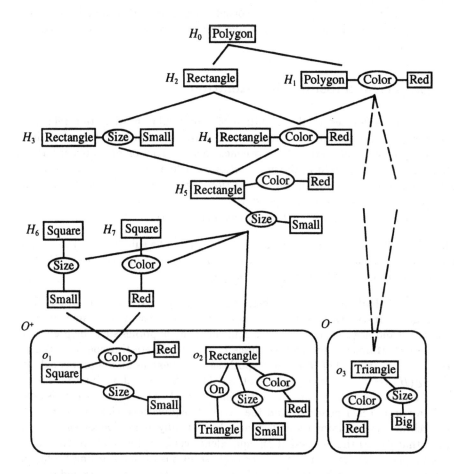

Fig. 1. A learning problem with two examples and one counter-example.

In the general case, there is no solution to this problem and we must search for approximate solutions. Therefore a search criterion is used to allow the choice of a *best* hypothesis, which will be proposed as a solution. In most learning systems, such a criterion uses two preference relations. The first relation compares hypothesis considering only the number of positive and negative instances that are subsumed by the hypothesis. Generally, hypothesis that subsume many positive instances and few negative instances are preferred. With such a relation the hypothesis H_2, H_3, H_4 and H_5 of Figure 1 will be regarded as equivalent.

The second relation is used to make a choice among hypothesis that are equivalent with respect to the first relation. In some cases (as in [12]) most specific hypothesis are preferred (H_5 is preferred to H_2, H_3 or H_4) and in other cases (see for instance [13] or [14]) most general or most simple ones are preferred (H_2 is preferred to H_3, H_4 and H_5).

In the rest of this paper we will consider learning with a search criterion depending only on the number of positive and negative instances subsumed by the hypothesis.

4   Empirical Subsumption Relation

This section is devoted to the definition of a particular subsumption relation, called empirical subsumption relation, which allows the representation of a minimal search space (considering the above mentioned restriction on search criteria). This relation will be used in the next section to define the notion of empirically conservative reduction and in the last section of this paper to define the notion of empirically ideal refinement operator.

Let us give the following definition as a preliminary :

Definition 2. Let $(S, \leq)$ be a quasi-ordered set and O a subset of S. The *extension* of an element s of S, denoted by $ext_O(s)$, is the set of elements of O that are smaller than s : $ext_O(s) = \{o \in O/o \leq s\}$.

When the evaluation of the hypothesis depends only on the number of positive and negative instances subsumed, all hypothesis that subsume the same number of positive and negative instances are equivalent (no hypothesis can be preferred to another). Moreover, given a hypothesis H, the only interesting specializations[1] (resp. generalizations) of H are the hypothesis whose extension is a proper subset of $ext_O(H)$ (resp. a proper superset of $ext_O(H)$). So, this allows us to define a new subsomption relation based on the inclusion of the hypothesis extensions. Such a relation is a preorder, and the set of its equivalence classes ordered by this relation is a minimal search space. Indeed, any reduction of the search space that causes the removal of an equivalence class of this relation will result in a proper reduction of the set of possible solutions of our learning problem. Conversely, this suggests that it could be possible to make proper reductions of the hypothesis language without any consequence on the expected learning results. This latter point will be more investigated in the following section by defining the empirically conservative reduction.

This new subsumption relation, that we call empirical subsumption relation, is defined as follows :

Definition 3. We call empirical subsumption relation relative to $(S, O, \leq)$, denoted $\leq_O$, the relation defined, for any couple (s_1, s_2) of elements of S, by : $s_1 \leq_O s_2$ if and only if $ext_O(s_1) \subseteq ext_O(s_2)$.

---

[1] specializations that improve the value of the criterion.

Notations. In the following, the empirical equivalence class of an element s of S relative to the subset O of S is denoted by $eq_O(s)$, and for any two elements s_1 and s_2 we write $s_1 \sim_O s_2$ if s_1 and s_2 are equivalent with respect to $\leq_O$.

The following example (based on the example of Figure 1) illustrates the definitions and notations given above :

Example 1. Concerning the extensions of the hypothesis of Figure 1, we have:

- $ext_O(H_0) = ext_O(H_1) = \{o_1, o_2, o_3\}$,
- $ext_O(H_2) = ext_O(H_3) = ext_O(H_4) = ext_O(H_5) = \{o_1, o_2\}$,
- $ext_O(H_6) = ext_O(H_7) = \{o_1\}$.

As a consequence, $\{H_0, H_1\}$, $\{H_2, H_3, H_4, H_5\}$ and $\{H_6, H_7\}$ are sets of empirically equivalent hypothesis. Moreover, each hypothesis of $\{H_0, H_1\}$ empirically subsumes each element of $\{H_2, H_3, H_4, H_5, H_6, H_7\}$ and each hypothesis of $\{H_0, H_1, H_2, H_3, H_4, H_5\}$ empirically subsumes H_6 and H_7.

Using the set inclusion properties we can easily prove that $\leq_O$ is reflexive and transitive. Since two elements with same extension can be different, the antisymmetry property is not fulfilled. Hence the empirical subsumption relation is a preorder.

Properties. Let $(S, \leq)$ be a quasi ordered set, O a subset of S and $\leq_O$ the empirical subsumption relation relative to $(S, O, \leq)$.

1. $\leq_O$ is a preorder.
2. if $s_1 \leq s_2$ then $s_1 \leq_O s_2$
3. if $s_1 \sim s_2$ then $s_1 \sim_O s_2$
4. if $s_1 <_O s_2$ and $s_1 \leq s_2$ then $s_1 < s_2$

Another desirable property that is not fulfilled in general is that, for any couple (e_1, e_2) of empirical equivalence classes such that $e_1 \leq_O e_2$ and for any element of e_2 there is an element of e_1 that is smaller with respect to $\leq$. Formally this property is defined as follows :

(P$_1$) $\forall(s_1, s_2) \in S \times S$, if $s_1 <_O s_2$ then $\exists s'_1 \in eq_O(s_1), s'_1 \leq s_2$

If property P$_1$ is satisfied and if we have a complete refinement operator for $(S, \leq)$ (see section 6) then we will be able to build a derivation from *any* element of e_2 to *an* element of e_1. A sufficient condition to fulfill this property is to provide an operator $\oplus$ on S satisfying the following properties :

1. $\forall(s_1, s_2) \in S \times S, s_1 \oplus s_2 \in S$
2. $\forall o \in S$, if $o \leq s_1$ and $o \leq s_2$ then $o \leq s_1 \oplus s_2$
3. $\forall(s_1, s_2) \in S \times S, s_1 \oplus s_2 \leq s_1$ and $s_1 \oplus s_2 \leq s_2$

When S is a set of logical formulas the logical conjunction fulfills these properties and the same apply with S_{nc}-graphs. On the other hand such an operator does not exist for S-graphs and the example of Figure 2 shows a case where property

P_1 is not satisfied. This figure shows two empirical equivalence classes e_1 and e_2 of the empirical subsumption relation relative to the set of S-graphs $O = \{o_1, o_2, o_3\}$. We can see on this example that the S-graph H_4 belonging to the empirical equivalence class e_1 has no specialization in the empirical equivalence class e_2 despite e_1 empirically subsumes e_2. This is due to the fact that there does not exist a S-graph H', such that $H_4 \geq_\pi H'$, $H' \pi$-subsumes o_1 and o_2 but does not π-subsume o_3. On the other hand, if we consider S_{nc}-graphs instead of S-graphs, we can build a S_{nc}-graph H' by addition of H_4 with any element of e_2. Furthermore, note that the S-graph H_3 of the figure leads to the same problem as H_4 does.

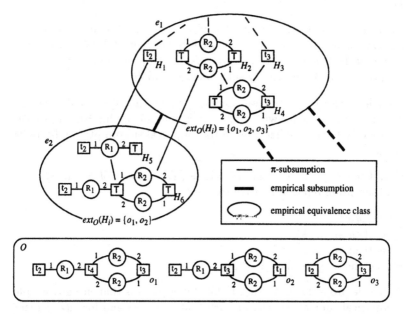

Fig. 2. Empirical subsumption relation relative to the set of S-graphs O (concept types t_1, t_2, t_3 and t_4 are pairwise incomparable).

5 Empirically Conservative Reduction

We propose here a new criterion that can be used to estimate the incidence of a restriction of the hypothesis language. As empirically equivalent formulas can not be distinguished from the learning point of view, the set of hypothesis to consider could be restricted without any incidence on the learning results. Note that such reductions are implicitly made by many learning algorithms and especially by all "lgg based learning algorithms" (see for example [11] or [1]). Indeed, such learning algorithms explore the search space by repeatedly computing the least general generalization (or lgg) of different subsets of the learning set. Since the lgg of any subset of the learning set is a most specific element of the corresponding empirical equivalence class (the class the lgg belongs to), these algorithms do not explore all the hypothesis set but only the most specific hypothesis of each empirical equivalence class. We investigate here which properties should be satisfied to make a "good" reduction. Note that we consider only reductions on the hypothesis set to explore and not on the subsumption relation.

We say, that a "good" search space reduction is a reduction that keeps at least one element in each empirical equivalence class and which does not reduce the comparability between empirical equivalence class. Such a good reduction is called an *empirically conservative reduction* and is define as follows :

Definition 4. Let $(S, \leq)$ be a quasi-ordered set and O a subset of S. Then a subset S_O of S is called an *empirically conservative reduction* of S with respect to O if

1. $\forall p \in S, eq_O(p) \cap S_O \neq \emptyset$
2. $\forall (p, q) \in S \times S$ if $p <_O q$ and $p < q$, then $\exists (p', q') \in S_O \times S_O, p' \in eq_O(p), q' \in eq_O(q)$ and $p' < q'$

The first condition means that any non empty empirical equivalence class of S must contains at least one element of S_O. The second condition means that for any pair of empirical equivalence classes (e_1, e_2) such that e_1 empirically subsumes e_2, if there exists an element of e_1 that subsumes an element of e_2, then the same must be true in S_O.

We give now some examples of reductions, illustrated on the preceding example (see Figure 2) :

Example 2. No reduction of e_1 to $\{H_4\}$ can be empirically conservative since no element of e_2 will be π-subsumed by an element of e_1. The same applies for reductions with $e_1 = \{H_2\}$ and $e_2 = \{H_5\}$. On the other hand, reductions with $e_1 = \{H_1\}, e_2 = \{H_5\}$ or $e_1 = \{H_2\}$ and $e_2 = \{H_6\}$ could be empirically conservative depending on reductions made on the other empirical equivalence class.

Concerning empirically conservative reductions for S-graph and S_{nc}-graphs, we can state the following results :

Let us first consider S_{nc}-graphs.

Theorem 5. *Let O and S_O be two sets of S_{nc}-graphs. If for any empirical equivalence class e, S_O contains the most specific element of e, then S_O is an empirical conservative reduction with respect to O.*

Proof. We proved in [3] that any set of S_{nc}-graphs has a unique and finite least general generalization (or lgg). Hence, for any empirical equivalence class e, the lgg of $ext_O(e)$ is the most specific element of e. Using the addition rule properties we can prove, that for any pair of empirical equivalence classes (e_1, e_2), if $e_2 \leq_O e_1$, then the lgg of $ext_O(e_1)$ π-subsumes the lgg of $ext_O(e_2)$. Hence, any reduction keeping at least the minimal element of any empirical equivalence class is an empirically conservative reduction. $\square$

Concerning S-graphs, the example of Figure 2 shows that this theorem does not hold for S-graphs.

6   Refinement Operators

In this section we take up the notion of ideal refinement operator defined in [7]. Then we show that nonexistence results of such operators stated by Van der Laag and Nienhuys-Cheng for clauses can be applied and extended for S-graphs and S_{nc}-graphs. Finally, using the notion of empirical conservative reduction we define the notion of empirically ideal refinement operator and show that such an operator can be defined for S_{nc}-graphs.

6.1   Ideal Refinement Operators

We give now the definition of ideal refinement operator proposed in [7]. To our opinion, the set of properties corresponding to this definition is a minimal one for learning. Van der Laag and Nienhuys-Cheng claim that an ideal refinement operator must be *complete*, *locally finite* and *proper*. An operator is proper if the refinements of an element computed by this operator are not equivalent to the refined element. Since, obtaining elements that are equivalent to a given element is not of any interest when solving a learning problem, such a property is interesting. Furthermore, note that the use of a non proper operator can even prevent a good pruning of the search space. An operator is locally finite if the set of one-step refinements is finite and computable for any element of the search space. Clearly this property is essential for any practical use. Finally, completeness ensures that any element can be obtained from any one of its generalizations (or its specializations in the case of an upward refinement operator).

Before defining what is an ideal refinement operator, let us give the definition of a refinement operator.

Definition 6. Let $(S, \leq)$ be a quasi ordered set. A refinement operator is a mapping on S :

- ρ is a *downward refinement operator* if $\forall s \in S, \rho(s) \subseteq \{s' \in S / s' \leq s\}$
- δ is an upward refinement operator if $\forall s \in S, \delta(s) \subseteq \{s' \in S / s \leq s'\}$
- The sets of *one-step refinements*, *n-steps refinements* and *refinements* of s are respectively defined as :
  - $\rho^1(s) = \rho(s)$
  - $\rho^n(s) = \{s' / \exists s'' \in \rho^{n-1}(s) \text{ et } s' \in \rho(s'')\}$
  - $\rho^*(s) = \rho^1(s) \cup \rho^2(s) \cup \ldots \cup \rho^i(s) \cup \ldots$

Now we can define what is an ideal refinement operator.

Definition 7. Let $(S, \leq)$ be a quasi ordered set. Then a refinement operator is ideal if it is locally finite, complete and proper.

- An operator is *locally finite* if $\forall s \in S, \rho(s)$ is finite and computable

- An operator is *complete* if $\forall s > r, \exists n \in N$ and $\exists r' \in \rho^n(s)$ such that $r' \sim r$
- An operator is *proper* if $\forall s \in S, \rho(s) \subseteq \{s' \in S / s > s'\}$

Although it is said "ideal" such an operator is not perfect. Note, for instance, that, given an ideal refinement operator ρ, any operator ρ' defined by $\rho'(s) = \rho^n(s)$ is ideal too. Hence, a property of minimality could enhance this set of properties.

6.2 Nonexistence Results

Van der Laag and Nienhuys-Cheng [7] have shown that ideal refinement operators do not exist for unrestricted sets of clauses ordered by θ-subsumption. In [3] we showed that these results can be applied to S-graphs sets ordered by π-subsumption. Furthermore, we proved that even with drastic restrictions of the graphs structure (S-chains), nonexistence results still hold. These results are mainly due to the existence of infinite uncovered chains. Existence of such chains means that between any two comparable and reduced S-graphs, there can be infinitely many reduced S-graphs, and thus that in the general case, the set of generalizations of an S-graphs set (i.e. the search space of a learning problem) is infinite. Another consequence concerning ideal refinement operators is that it becomes very difficult to ensure the existence of a *finite* derivation chain between any pair of comparable formulas.

6.3 Empirically Ideal Refinement Operators

We have shown in previous section, that ideal refinement operators do not exist for S-graphs and that structural restriction can not avoid this problem. This observation raises the following question :

Which restriction is needed to obtain existence results ?

Van der Laag and Nienhuys-Cheng propose in [5] to reduce the number of formulas to consider using a threshold on the size of the formulas. With such a limitation, the search space become finite and, as a consequence, derivation chains become finite too. Moreover, Van der Laag and Nienhuys-Cheng show that with an appropriate measure of size, comparability of formulas is not reduced : any two "small" comparable formulas remain comparable after the restriction. However, we raise the problem of the threshold choice : which consequences will have a particular choice ? which threshold to choose to have the least of undesirable consequences ? In this section, we address the second question. We show that for any learning set we can find a threshold, such that the corresponding restriction is an empirical conservative reduction. We then define the notion of empirical refinement operator. This definition use the notion of empirical conservative reduction with the idea that, when such reduction is possible we can use an "almost ideal" operator i.e. an operator that is not ideal but behaves as an ideal operator with respect to our learning framework.

Ideal Operator for Size Bounded Graphs. Before to define the notion of empirical ideal refinement operator, we give the definition of an ideal operator for size bounded S_{nc}-graphs. This operator was first defined for clauses by Van der Laag and Nienhuys-Cheng in [5]. As shown in [3], we can prove that the corresponding operator for S_{nc}-graphs satisfies the same properties (i.e. it is ideal for size bounded S_{nc}-graphs). But let us first define the size measure, called *newsize*, which is needed for the definition of the operator.

Definition 8. Let G be a S_{nc}-graph, R the set of r-vertex of G and r an element of R.

- $rsize(r) = degree(r)$ - (number of generic c-vertex which are neighbor of r).
- $maxsize(G) = \max_{r \in R}(rsize(r))$
- $newsize(G) = (maxsize(G), |R|)$

Properties. Let G and H be S_{nc}-graphs, R the set of r-vertex of G and (k, m) a couple of positive integers.

1. if $H \leq G$ then $maxsize(G) \leq maxsize(H)$
2. if $G \sim H$ then $maxsize(G) = maxsize(H)$
3. if $maxsize(H) < maxsize(G)$ then $H \not\leq G$
4. The number of S-graphs satisfying $newsize(G) \leq (k, m)$ (i.e. $maxsize(G) \leq k$ and $|R| \leq m$) is finite.

Now we define the ideal refinement operator for size bounded S_{nc}-graphs.

Definition 9 (cf. [3] according to [5]). Let G be a reduced S_{nc}-graph, such that $newsize(G) \leq (k, m)$ and $eq_{(k,m)}(K)$ be the set of S_{nc}-graphs with $newsize$ smaller or equal to (k, m) and equivalent to K. Then $H \in \rho_{rg}(k, m)(G)$ if H is reduced, $newsize(H) \leq (k, m)$ and one of the following condition holds :

1. ρ_{rg}^1 : $H < G$ and there exists $G' \in eq_{(k,m)}(G)$ and $H' \in eq_{(k,m)}(H)$ such that H' is obtained from G' by application of the join rule to itself (i.e. an internal join).
2. ρ_{rg}^2 : $H < G$ and there exists $G' \in eq_{(k,m)}(G)$ and $H' \in eq_{(k,m)}(H)$ such that H' is obtained from G' by an *elementary* restrict (replacement of a generic referent by an individual one or replacement of a type by one of its successors in the type lattice).
3. ρ_{rg}^3 : H is obtained from G by addition of a maximally general star graph (S-graph with an unique r-vertex) whose r-vertex is of type r, and such that G does not contain any occurrence of a r-vertex of type r.

We have proved in [3] that, for any (k, m), this refinement operator is ideal for S_{nc}-graphs with $newsize$ smaller than or equal to (k, m).

Empirically Ideal Operator. Since we have shown that there exists ideal refinement operators for size bounded S_{nc}-graphs, we raise the following question:
How to choose a "good" bound ?

In this section, we address this question using the notion of empirically conservative reduction of the search space. We say that if we have an operator that is ideal for an empirically conservative reduction of the search space, then this operator will behave like an ideal one. We call such operators *empirically ideal operators*. Then we propose a function that allows, for any set of S_{nc}-graphs, to compute a bound such that we can obtain an empirically conservative reduction of the search space and hence an empirically ideal operator.

An empirically ideal refinement operator is defined as follows :

Definition 10. Let ρ be a refinement operator defined on a quasi-ordered set $(S, \leq)$. Then ρ is *empirically ideal* if, for any subset O of S, there exists a subset S_O of S such that the two following conditions are satisfied :

- S_O is an empirically conservative reduction of S with respect to O.
- ρ is an ideal refinement operator for $(S_O, \leq)$.

We propose now a function, called *bound*, which for any set of S_{nc}-graphs compute a value allowing an empirically conservative reduction of the search space.

Definition 11. Let $O = (H_i)_{1 \leq i \leq n}$ be a set of S_{nc}-graphs and let us denote, for any $i, 1 \leq i \leq n, R_i(t_r)$ the set of r-vertex of type t_r of H_i. The *bound* function is defined as follows :

$$bound(O) = \left(\sum_{t_r \in T_r} \left(\max_{1 \leq i \leq n} (|\, R_i(t_r)\,|) \right)^n, \max_{1 \leq i \leq n} (maxsize(H_i)) \right)$$

Let us now define the refinement operator $\rho_{rg} bound_O$.

Definition 12. Let O be a set of S_{nc}-graphs and G a reduced S_{nc}-graph such that $newsize(G) \leq bound(O)$. Then $H \in \rho_{rg} bound_O(G)$ if $H \in \rho_{rg}(k, m)(G)$ and $(k, m) = bound(O)$.

We prove now that $\rho_{rg} bound_O$ is empirically ideal :

Theorem 13. $\rho_{rg} bound$ *is an empirically ideal refinement operator.*

Proof. Let O be a set of S_{nc}-graphs and S_O the set of S_{nc}-graphs K such that $newsize(K) \leq bound(O)$. Since $\rho_{rg}(k, m)$ is ideal for size bounded S_{nc}-graphs $\rho_{rg} bound$ is ideal for S_O. As shown in [3] we can prove that, for any set O of S_{nc}-graphs, the newsize of the lgg of the elements of O is smaller than or equal to $bound(O)$. Hence S_O contains at least the minimal element of any empirical equivalence class of O and using theorem 1 we conclude that S_O is an empirically conservative reduction of the set of S_{nc}-graphs. So for any set O we can build an empirically conservative reduction for which $\rho_{rg} bound$ is ideal and therefore $\rho_{rg} bound$ is an empirically ideal operator. □

7 Conclusion and Further Works

We have highlighted that the learning algorithms are essentially based on criteria that depend only on the number of positive and negative instances subsumed by the hypothesis. Therefore, reductions on the number of formulas of the search space can be made without any reduction of the quality of the expected learning results. Then we have defined a new criterion allowing to establish if a given reduction is such a "good" one, and have shown that such *empirically conservative reductions* could be made when hypothesis are represented using S_{nc}-graphs. However, we have shown that the same kind of reduction could not be made for S-graphs, although the problem of empirically conservative reduction's existence remains still open for this kind of conceptual graphs. Note that, this negative result is mainly due to the structure of the π-subsumption relation that has better properties in the case of S_{nc}-graphs than in the case of S-graphs. Hence, further works should study more precisely the structure of the π-subsumption relation and other conditions on which existence of empirically conservative reductions could be ensured.

Then we have define the notion of *empirically ideal refinement operator*. Such an operator is required to be ideal only on empirically conservative reductions of the hypothesis space. Hence, regarding the properties of empirically conservative reductions, the properties required for empirically ideal refinement operators ensure that such operators should behave like ideal ones concerning the expected learning results. Finally, we have applied these results to remedy to the problem of nonexistence of ideal operators. Namely, we have define a new refinement operator for S_{nc}-graphs that is empirically ideal. This new result is based on a theorem proved in [3], that give a bound on the size of the least general generalization of any set of S_{nc}-graphs (or S-graphs). We show in this paper that limiting the size of considered hypothesis by this bound, is a valid empirically conservative reduction.

However, this work does not provide an efficient refinement operator. Indeed, the size bound we have given is exponential according to the size of the learning set and thus prohibits its use in a learning algorithm on real world application. Furthermore, note that results from Muggleton and Feng [11] about least general generalization's size confirm our results. This suggests that it should not be possible to reduce this bound and that if better empirically conservative reductions (in term of search space size) do exist, such reductions should not keep the minimal elements of empirical equivalence classes in the search space. Thus better results on "lgg conservative reductions" should need proper reduction of the language. Note for instance, that for some restrictions of first order logic as the ij-determinate restriction proposed by Muggleton and Feng in [11], the lgg size bound can be reduced to a polynomial function of the learning set size.

References

1. Brezellec, P., Soldano, H.: ELENA: a bottom-up learning method. Tenth International Conference on Machine Learning. Amherst (Massachusetts, USA). Morgan Kaufmann. (1993) 9–16.

2. Champesme, M.: Apprentissage par detection de similarites utilisant le formalisme des graphes conceptuels. Phd Thesis, Paris XIII (1993).

3. Champesme, M.: Operateurs de raffinements ideaux pour les graphes conceptuels. Prepublication LIPN (1994).

4. Chein, M., Mugnier, Marie-Laure: Conceptual Graphs: fundamental notions. Revue d'Intelligence Artificielle, 6 (1992) 365–406.

5. Van der Laag, P.R.J., Nienhuys-Cheng, Shan-Hwei: Subsumption and refinement in model inference. ECML-93, European Conference on Machine Learning. Vienna, Austria. Springer-Verlag.(1993) 95–114.

6. Van der Laag, P.R.J., Nienhuys-Cheng, Shan-Hwei: Existence and Nonexistence of Complete Refinement Operators. ECML-94, European Conference on Machine Learning. Springer Verlag. (1994) 307–322.

7. Van der Laag, P.R.J., Nienhuys-Cheng, Shan-Hwei: A note on ideal refinement operators in inductive logic programming. ILP-94, Fourth International Workshop on Inductive Logic Programming. GMD. Bad Honnef/Bonn, Germany. (1994) 247–260.

8. Levinson, R.A.: APS: An architecture for experience-based knowledge acquisition. Workshop on Computational Architectures for Supporting Machine Learning and Knowledge Acquisition at Machine Learning Conference. Aberdeen (Scotland). Morgan Kaufmann. (1992).

9. Liquiere, M.: Apprentissage a partir d'objets structures: Conception et realisation. These de 3eme cycle, Montpellier. (1990).

10. Mineau G.W.: Acquisition d'objets structures destines a la classification symbolique. Premieres Journees Francophones d'Apprentissage et d'explicitation des Connaissances. Dourdan (France). (1992) 131–145.

11. Muggleton, S., Feng, C.: Efficient Induction of Logic Programs. In Muggleton S. (Eds.), Inductive Logic Programming. Academic Press. (1992) 281–298.

12. Plotkin, G.D.: A further note on inductive generalisation. Machine Intelligence, 6 (1971) 101–124.

13. Quinlan, J. R.: . Learning logical definitions from relations. Machine Learning, 5, (1990) 239–266.

14. Shapiro, E. Y.: Algorithmic program debugging. MIT Press. (1983).

15. Sowa, J.F.: Conceptual Structures, information processing in mind and machine. Addison Wesley. (1984).

16. Winston, P.H.: Learning structural descriptions from examples. In P.H. Winston (Eds.), The psychology of computer vision. McGraw Hill. (1975).

A New Parallelization of Subgraph Isomorphism Refinement for Classification and Retrieval of Conceptual Structures

James D. Roberts

Board of Computer Engineering
University of California, Santa Cruz CA 95064, USA

Abstract. Major applications of conceptual structures will require quick response times on extremely large knowledge bases. Although algorithmic developments have provided tremendous improvements in speed, we believe implementation on parallel processors will be needed to meet long-term needs. This paper presents a new parallelization of a subgraph isomorphism refinement algorithm for performing projection tests and retrieving conceptual structures (CS). The improved algorithm is faster, requires fewer processors, and is compatible with recent relation-based representations of CS.

The new parallelization takes advantage of the features of contemporary massively parallel machines by exploiting bit-parallelism in the data words. Processing numerous CS on a single parallel array using load balancing integrated with multi-level indexed search, it combines the strengths of prior parallel subgraph isomorphism parallelizations. It incorporates lattice codes of the concept-type hierarchy, forming all node candidate binding lists in parallel. Simulation results of the behavior of the refinement algorithm with parameterized synthetic data sets are presented.

1 Introduction

Testing for projections is the most time consuming portion of classifying a conceptual structure (CS) into a database of graphs. Classification speed is crucial for systems that must handle a large number of queries or inference operations, particularly for anticipated knowledge bases of hundreds of thousands of structures. Speed improvements can be achieved by reducing the number of necessary projection tests, and by reducing the time required for each projection. Techniques such as multilevel indexed search, as incorporated into the PEIRCE Workbench, tremendously speed structure retrieval by minimizing the number of projection tests required and in the future by re-using matching information across graphs in the hierarchy. We believe that even with these techniques parallel processing will be necessary for future large-scale applications. This paper presents a revised algorithm for reducing the time of projection tests through parallel processing.

For the purposes of this paper we will consider CS projection as equivalent to subgraph isomorphism (SI). Negation and nested contexts introduce important differences between SI and projection, but the operations of SI remain central. Other differences are already incorporated into our SI algorithm. Although having exponential time in the worst case, subgraph isomorphism has an empirical expected-case time of $O(n^4)$ bit operations (where n is the number of nodes in the larger of the two graphs) using the SI refinement algorithm developed by Ullmann [11]. The polynomial expected-case time means that parallelization can provide a substantial speedup. Our new algorithm exploits both the parallelism of multiple processors and the parallelism of the multiple bits in a processor's data word.

Several others have proposed or implemented parallel algorithms for subgraph isomorphism; here we summarize the three most relevant to our current work. Ullmann's original implementation used the bit-parallelism in the data word of a conventional serial processor to achieve $O(n^3)$ empirical time on graph isomorphism [11]. Implementations by Willett *et al.* on a data-parallel processor empirically show $O(n^2)$ time using n^2 single-bit processors, demonstrating an efficient parallelization of SI with substantial time improvement [12]. Lendaris has used an inherently parallel neural network approach to processing conceptual structures (CS), including join, simplify, and projection [5, 6]. His results demonstrate neural networks performing important filtering of candidate graphs which must then be further processed for projection. Our own prior work focuses on the computationally efficient approach of using multi-level indexed search to extensively prune the number of graphs to be tested then speeding each test through parallel processing. Our prior paper ([4]) discusses modifications to Ullmann's SI refinement algorithm to accommodate the specific requirements and improve performance in projection tests on CS, analyses the advantages of a parallelized multilevel indexed search over exhaustive comparison to all graphs in the knowledge base for a bounded number of processors, and describes parallel compilation (encoding) of the concept-type hierarchy.

This paper extends our prior work by revising the parallel SI algorithm to exploit multi-bit parallel processors, more typical of contemporary data-parallel machines. This allows processing numerous CS on a single, small parallel array. We thereby modify our prior "processor farm" method that uses multiple arrays for classification in a large database into an approach using a single array with work-load balancing. This paper also introduces using subsumption codes to avoid the bottleneck of consulting a subsumption table, rather than the less effective caching method we previously proposed. We also discuss graph storage and allocation, and present an algorithm for forming all SI node candidate binding lists in parallel. Additionally, this paper incorporates Levinson's recent work on a relation-based CS representation and summarizes refinement behavior on a parameterized synthetic data set.

Section 2 summarizes Ullmann's algorithm and prior parallelizations. This is followed by a brief overview of multilevel indexed search and lattice coding. Section 3 summarizes modifications to Ullmann's algorithm for conceptual structures, including Levinson's relation-based representation of CS with performance benefits. Section 4 highlights the new multi-bit parallelization and Sect. 5

presents the details. Revisions to the multilevel indexed search parallelization are discussed in Sect. 6. Parameterized simulations on synthetic data sets are presented in Sect. 7. Section 8 summarizes the results presented in this paper, including hardware implications, and suggests directions for future work.

2 Background

This section outlines the algorithms that form the basis of the parallelizations presented in the remainder of this paper and briefly summarizes our prior work. We first discuss Ullmann's serial SI refinement algorithm and its parallelization by Willett *et al.*. This is followed by a brief description of multilevel search and lattice coding as relevant to our new SI parallelization.

2.1 Subgraph Isomorphism Refinement

Ullmann's algorithm features a refinement technique to extensively prune the search tree in a backtracking algorithm [11], speeding SI tests. Given two graphs G_a and G_b we want to determine if G_a is a subgraph of G_b. Ullmann's refinement algorithm represents the graphs and the binding between nodes as three Boolean matrices: A the adjacency matrix of G_a, B the adjacency matrix of G_b, and M the 'match' matrix; a bit $m_{ij} \in M$ is 1 if node $i \in G_a$ is a candidate for binding to node $j \in G_b$ (otherwise m_{ij} is 0). Figure 1 diagrams the overall structure of the algorithm. First, graphs are selected for comparison; those selected could include all graphs (exhaustive comparison) [12], those which pass filtering criteria (single-level indexing) [12, 6], or those passing multi-level indexing (initial graph SI tests filtering subsequent graphs) [8, 4]. Second, the match matrix M is formed. Third, the backtracking refinement search is performed. The refinement procedure is called at each branch of the backtracking search tree, significantly pruning the number of branches explored.

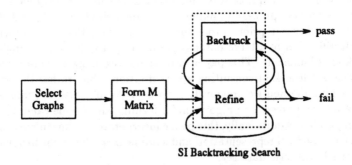

SI Backtracking Search

Fig. 1. Flow diagram of the refinement subgraph isomorphism algorithm

The initial match matrix is formed by iterating over all i and j and setting m_{ij} to 1 if the node type of i is the same as that of j and the arity of i is less than or equal to the arity of j, otherwise m_{ij} is set to 0, thus forming the set of all possible bindings. The M matrix is then refined to remove candidate bindings based on the topology of the graphs. This is not guaranteed to remove *all* invalid bindings. The refinement algorithm keeps a candidate binding m_{ij} iff $\exists(x)\, a_{ix} \wedge \neg(\exists(y) m_{xy} \wedge b_{yj})$. This logical criteria can be expressed by the matrix operations

$$M' = M \wedge \overline{A \times \overline{(M \times B)}} \ .$$

If at any time there is a row in M consisting of all zeros the refinement fails, as a node in G_a does not have a corresponding node in G_b. If all rows in M have a single one, the refinement succeeds and the match matrix corresponds to one of the possibly multiple subgraph isomorphisms. Otherwise, refinement iterates until the M matrix has not changed from the prior iteration. In the latter case the backtracking portion of Ullmann's algorithm determines if M represents a valid subisomorphism, checking that each row of M has exactly one 1 and that no column has more than a single 1. If so, we're done. Otherwise the backtracking portion recursively continues the search by arbitrarily selecting one binding for a node in G_a and calling refinement again. The full algorithm terminates when either an isomorphism has been found or there are no more bindings to try. Hughey *et al.* present examples of the SI refinement algorithm on conceptual structures [4].

The matrix operations are readily parallelizable with all candidate bindings (bits in the match matrix) processed simultaneously. Willett *et al.* present a parallelization for the DAP-610, an older parallel machine with 4096 single-bit processors [12]. Their coding optimizes the matrix expressions, eliminating the transpose for one of the Boolean matrix multiplications. They present three parallel implementations of Ullmann's algorithm. In the first, each processor stores 1 bit from each of the matrices and the parallel processor computes one SI test at a time. In the second, each processor stores the entire matrices and the processor performs up to 4096 SI tests simultaneously. In the third, a hybrid, graphs are processed using the second implementation until some number of graphs are completed, after which the remaining graphs are processed using the first implementation. Their experimental results on a small organic chemistry database show that the third method is clearly faster. Willett *et al.* propose a fourth method that uses the second method, adding new graphs as earlier graphs complete processing.

2.2 Multilevel Indexed Search

Multilevel indexed search (MIS) [8] can significantly reduce the number of SI tests necessary. As an independent surveys shows Levinson and Ellis' methodology to be the fastest known structure search pruning [2], it forms the framework and motivation for our parallelization efforts. Rather than providing a single level of indexing as in conventional databases, Levinson's Method III MIS creates a partial order over the knowledge base using the more-general-than relation. Objects are screened by predecessors in the poset and in turn screen successors.

Smaller and simpler objects prune out more time-consuming comparisons on larger objects. Method III empirically prunes by a factor of $O(N/\lg^2 N)$, where N is the number of objects in the knowledge base, and also supports conceptual clustering, generalization, and machine learning. We hypothesize that at any given point in the execution of MIS, $O(\lg N)$ graphs are available as candidates without risking unnecessary comparisons. Current work on multilevel indexed search explores ways of reducing the work in performing a SI test on a given graph by exploiting binding and other information from tests on its predecessors.

2.3 Lattice Codes

The transitive links in a poset can be compiled into codes for each element, avoiding both link chasing and large look-up tables for testing subsumption in the CS concept-type hierarchy. As a basic example, the poset is represented as an adjacency matrix and the reflexive-transitive closure of the matrix is computed. Each row of the resulting matrix is the code for its respective element in the partial order [1]. These simple codes are N bits long where N is the number of elements in the poset. More formally, the poset is plunged into a lattice and the lattice is then encoded; more advanced encoding algorithms approach the lower bound of $\Omega(\lg N)$ length codes for ideal poset topologies [1, 3]. Lattice codes can also be used to further prune comparisons in a Method III graph classification, provided the "query" graph is known to be in the knowledge base [3]. The time to compute subsumption, greatest lower bound, or least upper bound between two concept types or two graphs in the hierarchy is proportional to the length of the code.

3 Subgraph Isomorphism and Conceptual Structures

In this section we consider how Ullmann's SI refinement algorithm can be applied to CS projection, given the traditional node-based representation and a recently proposed edge-based representation.

3.1 Modifications to Ullmann's Algorithm for Node-Based Representations of CS

We make several modifications in applying SI refinement to conceptual structures [4]. First we must account for edge direction; Ullmann's algorithm is for undirected graphs and cannot be adapted to SI in DAGs. In CS, edge direction is determined by the relation type so that we treat relations as labeled edges and check argument numbers to verify direction in the backtracking portion of the algorithm. Treating relations as labeled edges rather than nodes also has a major speed advantage, as empirical times are $O(n^4)$ for serial and $O(n^2)$ for Willett *et al.*'s parallel implementations, where n is the number of nodes. Additionally, nodes may bind if concept node of G_a is subsumed by that of G_b. We now initially set m_{ij} to 1 iff $a_i \preceq b_j$ and |relations adjacent to a_i| $\leq$ |relations adjacent to b_j|. Additional criteria and filters, such as minimum loop length, can also be applied in forming the initial match matrix.

In testing projection between CS, there does not have to be a 1:1 mapping between the concept-type nodes and relations in G_a to those in G_b. This is readily handled by relaxing the mapping criteria applied in the backtracking portion of the algorithm.

3.2 A Relation-Based Representation for CS

Levinson's Universal Data Structure (UDS) combines the features and benefits of neural networks, semantic networks, relational databases, and conceptual structures in a single representational framework [7]. Its relation-based representation of graphs is of particular relevance to the new work presented in this paper. In UDS, conceptual structures are transformed such that the relations and their adjacent concept-types in the CS become nodes in the UDS graph. A node has the form *[relation, arg1, arg2,... argN]*, representing higher-order relations as a single node. Each node is in the UDS hierarchy, having a single subsumption code that can be used in forming candidate bindings in subgraph isomorphism tests.

The refinement subgraph isomorphism algorithm can be applied directly to the UDS representation without change, but with several benefits. Representing CS at the relational level usually reduces the the number of nodes for the SI algorithm and the number of refinement iterations are reduced due to the greater specificity of the nodes. Combined, these significantly speed SI tests and in turn classification in a database.

4 The New Multi-Bit Multi-Graph Parallel Algorithm

Having covered background information on the base algorithms and their adaptation to CS, we now present our new parallel implementation. Previous parallelizations, including our own prior work, assumed massively parallel machines with a single-bit processor such as the DAP-601 or the CM-2. Recent machines feature much wider data words in each processor, typically 32 bits. Since nearly all bit operations are independent of one another at any step of the refinement, it is possible to assign as many bits to a processor as it can handle simultaneously rather than place a single bit of each matrix in each processor. This increases speed by making much better use of the available resources and permits simultaneous comparison of many graphs even on a relatively small number of processors. This also reduces the cost of a machine by reducing the number of processors necessary.

In addition to converting from a single-bit to a multi-bit processor parallelization, our new parallelization covers graph storage, a parallel algorithm for forming the M match matrices, and revises the parallelization of Method III to account for a significant number of graphs being tested on a single parallel array. As noted, recent work on UDS can be directly incorporated. The overall structure is the same as in Fig. 1. We use Method III (or its successors) to select several graphs for testing, move them into processors, form the M matrices, then execute the SI refinement. In this paper we discuss only refinement in detail; the backtracking portion is assumed to be essentially the same as that of Willett *et al.*

The following descriptions involve three assumptions for the purpose of simplifying discussion. First, we assume the processors are connected in a 2 dimensional mesh topology, typical of most contemporary massively parallel machines. Second, we assume that the number of bits in the processor data word is at least as large as the number of nodes in a graph. Otherwise each row of the matrices must reside in multiple processors. Third, we assume that each column of the processor mesh has the same or slightly larger number of processors as there are nodes in the largest graph to be compared. Otherwise each column would be processing multiple graphs. Combining the second and third assumptions we are using n^2 processors (where n is the number of nodes in the largest graph) as with prior implementations, but here we are exploiting n-bit parallelism within each processor and computing n graph refinements simultaneously. The implementation details that arise when these conditions are not met can be dealt with in a straightforward manner.

The following sections cover our modification and parallelization of the refinement portion of Ullmann's SI algorithm. Refinement represents the bulk of execution time in SI, that in turn represents the bulk of time in Method III MIS classification.

5 Details of the Multi-Bit Refinement Algorithm

Adapting the SI refinement algorithm to contemporary multi-bit processors, we combine the features of Willett *et al.*'s methods and achieve additional benefits. In the new parallelization, we assign an entire row of each of the matrices to a processor and execute the refinement algorithm for many graphs on a single-data parallel array, giving a group of processors a new graph once they have completed their prior SI test. We exploit the bit-parallelism of b-bit processors so that a given SI test requires $1/b$ as many processors with no slowdown. Having fewer processors and the full matrix row in a single processor greatly reduces communication time. As with Willett *et al.*'s first method, having multiple processors per graph speeds refinement. As with their second method, having several graphs on a single parallel array minimizes wasted processors when many more processors are available than are needed for a single graph comparison. Although not yet implemented in our simulations, we propose a load-balancing approach similar to Willett *et al.*'s third and fourth method to avoid the poor processor utilization which would result if the implementation were to wait until all graphs finished before proceeding to the next group of graphs. We propose integrating this load balancing with the Method III queues, as described in Sect. 6.2. Finally, specific to CS, we use lattice codes of the concept-type hierarchy, performing a subsumption test on the codes in forming the M match matrix. This avoids the bottleneck of all processors consulting a single concept-type hierarchy and allows forming all rows of the match matrices of all graphs simultaneously.

5.1 Graph Storage and Assignment to Processors

The graph representations must first be moved from main memory into the processors. Although during computation each processor is assigned a single row of a graph's matrices (corresponding to a single node), each processor's memory holds the full adjacency matrix, node subsumption codes, and other data for a complete graph. Since only $O(\lg^2 N)$ graphs are actually tested, where N is the total number of graphs in the knowledge base, each processor stores a large number of graphs but processes only a tiny fraction of them. Graphs are assigned to processors for storage randomly, and the odds of a conflict in spreading the graph to nearby processors is quite small. If there is a conflict, the graph can be moved to another region of the processor array or held back until the nearby processors have completed the first graph they were assigned.

The following gives a high-level description of the process of moving graphs from memory to the processors' registers for the simpler case of forming the initial group of graphs to be processed. Here we present the case where the query graph G_b is known to be larger than the knowledge base graphs to be tested (a $G_a \preceq G_b$ test for Phase I of Levinson's Method III MIS) and $b \geq n$. Although other cases are somewhat more complex, simulation shows that this data retrieval phase is only a few percent of total refinement time.

1. Divide the array into equally sized subsections. Each section is 1 processor wide and $|G_b|$ processors tall.
2. Select one graph to be tested from among those stored in each array subsection that is in the MIS queue of graphs.
3. Distribute the selected graph to the other processors in the sub-array so that each processor receives one row of the A adjacency matrix, and the corresponding node lattice code and relation list. This is done for all selected graphs simultaneously.
4. Broadcast and store the rows of B in the processors, one row simultaneously for all active knowledge base graphs.

5.2 Forming the Match Matrix for Comparing Two CS

Figure 2 gives the pseudocode for forming the M matrices. The algorithm processes all nodes in all the selected knowledge base graphs simultaneously. In addition to this parallelism, the algorithm has the advantages that there is no communication between the processors and no consultation of a central concept-type subsumption table. Node lattice codes and relation labels for the query graph are broadcast from the controller to all processors.

As the refinement algorithm is called many times per SI test, and iterates several times per call, its time dominates over that of forming the M matrix even when the latter includes additional filters.

```
/***** FORM M MATRICES *****/

/** data in registers, each processor **/
code(n) Anode /* n-bit row of matrix A */
subsumptioncode acode /* subsumption code for Anode */
int arels /* number of Anode's edges (relation) */
code(m) Bnode /* m-bit row of matrix B */
code(m) Mrow /* corresponding row of match matrix */

/** Test Candidate Bindings **/
foreach nodex ∈ G_b
     Mrow(nodex) ← 1
     if |arels| ≰ |brels| /* brels broadcast from controller */
          Mrow(nodex) ← 0
     else if acode ≰ bcode /* bcode broadcast from controller */
          Mrow(nodex) ← 0
```

Fig. 2. Parallel algorithm for forming candidate binding (match) matrix

5.3 Multi-Bit Parallel Refinement

With the A, B, and M matrices established in the processors, the refinement algorithm itself simply executes the matrix operations with the termination conditions as given in Sect. 2.1. Details of the communications and bit-manipulations required in the Boolean matrix multiplications, including the transpositions, are machine dependent.

The allocation of one row of the matrices to a processor offers several advantages. With this layout, the "summing" portion of the Boolean vector-product, performed n times for each multiplication, is replaced by a zero-test on a single data word, avoiding time consuming inter-processor communication. The layout also allows testing if any row of a matrix is all zeros or if the matrix has changed since the last iteration in just one or two instructions. Further, with all rows of the matrices of a given graph allocated to a column of the parallel array, the Boolean matrix transpose requires only a regular, efficient broadcast of each row of the matrix to the other processors in the subarray, which then select the appropriate bits for their destination matrix row.

Since one graph can be significantly larger than the other, such as the query graph G_b in this section's discussion, many processors will have rows of the B matrix but not of A and M matrices, nor the intermediate matrices R and T. This does not significantly hurt performance, however, as all processors are active in the matrix operations during most of the steps.

The new algorithms provides a significant speedup for SI tests over prior parallelizations. Ullmann's original implementation used the bit-parallelism of a single processor to achieve $O(n^3)$ expected-case time (performing $O(n^4)$ single-bit operations). Willett *et al.*'s implementations use multiple single-bit processors to achieve $O(n^2)$ expected-case time. Combining bit and processor parallelism, the algorithm presented here performs $O(n^2/b)$ SI tests simultaneously in $O(n^2)$ expected-case time. Where $b \geq n$, the amortized expected-case time for a single subgraph isomorphism test is linear in the size of the graph. This has been observed in our limited number of experiments. In all cases the space-time product, including number of processors, bits per processor, and number of graphs tested simultaneously is $O(n^4)$, indicating efficient parallelizations.

The new multi-bit implementation of SI refinement significantly complicates implementation details over our prior method, however. In addition to the load balancing described in above, it is necessary to execute the backtracking portion of Ullmann's on multiple processors in the data parallel array as in Willett *et al.*'s methods two through four.

6 Implications for Classification into a CS Database

In addition to speeding SI tests, the multi-bit parallelization allows performing multiple tests simultaneously on a single data-parallel array. By incorporating multi-level indexed search (MIS) we can minimize, and in some cases eliminate, unnecessary graph comparisons.

6.1 Improvement in Multi-Level Indexed Over Exhaustive Search

The multi-bit implementation strengthens our prior analysis demonstrating the advantage of MIS over exhaustive parallel search [4]. With b-bit processors we now need $1/b$ as many processors, or we can process b times as many graphs simultaneously. Updating the prior analysis to the multi-bit implementation is straightforward. Except under conditions where nested parallelism (comparing multiple graphs and parallelizing the refinement matrix operations) is not possible, the new MIS implementation is typically faster than exhaustive search and a factor of b faster than the single-bit implementation. Taking the number of processors as the space measure, MIS always has a much better space-time product than exhaustive search. If memory is used as the space measure, exhaustive search has a space-time product a factor of $\lg N$ better as memory cost dominates over processor cost. However the million or more processors required[1] is impractical for the foreseeable future. With memory cost dominating, the parallel applications have better space-time products than their serial equivalents.

6.2 Modification to Prior Multi-Level Indexed Search Parallelization

It is necessary to reconsider the approach to parallelizing MIS. Previously, a parallel array of a thousand processors could handle only one to a few graphs; with

---

[1] A knowledge base with 64 K graphs averaging 16 nodes each would require a million processors for exhaustive parallel search, one with 1 M graphs averaging 32 nodes each would require 32 million processors.

a dozen or more graphs available at any stage in the MIS, comparisons would be farmed out to several processor arrays [4]. This processor-farm model implicitly included load balancing (each array requesting more work when it was done) and there were more graphs queued for comparison than the available processors could handle so that a graph need not be compared until it was certain that the comparison was necessary. On the minus side, communication was heavy as the full knowledge base graph and concept-type subsumption data had to be sent from wherever the MIS management was executing. Although having all graphs on a single parallel array significantly reduces communication outside the array and the use of concept-type subsumption codes greatly reduces communication within the array, the new implementation affects the implementation of the multilevel indexed search itself. With 32-bit processors, an array of 1024 processors could handle thirty-two 32-node graphs simultaneously, but at any given time the MIS algorithm would have identified only one or two dozen graphs requiring comparison. One must thereby choose between not using all the processors or speculatively comparing graphs.

Performing speculative graph comparisons using an earlier version of Phase I of Levinson's Method III appears to be the better choice. In the more recent implementations of Method III, the queue of candidate graphs begins with only the top of the poset (generalization hierarchy) and candidates are added as the poset is traversed, limiting not only the number of SI tests but also the number of graphs "visited." By initially enqueuing all graphs smaller than the query graph and then dequeuing them if a predecessor fails, the older version guarantees a large queue of candidate graphs. Although there will be some graph comparisons that are later determined to be unnecessary, this is expected to give a faster parallel implementation as far fewer processors would be idle. It should also be possible to integrate the MIS dequeuing with the refinement load balancing; if a currently executing graph is dequeued it can be abandoned and another graph started. The data parallel array can also handle much of the queue and other MIS management.

7   Simulation Results of Refinement Behavior

We have begun evaluating the new approach by simulating parallel execution of refinement on a serial workstation. A probabilistic analytical model for predicting the likelihood of a bit in the M match matrix changing from 1 to 0 (and in turn the probability of a SI test failing or succeeding) based on the densities of the A, B, and M matrices would be overly complex due to the dependencies between bits in the matrices. We have thereby run simulations of the refinement algorithm, emulating two data parallel machines: the proposed 16-bit MISC architecture [10] and the commercially available 32-bit MasPar MP-2 [9]. The simulations use parameterized synthetic data sets to study the effect of different graph characteristics on refinement behavior.

7.1 The Simulation

Our simulations go beyond the simple case described in Sect. 5, allowing graphs to have more nodes than bits in the processor data word (requiring multiple columns in the array), and graphs with fewer than half the processors in one column of the array (allowing multiple graphs in a single column). These add only a small number of inter-processor communications and additional operations.

Graphs in the knowledge base are generated as random adjacency matrices A with $|G_a|$ within a range of specified sizes. The probability of any bit in A being on is specified by the parameter $p(A)$. Ones on the diagonal are disallowed, prohibiting self-loops in the graphs, and the resulting graph is tested to see if it is connected. If an A does not correspond to a single connected graph, it is discarded and another is generated randomly. Query graphs are generated in the same way.

7.2 The Experiments

Our results include several experiments varying parameters as shown in Tab. 1. As before, $|G_x|$ is the size of a graph and $p(X)$ indicates the probability that a bit in the adjacency matrix of graph X is one. Both the "mid1" and "mid2" experiments have adjacency matrix densities corresponding to 1 or 2 edges per node; the others range from 2 to 9 edges per node. Figure 3 plots the portion of graphs remaining to be processed (neither failed nor unchanged) at each iteration averaged over 4 queries and Fig. 4 plots the cumulative number of graphs that have failed. All the examples show similar behavior in the number of iterations needed to complete most of the graphs, and only the "small" experiment shows a major difference in the total number of graphs that fail.

Experiment	$|G_a|$	$|G_b|$	$p(A)$	$p(B)$	$p(M)$
small	8	12	0.20	0.30	0.50
mid2	16	24	0.13	0.15	0.40
mid1	8-19	19-25	0.10	0.10	0.45
large	24-26	30-31	0.15	0.30	0.25

Table 1. SI refinement experiment conditions

In the context of a MIS knowledge base, the graphs selected for comparison are more likely to pass the SI than a random selection due to the filtering effect of the multilevel indexing. This is approximated in the experiments by the large values for $p(M)$.

Small changes to $p(A)$ and $p(B)$ from those of Tab. 1 have only a small effect on behavior. Reducing $p(M)$ to 0.20 results in all graphs failing within 1 or 2 iterations whereas increasing $p(M)$ to 0.60 for the "mid2" example results in as many as 17 iterations to complete all graphs and less than 50% failing refinement. Increasing $p(M)$ to 0.40 for the "large" experiment results in none

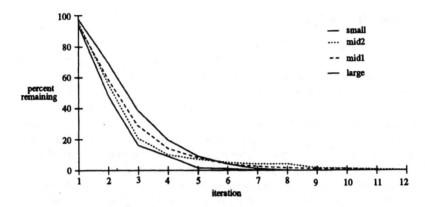

Fig. 3. SI refinement: graphs remaining

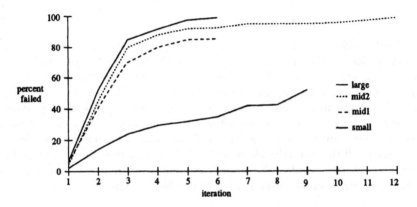

Fig. 4. SI refinement: cumulative graph fails

of the graphs failing with only 2 to 4 iterations required until all M matrices stop changing. The preliminary experiments of Tab. 1, Fig. 3 and Fig. 4 are intended to represent a middle ground of typical behavior.

Examining the simulation experiments individually, where processing continues until all graphs are complete, at any given time an average of 70% of the processors have completed their graph. Although 30% is not an unusually low processor utilization, it indicates that the load balancing effect of adding new graphs as others complete is an important issue. The simulations indicate that a *single* iteration of refinement requires some 1,000 instructions on a well-suited processor architecture whereas distributing graphs and forming the initial match matrix requires over 2,000 cycles. Combining these simulation results we conclude that it would be appropriate to bring in new graphs every 5 to 10 iterations of refinement depending on application characteristics.

Thus far we have discussed refinement iterations. Willett *et al.*'s experimental data for exhaustive search in a chemical database shows that nearly all SI tests are resolved in 3 to 7 calls to refinement (each of which includes numerous iterations) with a few (less than 1%) requiring 20–50 calls. By bringing in new graphs during refinement iterations, we effectively load balance at each call to refinement; the housekeeping required by this approach remains an open question at this stage in our research.

7.3 Hardware Implications

In the course of this work we have noted several characteristics of the refinement algorithm and MIS that have influenced our design of the MISC Machine, a proposed architecture for artificial intelligence, including CS processing. A comparison of simulated SI execution on the MISC and MasPar MP-2 not presented in this paper indicates that MISC would be an order of magnitude faster given equivalent hardware resources and implementation technology. The simulations also indicate that MISC with 1 K processors would be roughly 3 orders of magnitude faster than present workstations. As a result of these observations, we foresee small (256 to 1024 processor) MISC arrays configured as workstation coprocessor boards as a platform for future CS applications. The workstations could be networked into a distributed system for extremely large knowledge bases, multiple-domain applications, and other division of labor.

8 Conclusions

We have developed a new parallelization of Ullmann's subgraph isomorphism refinement algorithm featuring multiple bits per processor and multiple graphs per processor array. The new algorithm incorporates graph storage and allocation, the formation of the candidate binding matrices in parallel, and criteria specific to conceptual structures. The algorithm avoids the potential bottleneck of consulting a concept-type subsumption table by using subsumption codes as concept-type labels. We also present a revised parallelization of multi-level indexed search that exploits the characteristics of the new refinement algorithm. Results of the refinement algorithm's behavior on parameterized synthetic graphs is also included.

With these improvements the new algorithms provide a significant speedup over prior parallelizations, make efficient use of contemporary parallel processors, are superior to exhaustive parallel SI tests, and require only a small data parallel machine. The promise of $O(n \lg N)$ time using $O(n^2)$ processors of $O(n)$ bits each (where N is the number of graphs in the database and n is the number of nodes in a graph) in a "CS knowledge base machine" suggests exciting possibilities for large knowledge bases of conceptual structures.

Our broad goal for future work is the parallel implementation of a CS database classification system on an existing parallel machine. Key issues to be addressed along the way include improvements to Willett *et al.*'s data-parallel implementation of the backtracking portion of Ullmann's algorithm, load balancing to keep processors busy as some SI tests complete before others, and integration of the refinement SI algorithm with multi-level indexed search algorithms, including UDS. Work should also include further experiments with parameterized synthetic graph data sets and characterization of graphs in specific CS applications.

References

1. Hassan Ait-Kaci et al. Efficient implementation of lattice operations. *ACM Transactions on Programming Languages and Systems*, 11(1):115–146, January 1989.
2. Franz Baader, Bernhard Hollunder, Bernhard Nebel, Hans-Jurgen Profitlich, and Enrico Franconi. An empirical analysis of optimization techniques for terminological representation systems. In *Proceedings of the 3rd International Conference on Principles of Knowledge Representation and Reasoning*, Cambridge, MA, October 1992.
3. Gerard Ellis. Efficient retrieval from hierarchies of objects using lattice operations. In *Conceptual Graphs for Knowledge Representations*, pages 274–293, New York, 1993. Springer-Verlag.
4. Richard Hughey, Robert Levinson, and James D. Roberts. Issues in parallel hardware for graph retrieval. In *Proceedings of the 1st International Conference on Conceptual Structures*, pages 62–81, 1993.
5. George G. Lendaris. Representing conceptual graphs for parallel processing. In *Conceptual Graphs Workshop*, 1988.
6. George G. Lendaris. A neural-network approach to implementing conceptual graphs. In Timothy E. Nagel et al., editors, *Conceptual Structures, Current Research and Practice*, chapter 8, pages 155–188. Ellis Horwood, New York, 1992.
7. Robert Levinson. UDS: A universal data structure. In W. M. Tepfenhart, I. P. Dide, and J. F. Sowa, editors, *Conceptual Structures: Theory and Practice*, pages 230–250. Springer-Verlag, New York, 1994. Lecture Notes in AI 835.
8. Robert A. Levinson and Gerard Ellis. Multi-level hierarchical retrieval. *Knowledge-Based Systems Journal*, 5(3), September 1992.
9. John R. Nickolls. The design of the Maspar MP-1: A cost effective massively parallel computer. In *Proceedings of COMPCON*, pages 25–28, February 1990.
10. James D. Roberts et al. Hardware for PEIRCE. In *Proceedings of the International Workshop on PEIRCE: A Conceptual Graphs Workbench*, 1992,1993.
11. J. R. Ullmann. An algorithm for subgraph isomorphism. *Journal of the ACM*, 23(1):31–42, January 1976.
12. Peter Willett, Terence Wilson, and Stewart F. Reddaway. Atom-by-atom searching using massive parallelism: Implementation of the Ullmann subgraph isomorphism algorithm on the distributed array processor. *Journal of the Chemical Information Computation Society*, (31):225–233, 1991.

Transputer Network Implementation of a Parallel Projection Algorithm for Conceptual Graphs

Alfred Chan and Pavel Kocura

Department of Computer Studies
Loughborough University of Technology
Loughborough, England

Abstract : The design, implementation on a network of transputers, and testing of a parallel projection algorithm which differs from traditional methods is described. A new automated testbed generator which produces large, complex CGs knowledge bases for testing CGs systems, is introduced and discussed. Extensive tests, using automatically generated data, show that the algorithm is scalable and free from excessive communication overhead. A parallel version of the Peirce-logic-based theorem prover is suggested.

Keywords: Parallelism, Transputer, Subgraph Isomorphism, Conceptual Graph Processing, Projection.

Introduction

For a Conceptual Graph processor, graph projection is the most important operation. It is used for the 'abstract' matching of queries against the knowledge base, rules, or for the enforcement of semantic constraints.

Projection is commonly consider as being NP. It is an alarming experience to see how, with large graphs, the combinatorial nature of projection grinds the system to a halt. To alleviate this problem, and to explore the evident parallel nature of CGs, we have implemented a projection algorithm on a network of Transputers.

The paper deals with three main topics: the design of a novel subgraph isomorphism algorithm, the design and implementation of an Automated Testbed Generator for CGs processors, and the results of the performance tests of the parallel projection algorithm. It also suggests a parallel solution to Peirce-logic-based theorem proof.

1.0 Parallel Algorithm for Subgraph Isomorphism

1.1 Prior Works

The subgarph isomorphism problem has been vigorously researched because of its usefulness in many practical areas. One of the most prominent result was achieved by Ullman[5] with his subgraph isomorphism algorithm having a time complexity of $O(n^3)$. It is widely used by researchers and many researches carried out in the area of subgraph isomorphism for the past two decades have been based on Ullman's algorithm. We will briefly summarise those that are relevant to our work. Simulation of parallel subgraph isomorphism by Wipke et al.[7] demonstrated efficient speed up can be gained on a 5 to 25 processor's system. Implementation of Ullman algorithm on a massively parallel Distributed Array Processor (DAP) by Willet et al.[6] showed a time complexity of $O(n^2)$ running on a DAP with n^2 1-bit processing elements. His works demonstrated it is indeed possible to attain speed up by executing the subgraph isomorphism algorithm in parallel. Roberts et al.[3] has thoroughly discussed the application of fine-grain parallelism to the CG projection. His results showed an expected time complexity of $O(n^2)$ using $O(n^2)$ processing elements. His works also incorporated special data structure such as encoded Concept-Type Hierarchy to enhance the speed of the projection operation.

1.2 Data Representation

Input (in linear form) to our projection algorithm containing n-adic relations is transformed into its internal dyadic equivalent after parsing. We find that graph represented with only dyadic relation is more suitable for our algorithm computational wise. Multi-linked list structure is used as the internal representation for both graphs and lattices.

Data used by our algorithm is stored and processed in a format whereby concept with individual marker only occurs once. This is in accordance with the use of individual marker[4] which specified that a concept is unique. For this reason, concept nodes having common type label and individual marker are joined to reflect their uniqueness. As such, data stored is made up of a few very large connected graphs rather than many small disjoint graphs.

With this representation it is more efficient to enforce consistency among data. Changes to data such as addition or deletion of graphs can disrupt data consistency. In our representation there is no duplicated information as a result we only have to check the part where modification is applied. In the case of representing data with many disjoint graphs, every graph has to be checked to ensure changes are made to all of them.

1.2.1 Transputer and Parallel Processing

The transputer is based on the MIMD (Multiple Instruction Multiple Data) model, in which each transputer has its own private memory. Transputers are connected to each other through hardware links, forming a network, and solving problems collectively. This offers an advantage compared to shared memory architectures, in which an increase in the number of processors may lead to bus contention. Excessive inter-transputer communication, which could result in deadlocks, has to be prevented by the efficient distribution of data on the transputers, and by special routines which take care of inter-transputer traffic.

1.2.2 Parallelism Strategies and Limitations

There are two basic strategies for a parallel projection algorithm: A) the whole algorithm can be executed in parallel by running multiple copies of it concurrently, or B) only parts of the algorithm can be executed in parallel. The suitability of either option depends on the nature of the data, which determines the degree of parallelisation.

1.2.3 Transputer Network

The algorithm uses a transputer network configured into a pipeline structure. In the diagram below, transputers are represented as boxes, in which M = Master, S = Slave, the lines show transputer links, and arrows show data flow direction.

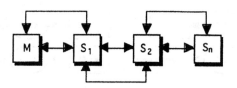

Fig 1.0 A Transputer Network

1.2.4 Parallel Projection Algorithm Design

Let us mention several essential points:

- Inputs into a projection are a) query (source) graphs, and b) the data (target) graphs.

- The size of a data graph, which can comprise the whole connected knowledge base is generally several orders of magnitude greater than that of a 'normal' query graph.

- The operation of distributing a data graph amongst transputers is not considered part of the Projection operation proper, and its overhead is excluded from the timings.

- Both data and query graph have only dyadic relations in its internal representation.

Considering the potential sizes of data graphs, strategy A is infeasible, because each slave transputer would have to have a copy of the whole knowledge base: the data graphs, the type and relational lattices, etc. If memory were not critical, then this option would be ideal, since each transputer would be processing its projection without having to co-operate with other slave transputers. The only communication overhead would be the time spent on distributing query graphs on slave transputers, and the collection of the projected graphs by the master transputer. Even so, this version of strategy A would fail if projections were performed serially, i.e. the query graph for next projection would depend on the results of the previous projection. Our algorithm adopts strategy B.

1.2.5 Pre-processing of Lattices and Data Graphs

According to strategy B, each slave transputer stores only partial data graphs. Thus, each slave will hold a unique section of the data graph, together with those parts of the lattices which contain inheritance links for the concepts and relations of that data graph section.[1]

During the pre-processing, the master transputer slices a data graph into tuples and distributes them over the transputer network. Each tuple consists of a relation and its source and sink concepts.

---

[1] We would like to thank here Dr. Dickson Lukose for contirbuting to the discussion of the design of our parallel CGs system during his stay with the Loughborough CGs group as a Leverhulme Fellow.

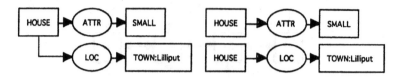

Fig 1.1 A Data Graph after sliced into tuples

Partitioning graphs into tuples before distribution ensures that a) each transputer will have an equal share of data, and b) data duplication will be minimised. The algorithm does not require the tuples to be reconstituted into a connected graph once they arrived at their allocated transputer. Before distribution, each concept and relation node is injected with a unique number (Data ID), which is then used during graph reconstruction.

After data distribution, the lattices are partitioned and distributed. Each transputer receives those lattice sections which correspond to the types and relations contained in the tuples stored in the given transputer's memory. The partitioning and distribution processes only occur once throughout the operation of the system.

1.2.6 First Stage

The algorithm is divided into two main, comprehensively parallelised processes. The first one projects a query graph against a data graph, the second process collates the matched tuples into a connected graph which answer the query.

1.2.7 Phase One - The Matching Process

1.2.7.1 The Query Graph

At the start of each projection, each node inside the query graph is allocated a unique number (Query ID). The number of outgoing and incoming links in each concept is recorded, to be used as one of a set of matching attributes. The query graph is then sliced into tuples and a copy of the set of sliced tuples is sent to each transputer for matching. The set of sliced tuples is rebuilt back into graph form in each transputer. The algorithm matches tuples in the query graph against the data graph tuples, using inheritance from its resident subsets of the type and relation lattices. A match succeeds only if all tuple attributes have been matched.

The matching attributes are concept label, referent, relation label, number of concept links, and the orientation of the concepts and of the relation within the tuple. A match succeeds only if all the following conditions are satisfied.

Matching Attributes in Query Tuple	Matching Attributes in KB Tuple	Matching Criteria
Concept Label *QCL*	Concept Label *KBCL*	*QCL* same type as *KBCL* or *QCL* is a supertype of *KBCL*
Relation Label *QRL*	Relation Label *KBRL*	*QRL* same type as *KBRL* or *QRL* is a supertype of *KBRL*
Referent *QRef*	Referent *KBRef*	*QRef* is a generic referent or *QRef* and *KBRef* share the same referent

Number of links to the Concept Node $QNosLinks$	Number of links to the Concept Node $KBNosLinks$	$QNosLinks \leq KBNosLinks$
Orientation inside the Tuple	Orientation inside the Tuple	Same Orientation

Table 1.0 Table of Tuple Matching Attributes

During a successful projection the query graph tuple is restricted, using the individual referents and types found in the matching data graph. The unique id numbers inside the data graph are carried over to the query tuples. The numbers injected into the query graph are not replaced, as each node in each tuple stores both numbers. After the transfer has been completed, before the slave continues on with the next match, the tuple is sent back to the master. When all the tuples allocated to a slave have been matched, the slave sends a termination token to the master.

1.2.7.2 Storing Matched Tuples

A skeleton structure is created in the master for classifying matched tuples. Inside the skeleton are containers for storing matched tuples. Each container only receives matched tuples with a specific token. This token obtained from the Query ID in the relation node of each matched tuple is used to assign it to its designated container. Upon receiving all matched tuples, it can be quickly determined if the projection fails or succeeds by checking for empty containers, since each container represents a specific part of the query graph. Phase two of the algorithm can proceeds if no container is empty.

1.2.8 Phase Two - The Graphs Construction Process

1.2.8.1 Distribution of Skeleton Structure

The second phase of the algorithm assembles matched tuples into query answer graphs. A potential valid graph can be formed by connecting a tuple picked from each skeleton structure container. For instance, if the skeleton structure has two containers with each container storing two matched tuples, then there is a potential of forming four projected graphs, providing the combination is valid. The skeleton structure is partitioned in such a way that no partitions will produce the same combination. The structures are delivered to the slaves for processing.

1.2.8.2 Graphs Construction

Set of tuples is picked from the partitioned skeleton structure. Before assembling them into graph, each slave first examines the set of tuples for duplication. The duplication test checks that no tuples share the same data relation ID. If that happens, it means that different tuples of the query graph have been matched to the same tuple in the data graph, and these would not form a valid graph. Thus, the next set of tuples is generated from the structure.

The number of tuples combinations generated from the partitioned structure can be expressed by $t_0 \times t_1 \times t_2 \times ... \times t_{n-1}$, where n is the number of container and t_i is the number of tuple in container i, for $0 \leq i \leq n-1$. It is combinatorial explosive to construct graphs by blindly trying out all tuples combinations. To remedy this, the query graph is referenced locally to obtain the structural information (Query ID) to assist the graph construction. With this information we know which tuple should be connected.

The construction process start by choosing a starting point in the query graph. A tuple is picked from the partitioned structure on the basis of the query ID located in the query graph at the starting point. The query graph is then traversed one tuple distance at a time to search for more tuple to connect. Tuples are joined if their concepts have the same type label, referent, and query and data IDs. The query and data IDs are used to prevent the creation of graph that do not exist in the data. Tuple found to have no adjacent connection is deleted from the partitioned structure to cut down the number of trials. Backtracking is used during traversal to select the tuples combinations in a systematic fashion.

If the number of connections made is the same as in the query graph, a valid graph has been constructed, which is then sent back to the master. After it has exhaustively tested all tuples combinations, the slave sends a termination token to the master. The parallel projection process is complete after all the slaves have sent their termination tokens.

1.2.9 Overview and Discussion of the Parallel Projection

The operations described above are implemented in parallel on a transputer network as individual processes. The following diagram gives an overview of the software configuration of the transputer network for the parallel projection algorithm.

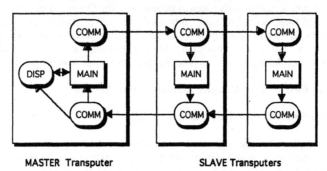

MASTER Transputer   SLAVE Transputers

Fig 1.2 Software Configuration of the Projection Algorithm in the transputer network.

During the projection operation, individual processes in the transputer, represented in Fig. 1.2 as boxes, run concurrently. Transputers communicate with each other, as shown by the lines joining the processes, and by their arrows, which show the direction of the flow of data. The "COMM" boxes are communication modules transferring data between master and slaves, the "DISP" box is the module receiving and displaying projected graphs. The "MAIN" boxes in both the master and slaves represent the projection algorithm. The only communication overheads of the parallel projection algorithm are associated with

a) the sending of the query graph and skeleton structures to the slave transputers, and

b) the collation of matched tuples and receiving of projected graphs from the slave transputers.

These overheads are minimal compared to the time spent on tuple matching and graph construction. Thus the algorithm is process-bound and should be scalable with the addition of more transputers. We will validate this claim in section 3 of this paper.

1.3. Algorithm Complexity

In this section we will work out the time complexity for our algorithm. Time complexity is useful as a mean to evaluate the algorithm performance. All the calculations are only estimate.

Phase I Time Complexity $\approx O\left(\frac{mn^2}{p}\right)$

n - number of data graph tuples
m - number of query graph tuples
n - time complexity for checking subsumption information in the lattice
p - number of transputers

Assuming m is several magnitudes smaller than n we will assume $m \approx \frac{n}{1000}$, we have a time complexity of $O\left(\frac{n^3}{1000p}\right)$ for Phase I.

Phase II Time Complexity $\approx O\left(\frac{rt^q}{p}\right)$

q - number of containers in the skeleton structure
t - average number of matched tuples in each skeleton
r - reduction to the search space from the information provided by the query graph
p - number of transputers

Number of all tuples combinations generated by the skeleton structure is $\approx t^q$ (This depends on how general the tuples are in the query graph during phase one tuples matching. Query graph with general types will have more matched tuples filling up the skeleton containers than query graph with specific tuples.)

Assuming $q \approx \frac{n}{1000}$, the Phase II time complexity is $\approx O\left(\frac{(rt)^{\frac{n}{1000}}}{p}\right)$

The overall time complexity is $\approx O\left(\frac{(rt)^{\frac{n}{1000}}}{p}\right)$

2.0 Automated Testbed Generator

To be of practical use, a CGs processor has to satisfy the requirement of correctness and efficiency. The implementation of an algorithm must do what is was designed to do (and nothing else), and do it as fast as technically possible. Rigorous testing of the developing system, and constant feedback for the implementors and designers, is essential. Because of the complex logical operations of the CGs processor, we cannot rely on time consuming manual tests. When estimating algorithm performance, it is necessary to realise that there are bound to be great differences between the CGs knowledge bases of different real-world domains: the 'shapes' and sizes of the lattices, numbers and complexity of definitions and canonical constraints, concept and relation types, individuals and rules. Thus, different versions of an algorithm for the same CGs function or operation might be suitable for different domains. This is particular true for Transputer-based systems, which can be reconfigured in different ways for different purposes.

At present there are no rigorous and realistically large CGs models of real-world domains, let alone the required variety, that would be needed for the testing of the whole 'family' of CGs processors developed by the Loughborough group. Thus we have designed and (to an operational level) implemented a universal, fully automated generator of test knowledge bases for CGs processor. The testbed generator, in its full form, has the following modules, which automatically generate a CGs knowledge base:

- Concept type lattice generator
- Concept canonical model generator
- Relation type definition generator
- Conformity table generator
- Query graph generator

- Concept type definition generator
- Relation type lattice generator
- Relation canonical model generator
- Data (fact) graph generator
- Logical rule generator

The system in its design, also includes a module for the automated evaluation and interpretation of timing results, as well as a module for the logical evaluation of the theorem proving operations.

2.1 Testbed Generator Implementation

Let us give some detail about those parts of the testbed generator which are used for the testing of the parallel projection algorithm. The data generator implemented was inspired by the example of systems used in the testing of conventional compliers[1,2]. The algorithm operates on the following data structures: concept type and relation type lattices, conformities and large, connected conceptual graphs, which contain the generated types, relations and conformities. These are generated by the lattice generator and conceptual graph generator modules, both of which use the random number generator module. A design diagram is shown below:

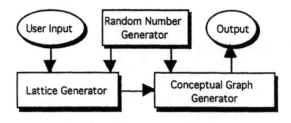

Fig 2.0 Structure of the Data Generator

2.2 Lattice Generator Module

The conceptual and relation type lattices are generated first, as they are needed by the other testbed generation operations, which place the data into graph nodes when conceptual graph is being compiled.

The lattice generator produces 'artificial' labels, each given a relative position in the lattice. The labels consist of a "C" (for concept types) or an "R" (for relation types), followed by a unique integer. The user interface of the testbed generator makes it possible for the user to interactively determine the size and other attributes of the lattices, e.g. the number of connections between types and the maximum number of types in each level in the lattice.

The actual structure of a lattice, i.e. the inheritance lines, is produced in the following way. First the generated labels are chained together to form a circular linked list. For each link in the lattice, two labels are picked at random from the list. Which labels will be selected is determined by the output of the random number generator. The chosen labels are connected according to their relative position in the lattice. Lattice construction terminates once the pre-set number of links in the lattice has been reached.

2.3 Conceptual Graph Generator Module

During the construction of graph, labels are picked at random from the type and relation lattice. The structures of a conceptual graph are given by the BNF grammar that defines the syntax rules to which graph has to conform [4]. The syntactical checking is done by going from the top of the parse tree and testing whether it can parse the syntax, using all the available rules. The same routine can be used for generating a syntactically correct graph. Instead of choosing which path to follow, based on the syntax as in normal parsing, a path can be chosen at random. The system does this when generating a new graph. As the algorithm is walking through the parse tree, the random number generator suggests a choice every time a branch appears. Some paths through the tree might lead to premature termination, thus generating a graph smaller than required. To avoid this, the terminating paths are shut off until the graph has reached the size pre-set by the user. Thus the generated graph will not be smaller than the preset size, but it may be bigger than required, since the graph generator may still be a long way from termination when the pre-set size has been reached.

2.4 Discussion

The testbed generator produces syntactically correct lattices and graphs. At the present stage it does not generate graphs which comply with arbitrary semantic constraints. Semantically correct graphs will be possible when the system is able to generate a set of semantic constraints (canonical models) for each entry in the lattices. This operation is still being developed. However, as it is, the data generator provides adequate test data for the measurement of the performance of the parallel projection algorithm.

3.0 Testing

3.1 Test Data

To assess the actual performance of the new algorithm, we simulated a hypothetical user querying a CGs knowledge base. In real-life applications, the user may ask queries based on some partial knowledge of the domain, and the system will be required to fill in the gaps. Such queries will consist of relatively small, partially instantiated conceptual graphs, whereas the system's knowledge base will consist of a relatively small number of relatively large, (almost) fully instantiated graphs. In the spirits of this scenario, the testbed generators produce a query graph with 11 tuples, a data graph with 1507 tuples, and lattices of 40 conceptual and relational types, In the test, while the system was timing the whole process, this query graph was projected onto the data graph. To see how the algorithm scales up, the number of transputers in the network was varied. A query graph used in the tests is shown below. To save space, we do not show the data graph.

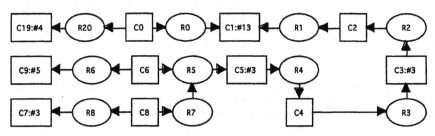

Fig 3.0 A Query Graph

3.2 Timings

At the start up of the system data graphs are partitioned and distributed across the transputer network. This is performed only once throughout the operation of the system and is considered as the preprocessing stage of the system. As a result of this the authors exclude it from timing consideration. The authors felt it is more appropriate to measure the actual time used during a projection operation.

Another point to note concern the exclusion of timing for parsing query graph. The parsing time is the same when running on a single transputer or multiple transputers, therefore it is excluded from timing. Measurements were compiled from the run times of the projection operation performed on a varying number of transputers. The timing proper of the algorithms starts once the query graph has been parsed and stored in the memory, but before it has been sliced into tuples. The timing terminates once all projected graphs have been received from the transputer network. In this particular test, the projection algorithm found in the data graph six projection images of the query graph. The test was repeated for different numbers of transputer. The results are in the table below, which shows the total query time as the sum of the tuple matching and graph construction times

Transputers	Tuples Match Time (s)	Graph Construction Time (s)	Total Query Time (s)
1	4.353216	268.309888	272.63104
4	1.220480	68.000000	69.220480
8	0.662912	46.396224	47.059136
12	0.571072	35.821440	36.392512
16	0.371776	36.789760	37.161536
20	0.397440	24.241920	24.639360
24	0.371712	24.364928	24.736640
28	0.298624	24.120704	24.419328

Table(1) Timing of projection vs the number of transputer.

3.3 Speed Up Factor

The speed up of the algorithm is calculated by dividing the projection time of one transputer by the projection time of the actual number of transputers used in a particular run. The following is the speed up equation

$$S_p = \frac{T_1}{T_p}$$

where S_p - speedup factor, T_1 - projection time on one transputer, T_p - projection time on p transputers. By converting the previous timing results into speedup factors, we can examine the scalability of the projection algorithm.

Transputers	Speed Up (Tuples Matching)	Speed Up (Graph Construction)	Speed Up (Whole Operation)
1	1	1	1
4	3.567	3.946	3.939
8	6.567	5.783	5.793
12	7.623	7.490	7.491
16	11.709	7.293	7.336
20	10.953	11.068	11.065
24	11.743	11.012	11.021
28	14.578	11.1236	11.165

Table(2) Speed Up Factors of the Projection Algorithm

Speed Up Factor of Tuples Matching   Speed Up Factor of Graph Construction

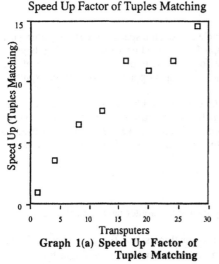

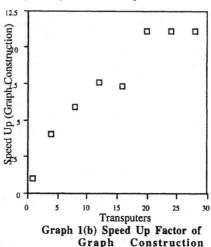

Graph 1(a) Speed Up Factor of
Tuples Matching

Graph 1(b) Speed Up Factor of
Graph Construction

Speed Up Factor of the Projection Algorithm

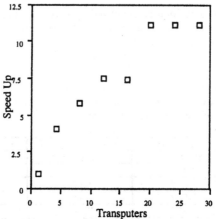

Graph 1(c) Speed Up Factor of the Projection Algorithm

3.4 Conclusion and Discussion

The discussion is carried out in two parts resolve around the performance of the algorithm and comparisons to prior works.

3.4.1 Run Time Results

Looking at the speed up graph 1(c) we can clearly see the speed up started levelling off after 16 transputers. For tuples matching, graph 1(a) showed near linear speed up beyond the number of available transputer we used for the test is possible. Phase one of the algorithm is therefore do not contribute to the speed up stoppage beyond 16 transputers. Graph 1(b) confirmed this was the case.

The limited speed up can be attribute to the uneven load balancing in phase two - the graph construction stage. The partitioned skeleton structure distributed to each slave generated roughly an equal number of tuples combinations nevertheless the time used to check the tuples combination generated from each partitioned skeleton structure is not constant.

3.4.1.1 Connection Strategies

The reason for using the pipeline topology for the transputer network is because of its simplicity.

The communication cost incurred in our algorithm is $p \times T_{comm1}$ where p is the number of transputer and T_{comm1} is the communication time in a single transputer. Improvement to the performance of the algorithm can be improved by using other connection strategy such as binary tree, fat tree or hypercube. For instance, with binary tree, the communication costs reduce to $(1 + \log p) \times T_{comm1}$.

As the communication times only make up a small percentage of the total projection time it is not our main concern at the moment. A better communication scheme will be incorporated in the near future.

3.4.1.2 Comparison to Prior Work

The work done by Roberts et al. is the closet to us where useful comparisons can be made. Their work reported a time complexity of $O(n^2)$, running on $O(n^2)$ processing elements. n is the number of nodes in the query graph

Our algorithm has an estimated time complexity of $O\left(\dfrac{(rt)^{\frac{n}{1000}}}{p}\right)$ running on p transputers. (see section 1.4)

3.4.2 Conclusion

We have demonstrated in practice that our parallel projection algorithm is scalable up to 16 transputers with respect to the data set we used. However due to the in-balance of load generated in the second part of our algorithm further work is required to rectify this situation. A new dynamic load balancing scheme is currently begin devised to redistribute the load to push up the speed up limit.

The parallel projection algorithm's performance is inextricably tied with the nature of the data it is processing. More works will be performed in this area to quantify the relationships between different data set and the algorithm perofmrnace.

Recall earlier that our representation of data consisted of a few massively large graphs, several advantages have been gained by not modelling the parallel algorithm on Ullman's algorithm. For example, we avoided using a matrix as the underlying structure for storing graphs which allow us to modify and use the memory more flexibly and efficiently. Thus changes to data can be performed on the fly by modifying only those parts that need to be changed, as opposed to rebuilding the whole data structure.

4 Future Work

We will be working on a detail analysis of the performance of the algorithm and of its response to different kinds of knowledge bases, as produced by the data generator. Also we will be looking into using transputer memory as 'unlimited' primary storage for very large knowledge bases.

The next step in the development of the LUT Transputer-based CGs processor is the detailed study of the parallel aspects of Peirce Logic inference and theorem proving. This is briefly summarised below.

4.1 The application of parallelism to the Peirce Logic Inference

We have demonstrated that CGs operations can be speeded up by using our new parallel projection algorithm, which makes it possible for the system to query realistically large CGs knowledge bases.

However, inference in Peirce Logic, or in any other type of logic, is a qualitatively different operation. Conventional implementations of theorem provers generally tend to be slow and unsuitable for practical use on real-world domains. We are attempting to remedy this problem by parallelising the inference process. However, although the incorporation of the parallel projection algorithm into an existing Peirce-logic-based CGs processor and theorem prover, implemented at Loughborough by Heaton, will cut down the inference time but not enough. The theorem prover will still be executing its Alpha/Beta rules [6] in a sequential fashion. Ideally we would like it to process rules in parallel.

The Loughborough CGs theorem prover works from a set of graphs {graphs} towards the empty set { }. A proof will usually involve large number of rules and generate sets of intermediate results. An example is illustrated in Fig 3.5.1 with Graph0 representing the set of graphs to be proved and Graph11, 12 and 13 as the results.

Parallelism can be introduced by processing intermediate sets of graphs simultaneously. To achieve this, the proof mechanism must proceed breadth-first, i.e. each inference step will generate all intermediate sets of graphs offering a high probability of proof completion. This sets of graphs is sent to available transputers for processing. Sets of graphs that have already been processed or currently being processed are also sent to prevent different transputers from doing the same inference steps or repeating inferred steps. During inference transputers will feed/consume graphs to/form each other. Transputers that have finished a proof act as consumers. Transputers with more than one set of graphs to process act as feeders. To use the new projection algorithm, each transputer used for inference must be connected to a projection-dedicated transputer network, as shown in the fig 3.5.2. M - Master Transputer, Tp - Slave Transputer used for Projection, TI - Slave Transputer used for Inference.

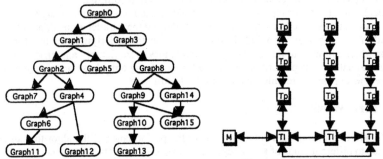

Fig 3.5.1 An Inference Example   Fig 3.5.2 Configuration of the transputer network for Parallel Inference

Acknowledgements

We thank Southampton University High Performance Computing Centre for giving us access to their transputers system for this research.

References

1. A. Celentano, Compiler testing using a sentence generator, *Software-Practice and Experience, Vol., 10, 897-918(1980).*

2. K. V. Hanford, Automatic generation of test cases, *IBM system Journal, No. 4, 1970.*

3. J. D. Roberts, R. Levinson and R. Hughey. Issues in Parallel Hardware for Conceptual Graph Retrieval, *ICCS' 93.*

4. J. F. Sowa, Conceptual Structures - Information Processing in Mind and Machine, *Addison-Wesley Publishing Company, Inc. 1984.*

5. J. R. Ullmann. An algorithm for subgraph isomorphism. *Journal of the ACM, 16:31-42, 1976.*

6. P. Willett and T. Wilson. Atom-by-Atom Searching Using Massive Parallelism. Implementation of the Ullman Subgraph Isomorphism Algorithm on the Distributed Array Processor, *J. Inf. Comput. Sci. 1991, 31, 225-233.*

7. W. T. Wipke and D. Rogers. Rapid Subgraph Search Using Parallelism, *J. Chem. Inf. Comput. Sci. 1984, 24, 255-262.*

Spanning Tree Representations of Graphs and Orders in Conceptual Structures

Andrew Fall

School of Computing Science, Simon Fraser University
Burnaby, B.C., V5A-1S6, Canada
electronic mail: fall@cs.sfu.ca

Abstract. Graphs and partial orders are fundamental to conceptual structures theory. Conceptual graphs are used in the knowledge base and canonical basis as well as for type and conceptual relation definitions. Partial orders are used to specify the associations among types and relations, as well as graphs in the generalization hierarchy. The speed with which basic operations on these structures can be achieved will have a pronounced effect on efficiency. In this paper we explore the use of spanning trees as underlying representations of graphs and partial orders, and their effect on storage and operational efficiency. The simple structure of trees can be exploited to improve matching and joins of graphs (through normalization), to refine search algorithms on orders, and for taxonomic encoding. Although this inquisition is preliminary in nature, we provide some useful techniques for gaining efficiency through the use of spanning trees, including an implementation with a special form of feature term.

1 Introduction

Spanning trees are a valuable tool for improving the operational efficiency of graphs and partial orders in conceptual structures. In this preliminary investigation, we explore their use for representing conceptual graphs, the type and relation hierarchies, and the generalization hierarchy, as well as how they may improve operations on these structures. We only deal with *atomic* conceptual graphs in which all relations are both dyadic and invertible. Atomic conceptual graphs contain no logical connectives (i.e. they are connected), no logical quantifiers (other than the implicit existential), and no nesting (i.e. there is only one context) [4, 5]. The *inverse* of a dyadic conceptual relation R is a relation R^{-1} which is semantically identical to R with the direction of the arrows reversed. For example, the inverses of AGNT and PARENT are AGNT_OF and CHILD, respectively. Similar assumptions have been made in [5, 10, 11, 14].

We first discuss the notions of cardinality constraints and functional relations. Although cardinality can be expressed using sets, actors or complex nesting of contexts, it is important to have the ability to express such constraints simply and declaratively. Graph normalization techniques introduced in [10] are expanded upon in section 3 to prepare for constructing the *spanning tree normal form* that we introduce in section 4. Of particular importance to operational efficiency is the elucidation of functional relations in graphs. In section 5 we present *sparse terms* as a flexible and efficient implementation of spanning trees in which unification can partially automate CG operations. We then adapt the canonical formation rules and other operations to our spanning tree representation. After briefly discussing the use of spanning trees for representing the type and relation hierarchies, we explore their use in the generalization hierarchy to specify a *generalization hierarchy normal form*, to enhance search operations such as matching and retrieval, and to efficiently perform a topological traversal.

2 Cardinality Constraints

Although some conceptual relations are functional in character, CG theory provides no simple way to represent these and other forms of cardinality constraints declaratively, without resorting to the use of actors, sets or complex nesting of contexts. Actors and dataflow graphs provide a powerful means to represent functional dependencies among concepts, but they imply computation of dependent concepts from independent concepts. Sets provide this capability within concepts, but do not restrict the number of relations of a particular type, which can be a valuable constraint for normalization and matching. For example, the canonical graph: [EAT]→(AGNT)→[ANIMATE] does not tell us whether an act of eating must have exactly one agent or may have multiple agents (i.e. if AGNT is a *functional* relation of EAT). Another example is: [PERSON]→(SPOUSE)→[PERSON], which says that the spouse of a person must be person, but does not constrain a person to have at most one spouse. For illustration, we assume that both of these cases are functional.

Definition 1. A *cardinality constraint*, @n ($n \in \mathcal{Z}^+$), between a concept c and a relation r states that at most n relations of type r may be connected to c.

A cardinality constraint is denoted on the arc between the concept and the relation. Thus, the above example becomes: [EAT]-@1→(AGNT)→[ANIMATE]. Restricting a relation to one occurrence for a concept (i.e. $n = 1$) is a *functional cardinality constraint*, and it is these constraints that we focus on. The connection to logic is simple: if the variable representing the independent concept appears in two instances of the relation, then the variables representing the dependent concepts must be equal. This provides a sort of uniqueness constraint. Our example translates to: $\exists x \exists y$ (EAT(x) $\wedge$ ANIMATE(y) $\wedge$ AGNT(x, y) $\wedge \forall z$, AGNT(x, z) $\supset z = y$). We do not suggest that all functional dependencies can or should be expressed in this way. Rather, we feel that by notating functional relations, normal forms for CGs will be more distinct and easier to determine.

Cardinality constraints blend well with aggregation techniques and set cardinality [9, 13]. For *set coercion*, a cardinality constraint can be moved into the set notation. On expansion, the set cardinality can be moved out a cardinality constraint. In order to ensure set joins, we make concept sets functional. As an example, for: [DANCE]→(AGNT)→[PERSON: Liz], set coercion on PERSON results in: [DANCE]-@1→(AGNT)→[PERSON: {Liz}], whereas set expansion on: [DANCE]-@1→(AGNT)→[PERSON: {Liz,Kirby}@2] results in: [PERSON:Liz]←(AGNT)←@2-[DANCE]-@2→(AGNT)→[PERSON:Kirby].

3   Normalization

Normalization is important to enhance the similarity among conceptual graphs and can be achieved via transformation rules [10]. Of interest in this paper are rules regarding privileged relations, symmetry and elimination of redundancy. We have assumed that all relations are invertible so, for e.g., the inverse of WORKS_FOR is EMPLOYS, whereas the inverse of SPOUSE is itself (i.e. it is symmetric). We later show how our proposed representation automatically performs some simplification, reducing redundancy that can arise during joins.

Explicitly representing functional relations can be exploited to determine a precedence between a relation R and its inverse R^{-1}. Priority is given to functional relations. Thus, assuming a world in which a person has at most one nationality, we would prefer: [PERSON]-@1→(CITIZENSHIP)→[COUNTRY] to: [COUNTRY]→(CITIZEN)-@1→[PERSON]. If both R and R^{-1} are functional, we incorporate both (i.e. we perform *symmetry completion* [10]). By doing this, we can traverse all functional relations in the direction of their arcs. If neither R nor R^{-1} are functional, other preference schemes need to be specified.

Normalization will also incorporate selectional constraints related to the graph, particularly those which add functional relations between concepts. To illustrate, consider the well-known example: *a cat sitting on a mat*. The following figure shows a normalized version of the conceptual graph, in which the concept SIT imposes the selectional constraint that it has exactly one agent.

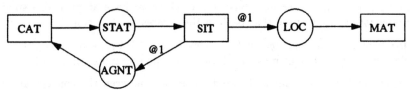

Fig. 1. Normalizing "a cat sitting on a mat"

4   Spanning Tree Normal Form

It is easy to specify a spanning tree for any conceptual graph, with coreference linking identical concepts as in the linear form. Any traversal of a graph that visits every concept and relation defines a spanning tree: the first node visited is the root and cycles are broken by introducing coreference. Our goal is to specify a *spanning tree normal form* (STNF) which can be used to improve the efficiency of CG operations as well as to organize and search the generalization hierarchy. STNF exposes functional relations in graphs and permits an implementation of CGs using feature terms. In [14] there is also a proposal for a normal form which is a spanning tree, but the tree is determined in an ad hoc manner (alphabetical order is used to select the root and relations to expand partial trees).

Definition 2. A *spanning tree T* for a conceptual graph G is a connected acyclic subgraph of G containing all the concepts of G (but not necessarily all the relations). To each spanning tree, one concept is designated the *root*.

In the linear form [13], concepts and relations form the nodes of a spanning tree, and arcs are labeled with directional arrows. For STNF, only concepts are nodes while relations are arc labels. The direction of arcs is implicitly downward. Although this format is suitable for binary relations, which form the majority of conceptual relations [12], it may be possible to accommodate monadic and higher-order relations, but we do not explore this here. We assume that our graph is normalized as described in section 3 and that we have linear extensions τ and ρ of the type and relation hierarchies, respectively. Since some graphs may require multiple root elements, we actually construct a spanning forest. We maintain the individual trees in a list ordered by the type of the root concepts (according to τ). When drawing forests, we add an untyped dummy root to connect the trees together.

We give below an algorithm that takes as input a normalized conceptual graph G, and outputs a spanning forest F that represents G in STNF. The concepts and relations of G are the ordered lists C and R, respectively. Each node of in the forest is a concept c to which a (possibly empty) list of children is associated (via $children(c)$). Each child contains a pair: the child concept and the connecting relation. The root of the tree containing a concept c is obtained by calling $tree(c, F)$.

Algorithm 1 $STNF(input: G = < C, R >; output: F)$

1. $F := C$
2. for each concept $c \in C, children(c) := \emptyset$
3. for each relation $r(c_i, c_j) \in R$ (taken in order)
4.   if $(tree(c_j, F) = c_j \ AND \ tree(c_i, F) \neq tree(c_j, F))$ then
5. $children(c_i) := children(c_i) \cup \{< r, c_j >\}$
6. $F := F - \{c_j\}$
7.   else
8. $children(c_i) := children(c_i) \cup \{< r, coref(c_j) >\}$
9. end

First, we start with a forest consisting of each concept in the graph G as a tree (lines 1 and 2). We consider relations one at a time and update the forest as necessary. A node is always placed below the entering concept c_i, labeled with the relation type. If the exiting concept, c_j, is the root of a different tree in the forest from c_i simply connect this tree below c_i (lines 5 and 6). We do this by adding the relation, concept pair to the children list of c_i and removing the tree rooted at c_j from the forest. If, however, c_j is not a root or is in the same tree as c_i, the node below c_i will contain a coreference label linking to c_j (line 8). Once we have visited all relations, we have a spanning forest for our graph. The time complexity of this algorithm is near linear in the number of concepts and relations in the input graph if the $tree$ function is implemented using a union-find algorithm.

The order in which we visit relations (line 3) is important. We consider all functional relations, before any non-functional ones. Within these groups, the order depends on the types of the relation and two incident concepts. The order of precedence is the relation, followed by the entering concept and lastly the exiting concept. Exploring the consequences of choosing different precedence orderings is a topic for further research. It may still be possible for there to be two or more arcs with precisely the same relation and incident concept types. In this case, contextual information may be needed for selection. In this preliminary analysis, we simply select one arbitrarily, and this is the only place where non-uniqueness can enter into the process. Thus, our construction computes a spanning tree normal form which is nearly unique for normalized graphs. As an example of this construction, Fig. 2 shows the STNF of the graph in Fig. 1. Note that both AGNT and LOC are functional relations of SIT. The last relation visited is STAT, which is added using coreference. In diagrams, we notate functional relations using thick lines and non-functional ones with thin lines.

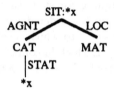

Fig. 2. Spanning tree normal form

Another well-known example, with a cycle, is: *a monkey eating a walnut using the walnut's shell as a spoon.* Figure 3 shows the normalized graph as well as its STNF. For illustrative purposes, we assume that an entity can only be (intransitively) a part of at most one other entity, and that an instance of eating has one agent and one object. Thus the relation PART is inverted to PART_OF. We assume that the linear ordering of relations is AGNT < OBJ < PART_OF < INST < MATR. We first add MONKEY and WALNUT as children of EAT, then a coreference link to WALNUT as a child of SHELL, and finally we add the non-functional relations INST and MATR in the tree rooted at EAT.

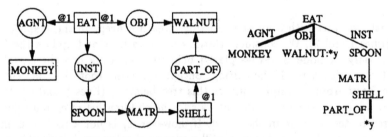

Fig. 3. A cyclic graph and a tree representation

For a more complicated example, consider the statement: *a woman eating a dinner cooked by her husband*, which is shown in Figure 4. In this case, we end up with two trees since both EAT and COOK only have exiting relations in the normalized form. Assuming the types are ordered by COOK < EAT < WOMAN < MAN, we obtain the STNF as shown.

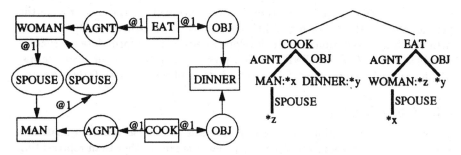

Fig. 4. A woman eating a dinner cooked by her husband

4.1 Pivoting

Given a graph in STNF, we may need a certain concept to be the root of one of the trees in the forest in order to perform graph matching, to obtain different viewpoints of a graph, or to further normalize the spanning tree for storage in the knowledge base. We call this process *pivoting*. Although there are several approaches to pivoting, we have chosen one that is particularly simple, yet useful for organizing the knowledge base. We call the node of a concept in a spanning forest that maintains the type information (and possibly has a subtree) the *dominant* node. All other, corefering nodes are called *subordinate*. Basically, to pivot a concept which is not already a root is accomplished by replacing the dominant node for the concept by a subordinate node and adding the subtree rooted at this node as a top level tree in the forest. Pivoting can easily be carried out, as shown in the following figure which shows pivoting of the STNF form of the graph in Fig. 3 on the concepts "WALNUT" and "SHELL".

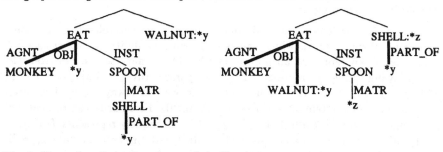

Fig. 5. Examples of pivoting the graph in Figure 3

5 Spanning Tree Implementations

One of the advantages of using spanning tree representations is that there is a flexible logic programming implementation which can efficiently compute unification and anti-unification, as well as incorporate inheritance. Some of the limitations of Prolog terms have led to a variety of extensions, such as rational trees, psi-terms [2] and sparse terms [7]. In their minimal form, sparse terms have proven to be useful for taxonomic encoding of partial orders and lattices, which we describe briefly in section 7. More expressive variations provide the functionality of the psi-terms of LIFE [2] and feature structures [3]. The ease with which new variations can be developed permits us to formulate a structure ideally suited to CGs in STNF. We first present the basic form of sparse terms, which are used for encoding, followed by a description of *sparse feature terms*.

5.1 Sparse Terms

Sparse matrix representation provides the basis for sparse terms. A matrix is sparse if the ratio of non-zero to zero entries is small. Rather than representing every entry, only the non-zero entries with their coordinates are maintained, providing an overall space savings. A logical term is sparse if the ratio of specified or coreferenced positions (i.e. predicates, functors, atoms, named variables) to anonymous variables is small. Anonymous variables only reserve positions and do not contribute to the information content of a term. A space savings can be realized by representing only the specified and coreferenced positions, along with their position index. Consider the term: $a(b(_{-1}, c, d, _{-2}), _{-3}, _{-4}, e(_{-5}, f(_{-6}, _{-7}), _{-8}))^1$. The ordinary and sparse forms are shown below in a rooted graph notation. The sparse term can be represented linearly as: $a - [1.b - [2.c, 3.d,], 4.e - [2.f]]$, where the lists are ordered according to increasing index.

Fig. 6. Sparse logical terms

In addition to the benefits of eliminating anonymous variables, there are some properties of sparse terms which endow them with flexibility and conciseness. A sparse term can represent an infinite number of ordinary terms. For example, the following terms are represented by the sparse term $f - [1.a]$: $f(a)$, $f(a, _)$, $f(a, _, _)$, $f(a(_), _), \ldots$ Arity is not bound, so we can always append an arbitrary number of anonymous variables as arguments. Fixed arity can be specified following the symbol it affects (e.g. $f/2 - [1.a/0]$ uniquely represents $f(a, _)$).

Positions in terms can be specified as filled, but the actual symbol (predicate, atom or functor) occupying the position can be left unspecified. The term $f(_, b(_, _), _, c(d, _, e), _)$ can thus be represented as $[2, 4.[1, 3]]$. Of course, unification can only fail if there are different symbols at the same location in both terms. By representing a minimal amount of information, functorless terms are particularly useful for reducing storage space in taxonomic encoding [7].

Another distinction from Prolog terms is the ability to index with feature labels rather than just numerically. Since we explicitly store the indices, this is trivial to incorporate and it provides an elegant yet efficient implementation of attribute-value matrices as used in several grammar formalisms (see [3]).

---

[1] The anonymous variables have been subscripted for clarity

5.2 Sparse Feature Terms

It is simple to derive variations of sparse terms which incorporate the functionality required for a particular domain. We develop one variation which incorporates coreference, inheritance and possible duplicity of features. Such terms are a variation of *order-sorted feature terms* (or just feature terms) and are amenable to the representation of conceptual graphs as spanning trees.

To represent arbitrary graphs, we need the capability for nodes (i.e. positions of a term) to corefer. The equality constraints imposed by Prolog variables are insufficient for this purpose because (i) only leaves of a term can corefer, and (ii) after instantiation, the coreference is lost (it is maintained implicitly by identity between the subterms at these positions and additional variables introduced in the instantiation). Even though implementations may maintain this identity using, for example, pointers in order to avoid duplicating the subterms, the surface form loses the explicit coreference. To provide *persistent* coreference, we use a mechanism as in LIFE [2], where coreference is specified using labels that do not disappear upon instantiation. The common subterm is maintained at one of the corefering positions (the dominant position) along with the label, while the rest of the positions contain only the label. As an example, the following figure shows a feature term and its corresponding graph.

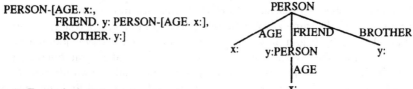

Fig. 7. Example feature term

We also provide an automatic inheritance mechanism, as supported in LIFE [2]. This requires the presence of a type hierarchy from which symbols are taken. When unifying two different symbols, instead of failing (as in Prolog), we use the greatest lower bound (or maximal common subtype) of these symbols in the hierarchy. Unification only fails if the result is ⊥ (absurd). Thus dog-[colour.brown] unified with pet-[name.rex] results in pet_dog-[colour.brown, name.rex].

To represent non-functional relations, we allow duplication of features. To avoid duplicating the feature label, we keep all features of the same type together as a set. For e.g., suppose the act of eating may have multiple utensils, then: EAT-[AGNT.PERSON, OBJ.PASTA, INST.{FORK, SPOON}] represents: *a person eating pasta with a fork and spoon.* Note that this set notation is different from aggregation and set coercion. This distinction affects unification: unique features (functional relations) are treated as before, but we can only take the union of non-unique features (non-functional relations) since there is no constraint for matching elements in one set with those in the other. If the feature is unique in one term, but not in the other, then we unify all occurrences together.

A final modification to deal with individuals and conformity checking, for representing CGs, can easily be integrated into the unification algorithm.

6 Conceptual Graph Operations

We now demonstrate how the canonical formation rules can be performed on graphs in STNF, and in particular how unification can be exploited to efficiently implement these rules by observing the constraints imposed by functional relations. We focus our attention on restrict, join and simplify.

We can characterize restrict and join using feature term unification, which provides the capability to extend both of these operations. *Restricting* a concept c in a graph G to a type t can be viewed as joining c with a graph R consisting only of one concept of type t. If the concept we wish to restrict is a root of the STNF of G, then we can achieve this simply by unifying the terms representing R and G. The inheritance of our feature term unification automates the restriction. If the concept we wish to restrict is not a root, we can perform this unification once we find the subtree rooted at this node. Locating nodes within a graph in STNF may be accomplished by tree traversal or we may incorporate more efficient techniques, such indexing, into the implementation. The automatic inheritance offered by feature terms indicates a useful extension to the restrict operation. Suppose we wish to restrict a concept c to be of a certain type t which is not a subtype of $type(c)$. Through unification, the type of c will be restricted to the maximal common subtype $type(c) \cap t$.

There are several forms of *join* operations that we distinguish. In [4] a distinction is made between *internal* and *external* joins. An internal join identifies two concepts within the same graph and an external join identifies concepts between two graphs. A *simple* join only joins two concepts and a *maximal* join maximizes the extent of a join. We show how functional relations affect join operations.

A simple, external join of two concepts can be easily accomplished. As for restrict, the concept types need not match as unification will find the maximal common subtype of the two types. Any join may result in additional joins due to functional relations, and so a simple join may not be possible. The scope of a *minimal* join in a graph are those concepts in the two graphs for which there is a sequence of relations from the joined concepts, where each relation is functional in at least one of the graphs. Thus joining two concepts will propagate through the functional relations of the two graphs. This is automated through the unification of sparse feature terms. To illustrate, consider: *a married woman who has a child*, and: *a parent of a student, who is married to a student*. We represent these using feature terms as x:WOMAN-[SPOUSE.MAN-[SPOUSE.x:], CHLD.{CHILD}] and x:PARENT-[SPOUSE.STUDENT-[SPOUSE.x:], CHLD.{STUDENT}]. By performing a minimal join on WOMAN and PARENT, we obtain: x:MOTHER-[SPOUSE.MALE_STUDENT-[SPOUSE.x:], CHLD.{CHILD,STUDENT}]. Because SPOUSE is functional, the join propagates over this relation. For CHLD, however, we have no information to say that these children may or may not be the same, so we simply take the union of the two sets. The two graphs in this example are shown in Figure 8.

In order to maintain STNF, we must determine how to update the combined forests. Algorithm 2 computes a minimal external join. The inputs are two concepts to join, $Concept_1$ and $Concept_2$, and their containing STNF forests F_1

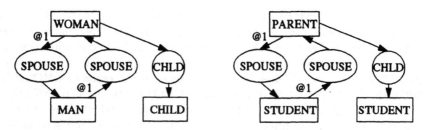

Fig. 8. Two graphs to join

and F_2. The output is a STNF forest representing the result of the join. The first step is to determine the dominant trees within F_1 and F_2 that contain the concepts to join (lines 1 and 2), and to set the output forest to the union of the trees in the input forests (line 3). If the first concept is the root of its containing tree T_1, then we unify this tree with the subtree rooted at the second concept (line 5). We then remove the now redundant tree T_1 from the output forest (line 6). These steps are valid whether or not $Concept_2$ is also a root. The predicate *unify* denotes the feature term unification algorithm. If this unification fails (i.e. the result is $\bot$), the entire algorithm will fail. If it succeeds, then the unified terms will corefer, and the first term will be dominant. Recall that for a set of corefering positions, one position is dominant (i.e. it holds the actual subterm information), and the rest are subordinate. Unification merges two coreference classes; the dominate position of the first term in the *unify* predicate will be the dominant position of the resulting coreference class.

If $Concept_2$ is the root of its containing tree T_2, but not $Concept_1$, then we unify this tree with the subtree rooted at the first concept and remove the tree T_2 from the output forest (lines 7-9). If neither are roots, then we consider the relation which connects to their parent concept in the trees. Preference is determined as in STNF construction, according to the relative order of the relation types and incident concept types. This is specified by calling *preferred*, but we do not elaborate further on this function. In either case, the subtrees rooted at these concepts are unified, and the favored concept is dominant (lines 10-13).

Algorithm 2 *Min_External_Join(input: $Concept_1, Concept_2, F_1, F_2$; output:F)*

1. $T_1 := tree(Concept_1, F_1)$
2. $T_2 := tree(Concept_2, F_2)$
3. $F := F_1 \cup F_2$
4. *if* $(Concept_1 = T_1)$ *then*
5. $unify(Concept_2, T_1)$
6. $F := F - \{T_1\}$
7. *else if* $(Concept_2 = T_2)$ *then*
8. $unify(Concept_1, T_2)$
9. $F := F - \{T_2\}$
10. *else if* $Concept_1 = preferred(Concept_1, Concept_2)$ *then*
11. $unify(Concept_1, Concept_2)$
12. *else*
13. $unify(Concept_2, Concept_1)$
14. *end*

For a simple, internal join, both concepts may be in the same tree. Here precedence is determined as before, except if one concept is above the other in the tree. In this case, the higher concept has precedence over the lower one. The algorithm for computing a minimal internal join is otherwise the same as the above algorithm. Joining the two children in the previous example results in x:MOTHER-[SPOUSE.MALE_STUDENT-[SPOUSE.x:], CHLD.{YOUNG_STUDENT}].

Since minimal joins observe the constraints of functional relations, any maximal join must be an extension of a minimal join. When constructing a maximal join, each simple extension may propagate as a result of functional relations. Thus, a maximal join may be characterized as a sequence of minimal joins. Performing a maximal join can either be accomplished directly, or by executing this sequence of minimal joins. The restrictions imposed by functional relations reduce the number of possible maximal joins, and so improve the efficiency of enumerating and determining maximal joins. The above two examples of simple external and internal joins result in a maximal join of the two initial graphs.

By representing functional relations and implementing graphs in STNF using sparse feature terms, the *simplify* operation is often performed automatically. Since functional relations enforce uniqueness of certain relations, duplicity of such relations is impossible. The only way redundancy can enter into a graph in STNF as a result of a join is on the non-functional relations. Due to the way non-functional relations are stored in feature terms, however, this possibility is further reduced (since we can easily eliminate duplicate coreference labels in the set of one relation). Since normalization selects a preferred direction to traverse relations, explicit simplification is only required for the case of relations which are both symmetric and non-functional (e.g. NEAR).

The operations of projection and matching may be improved for graphs in STNF, although detailed discussion of this is beyond the scope of this paper. The exposure of functional relations for graphs in STNF imposes important restrictions which can be used to guide the search for valid projections and matchings, and reduce the number of possibilities that must be considered. Of course, the worst-case behaviour of these operations is unaffected by STNF, but the average-case behaviour may be substantially improved.

7   Representing the Type and Relation Hierarchies

There has been much research into encoding lattices and partial orders for the efficient implementation of operations such as *greatest lower bound, least upper bound* and *subsumption checking* (e.g. [1, 8]). Although the details of taxonomic encoding are beyond the scope of this paper, for completeness we mention the use of spanning trees in encoding. The basic notion of encoding involves the association of a code with each element in the partial order. Taxonomic operations are then performed using the codes rather than searching in the original order. The goal of encoding is for the improved time of computing operations to outweigh the space requirements. Possible implementations of codes include bit-vectors, integer-vectors and logical terms. We have found the minimal *functorless* form of sparse terms to be useful for implementing encodings [7], particularly for very large, wide orders in which each element has relatively few ancestors (such as may be expected for the type hierarchy).

8 Representing the Generalization Hierarchy

A CG database contains of some of the (infinitely many) canonical graphs which can be obtained from the canonical basis B using the canonical formation rules. The generalization hierarchy organizes graphs into a partially ordered set of equivalence classes [5, 11], where each graph in a class is canonically derivable from all others in the class, and one class subsumes another if each graph in the latter is derivable from each graph in the former. The generalization hierarchy consists of both the canonical basis (which represents things that could exist) and the database graphs (which represent things that do exist). Although B may not form an anti-chain, there is a subset B_0 of B which forms the initial level, or co-atoms, of the generalization hierarchy. Our goal is to use STNF to assist in the organization and search of this hierarchy. The advantages of explicitly maintaining the generalization hierarchy are described more fully in [6]. This hierarchy can be encoded so that many operations among graphs in the hierarchy can be performed efficiently using only lattice operations, avoiding matching altogether. In our case, we maintain the full hierarchy, but mark one parent of each graph as dominant, to identify a spanning tree. Since graphs are represented in STNF, operations may be more efficient.

We first describe the process of constructing the spanning tree for the generalization hierarchy incrementally, leading to another normal form. We start with an empty generalization hierarchy consisting only of $[\top]$ and $[\bot]$. We also need to order the children of any element, so we define a total order on graphs (perhaps based on the linear extensions of the type and relation hierarchies, and the form and content of the graphs). The actual method used to specify this ordering is not important to the following discussion.

Suppose we have a generalization hierarchy organized with an underlying spanning tree T_G and we wish to add a graph Q which is in STNF. We essentially use the algorithm of [6] to search the hierarchy and find the immediate predecessors (IP) and immediate successors (IS) of Q. We store all graphs so that every STNF graph G is a simple specialization of its parent G' in T_G. That is, G and G' are unifiable and the term for G' subsumes the term for G. This cannot be achieved for all ancestors of G, but if it holds for all ancestors in T_G (i.e. the graphs on the path from G to the root $[\top]$), then we can improve search and matching operations. The position of Q in T_G will be below the leftmost IP.

As we find each predecessor C of Q in T_G, we modify the form of Q. Since both C and Q are in STNF, the spanning trees in C will be contained in the trees of Q (modulo symmetric relations and coreference). For each tree of C whose root is not a root of Q, we must pivot. Pivoting does not destroy the STNF properties, but creates additional trees, so we essentially flatten Q until C is more evident in its forest. When all the ancestors of Q in T_G have been processed, Q will be in *generalization hierarchy normal form* (GHNF). The advantage of a storing graphs in GHNF is that if we have graphs Q and Q' for which Q subsumes Q' in T_G, then Q and Q' are unifiable. That is, not only is Q' a specialization of Q, the feature terms representing Q and Q' are related by term subsumption.

8.1 Depth-first Topological Traversals

The spanning tree T_G underlying the generalization hierarchy can be viewed as representing a left-to-right (LR) depth first (DF) traversal of the generalization hierarchy. We show here a relation between LR-DF traversals and DF topological traversals, where a topological traversal is any traversal that obeys the topological property that a node cannot be visited until all of its parents have been visited. In [6], the advantages of searching the hierarchy for IP and IS topologically are described.

We make the distinction between breadth first (BF) and depth first topological traversals. In BF traversals, we visit nodes by level. The level in an ordinary BF traversal is the length of the shortest path to the root, since we place an element in the search queue when it is first accessible. The level for a topological BF traversal, however, is the length of the longest path to the root because we place an element in the search queue only when last accessible (when the last parent has been visited). DF traversals, on the other hand, select the next candidate node to visit with the longest leftmost path to the root (in a LR traversal), where conflicts are resolved by choosing the leftmost element. For ordinary DF traversal, a candidate is any unvisited node which is connected by an arc to the tree traversed so far. When observing the topological property, however, the only candidate nodes are those whose parents have all been visited.

It should be clear that BF and DF topological traversals are implemented differently (using a queue in the former and a stack in the latter) and may visit nodes in different orders. The proposal in [6] performs a BF topological search of the generalization hierarchy to perform updates and retrievals. We feel that it is interesting to explore DF topological searches for several reasons. First, such a search would result in finding the first member of IP earlier than a BF topological search. Second, we show how the spanning tree T_G can be used to perform a DF topological traversal without needing to mark elements as visited. Third, we can utilize GHNF more fully to improve the efficiency of graph comparisons.

Although we cannot use the LR-DF traversal suggested by T_G in the search algorithm, there is an interesting connection between DF traversals and DF topological traversals. If T_G represents a LR-DF traversal of a hierarchy P, then a right-to-left (RL) DF traversal of T_G is a RL-DF topological traversal of P.

Theorem 3. *Suppose G is a rooted directed acyclic graph and T_G is the tree resulting from a LR-DF traversal of G. Then a RL-DF traversal of T_G is a RL-DF topological traversal of G.*

Proof. Consider any point in a traversal of T_G. Suppose the next node to visit, v, with parent p in T_G, has an unvisited parent p_1. Since p_1 is unvisited, it must be to the left of p in T_G, but then during the initial DF traversal p_1 would have been visited before p, and so v would be below p_1 not p in T_G.

Thus, simply by following a RL-DF traversal of T_G we can perform a DF topological traversal of the ordered set without the overhead of checking when all parents have been visited.

In order to fully utilize the spanning tree structure of the generalization hierarchy and the GHNF form of graphs, we describe a modification of the search algorithm of [6]. The problem is to to find the immediate predecessors (IP) and then the immediate successors (IS) of a graph Q, which may or may not be in the hierarchy. We assume that after a comparison between Q and a graph u in which $u > Q$, it is desirable to compare the children of u with Q so that we can benefit from the result of the match (while still obeying the topological property). By following the depth first topological traversal described above, this can be achieved with very little effort: we don't even need to mark elements as visited. By marking only those which successfully match Q, we can perform the search with a minimum amount of administration. Furthermore, since graphs are in GHNF, we will successively compare graphs whose GHNF forms most closely match until a subtree is traversed or until a graph is found which doesn't match Q. Another advantage of this approach is that by performing a DF topological search, the focus (as described in [6]) becomes restricted more quickly, providing a more constrained target for guiding the search.

9 Conclusion

We have explored the use of spanning tree representations of graphs and partial orders in conceptual structures. We first proposed a means of declaratively representing cardinality constraints without the use of sets or dataflow graphs. Of particular interest are functional relations, which restrict the number of occurrences of a particular relation type to one. These constraints are important for improving the efficiency of matching and join operations. We extended and refined CG normalization, as introduced in [10], through the use of functional relations. We proposed a spanning tree representation of conceptual graphs, leading to a spanning tree normal form (STNF) which is based on semantic content and is less ad hoc than some previous proposals. Graphs represented in STNF have a natural and elegant implementation using a variation of sparse logical terms [7] called sparse feature terms, which are customized for the specific requirements of conceptual graphs. We described how the canonical formation rules can be effected using graphs in STNF and implemented using feature terms, providing a scheme in which these operations can benefit from the efficiency of feature term unification. In particular, functional relations expand the scope of joins and matches, reducing the space of possibilities for maximal joins and projections. The type and relation hierarchies are both partial orders, which can be encoded to improve the efficiency of operations such as computing greatest lower bounds and checking subsumption. Particularly efficient encodings are possible using the basic form of sparse term. Finally, we showed how identifying an underlying spanning tree for the generalization hierarchy can benefit both storage and traversals. A spanning tree can assist in a further refinement of STNF to generalization hierarchy normal form (GHNF) in which all graphs on the same path to the root are unifiable. Furthermore, by traversing this left-to-right depth first tree in a right-to-left depth first manner, we achieve a depth first topological traversal which can be used as an alternative search procedure of [6]. An advantage of this search, in addition to its efficiency and simplicity, is that graphs which are closely related have a higher chance of being compared successively, so we can take advantage of the results of previous matches.

This work presents a preliminary inquest into the use of spanning trees to improve the storage and operational efficiency of graphs and partial orders. We expect that many improvements are possible, and due to the largely independent nature of our proposals, they may be selectively adopted or enhanced.

References

1. H. Aït-Kaci, R. Boyer, P. Lincoln and R. Nasr, Efficient Implementation of Lattice Operations. *ACM Transactions on Programming Languages.* 11(1): 115-146. 1989.
2. H. Aït-Kaci and A. Podelski. Towards a Meaning of Life. Journal of Logic Programming. 16(3/4): 195. 1993.
3. B. Carpenter. The Logic of Typed Feature Structures. Cambridge University Press, London, England. 1992.
4. M. Chein and M. Mugnier. Specialization: Where do the Difficulties Occur? Proc. of *7th Annual Workshop.* Las Cruces, NM. H. Pfeiffer and T. Nagle (Eds.), 1992.
5. G. Ellis. Compiled Hierarchical Retrieval. Conceptual Structures: Current Research and Practice. T. Nagle et al (Eds.). Ellis Horwood, New York, 1992.
6. G. Ellis. Efficient Retrieval from Hierarchies of Objects using Lattice Operations. Proc. of *First Int'l Conf. on Conceptual Structures.* Quebec City, Canada, 1993.
7. A. Fall. Sparse Logical Terms. To appear in *Applied Mathematics Letters*, 1995.
8. A. Fall. The Foundations of Taxonomic Encoding. Submitted to *ACM Transactions on Programming Languages and Systems.* Also available as Simon Fraser University Technical Report SFU CSS TR 94-20.
9. D. Gardiner, B. Tjan and J. Slagle. Extending Conceptual Structures: Representation Issues and Reasoning Operations. Conceptual Structures: Current Research and Practice. T. Nagle et al (Eds.). Ellis Horwood, New York, 1992.
10. G. Mineau. Normalizing Conceptual Graphs. Conceptual Structures: Current Research and Practice. T. Nagle et al (Eds.). Ellis Horwood, New York, 1992.
11. M. Mugnier and M. Chein. Polynomial Algorithms for Projections and Matching. Proc. of *7th Workshop.* Las Cruces, NM. H. Pfeiffer and T. Nagle (Eds.), 1992.
12. L. Roberts, R. Levinson and R. Hughey. Issues in Parallel Hardware for Graph Retrieval. Proc. of *First Int'l Conf. on Conceptual Structures*, Theory and Applications. Quebec City, Canada, 1993.
13. J. Sowa. Conceptual Structures: Information Processing in Mind and Machine. Addison-Wesley, 1984.
14. G. Yang, Y. Choi, J. Oh. CGMA: A Novel Conceptual Graph Matching Algorithm. Proc. of *7th Workshop.* Las Cruces, NM. H. Pfeiffer and T. Nagle (Eds), 1992.

An Implementation Model for Contexts and Negation in Conceptual Graphs

John Esch & Robert Levinson

Unisys Government Systems Group
P.O. Box 64525   U1T23
St. Paul, MN 55164
(612) 456-3947
esch@email.sp.paramax.com
Department of Computer and Information Sciences
225 Applied Sciences Building
University of California
Santa Cruz, CA 95064
(408) 459-2087
levinson@cse.ucsc.edu

Abstract. An implementation model for a retrieval and inference system based on the theory of conceptual graphs is presented. Several hard issues related to the full implementation of the theory are taken up and solutions presented. The solutions attempt to exploit existing but not fully recognized symmetries in CG theory. These symmetries include those between formation and inference rules, AND and OR, positive and negative, copy and restrict, general and specific, etc. Topics taken up include the implementation of Sowa's formation rules, the storage of a conceptual graph hierarchy involving contexts and negation as a conjunctive normal form (CNF) lattice, the extension of existing retrieval algorithms, such as Levinson's Method III and UDS, to handle complex referents and nested contexts, the checking of consistency, and the definition of Peirce's inference rules in terms of formation rules. A distinction is made between syntactic implication and semantic implication. The issues tackled in the paper lay the foundation for a full scale graph-based first-order logic theorem prover.

KEYWORDS: Contexts, Negation, Conceptual Graphs, Consistency, Inference, Retrieval, Knowledge Representation

1   Introduction

At a given level of a hierarchy, a particular system can be seen as an outside to systems below it, and as an inside to systems above it; thus, the status (i.e., the mark of distinction) of a given system changes as one passes through its level, in either the upward or downward direction. The choice of considering the level above or below corresponds to a choice of treating the given system as autonomous or controlled (constrained) [11].

Is your conceptual graphs database consistent? Does it store and process nested context graphs? How is negation handled? Are complex referents too complex to use? Does it properly map the mathematical relationship of Peirce's rules to Sowa's [21, 22, 23] formation rules? Although conceptual graph (CG) theory has been discussed in depth for the past ten years, and several implementations have been developed [5], these full implementation questions have largely been neglected.

In this paper we address some of these omissions. In particular, we give implementation methods for handling negation, arbitrarily nested contexts, the canonical formation rules, and lambda abstraction. Although, we don't give a full scale inference system, we do show how Peirce's inference rules can be incorporated within the same scheme, and how inference might proceed. Along these lines we give a mechanism for ensuring consistency (no logical contradictions) in a CG database. Some of the major contributing notions include:

1. The storing of the CG generalization hierarchy as a CNF lattice that exploits the duality between Boolean AND and OR for storage and retrieval efficiency.
2. The extension of previous implementation models [3, 18] that exploit containment links to avoid repeated storage of structures, to store nested contexts.
3. The association of labels (codes) with nodes and links that indicate whether the positive or negated version is being asserted or inferred.
4. The representation of a CG hierarchy as a hierarchy over equivalence classes of CGs rather than over individual structures as is discussed in [24].
5. A more elegant statement of the canonical formation rules that more clearly distinguishes restriction and join.
6. The treatment of contexts and complex referents in the same unifying manner.
7. The extension of existing retrieval algorithms for exploring the generalization hierarchy to cover the full variety of CGs [16, 3, 4].
8. A formalization of Peirce's inference rules in terms of formation rules.

In Section 2, Formation Rules, we restate the basic rules for transforming one CG to another. These are foundational because the generalization hierarchy/lattice is based on the partial ordering relation induced by the projection operator. For one graph to project into another, it must be possible to derive the second by applying a sequence of formation rules to the first.

Section 3, Lattice Terminology, explains the basic terminology needed to understand and use lattices to store the CG database. Section 4, Adding Contexts, adds complex referents which include contexts. Previous work defining a Universal Data Structure (UDS) to store the CG database [18] did not include complex referents with icons, indexes, and symbols. This section shows how various forms of them, e.g. names, variables, literals, and graphs for contexts, can be added to the generalization lattice. Section 5 is on Adding Negation. It describes how to extend the generalization hierarchy/lattice to cover the negative graphs that were not included in previous work either. In Section 6, Extended Algorithms, the basic structure of the previous Method III algorithm [4, 15] is extended, without increasing the complexity, to cover complex referents, such as names, variables, contexts, and negative graphs. Section 7 covers Peirce Inference Rules. With conjunction and negation it is possible to represent all logical expressions. Peirce defined a compact system of logic that John Sowa applied to CGs. This section restates Peirce's rules and shows how they are handled with the extended generalization hierarchy/lattice. And Section 8 concludes by giving future directions.

2 Formation Rules

The formation rules are a foundational part of CG theory because they define the partial ordering relation used to build the syntactic generalization hierarchy. They were originally defined by Sowa in 1984 [21] and refined in his latest book [23]. We have streamlined and clarified those definitions by defining three kinds of operations on conceptual graphs, equivalence preserving, specialization, and generalization. Each of the formation rules has a mutual inverse: Simplify with Copy, Restrict with Join, and Unrestrict with Detach. Some work in the same spirit is taken up in [2].

In the following, each of the formation rules is defined in terms of more primitive individual concept and relation operations. The notation type(r1) $\leq$ type(r2) is used. It means that r1 is a subtype of r2 and hence more specialized. We also assume that a relation subtype has the the same corresponding arcs as the relation type.

2.1 Equivalence Formation Rules

The ϕ function maps a CG to an equivalent predicate logic expression. So two CGs, mapped by ϕ to logically equivalent expressions, have the same meaning; i.e., they are in the same equivalence class. Consequently, transformation rules, where the initial and final CGs are in the same equivalence class, are called equivalence rules.

We define Relation Simplify, its inverse Relation Copy, Concept Simplify, its inverse Concept Copy, and use them to define Simplify & Copy equivalence rules.

Definition: Relation Simplify. If r1 and r2 are relations, type(r1) $\leq$ type(r2), and each pair of corresponding arcs connects to the same concept or coreferent concepts in the same context, then r2 and its arcs may be deleted.

It can be shown that the initial and final CGs are in the same equivalence class as follows. First, ϕ maps relations to a predicate corresponding to the relation's type. Second, ϕ maps the predicate's arguments to the referents of the concepts to which the relation is connected. Third, since each corresponding pair of arcs connect to the same or coreferent concepts, the corresponding pair of arguments will be the same. Thus, the predicates have the same arguments. Fourth, the predicate for r1 implies that for r2. And fifth, since type(r1) $\leq$ type(r2) and predicate(r1) => predicate(r2), we can delete r2.

Definition: Relation Copy. If relation r1 has arcs connected to concepts c1, c2, ... ,cn, then relation r2 may be added where both type(r1) $\leq$ type(r2) and corresponding arcs also connected to c1, c2, ..., cn or concepts coreferent with them in the same context. Note that r2 doesn't have to be connected to the identical concepts, only corresponding coreferent ones in the same context.

Definition: Concept Simplify. If both c1 and c2 are coreferent concepts in the same context[1] and type(c1) $\leq$ type(c2), then concept c2 may be deleted and all relation arcs and coreferent links that were connected to c2 are connected to c1.

This works for the same reasons it works in relation simplify. The function ϕ maps concepts c1 and c2 to monadic predicates corresponding to their type labels. Since they are coreferent, these predicates will have the same argument and, since type(c1) $\leq$ type(c2), the more general predicate is implied and can be deleted.

Definition: Concept Copy. If type(c1) $\leq$ type(c2), any concept c1 may be split into c1 and a coreferent concept c2 in the same context with relation arcs and coreferent links that were connected to c1 connected to either c1 or c2. This is allowed because the conformance operator :: is transitive; that is, if referent x conforms to type c1, written c1::x, then x conforms to any supertype of c1, i.e. c2::x.

Next we combine these primitive operations to define the simplify and copy formation rules as a sequence of the corresponding primitive operations.

Definition: Simplify. The composition of a sequence of relation simplify and/or concept simplify operations on the same graph.

Definition: Copy. The composition of a sequence of relation copy and/or concept copy operations on the same graph.

Simplify and Copy are inverses; the composition sequence is just reversed. They are equivalence rules because they are really only syntactic changes to the graph. They do not change the meaning of the beginning graph, the final graph, or any graph in between because ϕ maps all to equivalent logical expressions.

2.2 Specialization Rules

There are two kinds of specialization, restricting the type or referent, or making sets of concepts coreferent. Restriction makes the type or referent of concepts more specialized. Again, the full restrict rule is defined in terms of a more primitive one.

Definition: Concept Restrict. In general, a concept may be restricted by 1) replacing its type with a subtype, 2) replacing a blank or generic referent with a name, variable, or individual referent, 3) replacing a name with a variable or individual referent, 4) replacing a variable with an individual referent, or 5) adding another name, variable, or individual referent.[2]

Definition: Restrict. A sequence of one or more concept restrict operations on the same graph.

Where restrict operates on the referents of a graph, join operates on the lines of identity structure of the graph. To join two concept nodes is to make them coreferent. They do not have to be physically merged, that's a syntactic operation, a

---

[1] They could be coreferent either because they have the same referent or because they have a coreferent link.

[2] The section on Adding Contexts goes into more detail.

concept simplify. The inverse of join (detach) is not to split two concepts leaving them coreferent (that's a concept copy), it is to sever the coreferent link. Consequently, the real effect of the join and detach operations are to make or unmake coreference links. The join formation rule is defined in terms of the more primitive concept join operation.

Definition: Concept Join. If two concepts in the same context are not coreferent, make them coreferent. If either of the two concepts were part of a larger line of identity, then all concepts in the two lines of identity become part of the same line of identity.[3][4]

Definition: Join. A sequence of one or more concept joins creating, extending, or merging one or more lines of identity in the same context.

2.3 Generalization Rules

The inverse of specialization is generalization. Consequently, the inverse of each of the specialization rules gives a generalization rule. The inverse of restrict is unrestrict, and the inverse of join is detach. The Opposite of restricting is to broaden, to make the type or referent of concepts more general.[5] Again, the full unrestrict rule is defined in terms of a more primitive one.

Definition: Concept Unrestrict. A concept may be unrestricted (broadened) by 1) replacing its type with a supertype, 2) replacing an individual referent with a variable, name, or blank (generic referent), 3) replacing a variable referent with a name or blank, or 4) replacing a name referent with a blank.

Definition: Unrestrict. A sequence of one or more concept unrestrict (broaden) operations on the same graph. Since join makes concepts coreferent, to be the inverse of join, detach must make them not be coreferent.

Definition: Concept Detach. Making a subset of a line of identity's concepts part of a new, distinct line of identity. The simplest case is to make two concepts in the same context, that are not coreferent with any other concepts, to be not coreferent.

The full detach rule, defined next, is the operation of severing possibly multiple coreferent links either by unlinking or by changing referent variables to be distinct. It's like copying whatever relations one wants, grabbing a handful of relations, pulling them to another part of the page, concept copying all concepts connected to those relations, and concept detaching all the copied concepts.

Definition: Detach. A sequence of one or more concept detach operations which partitions the corresponding lines of identity of one graph into distinct lines.

---

[3] This definition specifically does not required the types and referents to be the same. If the referents were the same, they would already be coreferent, and the join would really be a concept simplify. The real act of joining is to make them coreferent.

[4] As for types, consider the definition of a pet-cat, $TYPE\ pet - cat(pc)\ IS$ $[PET : *pc]...[CAT : *pc]$. This is the normal way to specify multiple supertypes. To do it, one has to make concepts with different type labels coreferent. All that doing so implies is that the referent of the two concepts conforms to two types. At the implementation level, all types are converted to bitcodes for fast comparison. [4].

[5] "Unrestrict" is the currently used term; however, in the future, the use of the term "broaden" might be more appropriate.

2.4 Summary of Formation Rules

Detach and Join are inverses of each other. Simplifying two already coreferent concepts does not change the semantics, the mapping to logic is the same. For a specialization to occur, the two concepts couldn't have been coreferent before the join. The old definition emphasized the physical aspect of merging the two concept nodes. However, the real act of joining was to make them concepts of the same referent.

The CG formation rules can be simplified and clarified by considering simplify and copy to be forming syntactic variants, or equivalence rules. Join and detach are inverses that make and break coreference links, a form of specialization and generalization, respectively. And restrict and unrestrict (broaden) are inverses that specialize and generalize the types and referents of concepts.

3 Lattice Terminology

The CG hierarchy is stored in the form of a lattice based on the formation rules described above. The basic algorithms depend on the mathematical properties of lattices. Techniques for improving algorithm efficiency embed this lattice in a bit encoded lattice where operations are much more efficient [4]. In this section we give a few definitions and apply them to the CG lattice. The sections following this one, on adding contexts and negation, extend the use of the lattice structure.

A partially ordered set or <u>poset</u> $\langle P, \geq \rangle$ is a set P and a partial ordering relation $\geq$ over P. The least upper bound, LUB, of elements x and $y \in P$ is written as LUB(x,y). The greatest lower bound, GLB, of elements x and $y \in P$ is written as GLB(x,y). The LUB and GLB of subset $S \subseteq P$ is written as LUB(S) and GLB(S).

Note that the LUB (GLB) may not exist for two reasons: there are no common bounds, or there is not a unique least (greatest) bound. However, when they both exist we have a <u>lattice</u>, i.e. a non-empty poset $\langle P, \geq \rangle$ where LUB(x,y) and GLB(x,y) exist $\forall x,y \in P$. A <u>complete lattice</u> $L\langle P, \geq \rangle$ is a lattice where LUB(S) and GLB(S) exist $\forall S \subseteq P$. In particular, for $L\langle P, \geq \rangle$ to be complete it must have both a unique <u>top</u> element $\top \stackrel{d}{=}$ LUB(P) and unique <u>bottom</u> element $\bot \stackrel{d}{=}$ GLB(P).

The complete CG generalization lattice is based on the subsumption relation $\geq$ over the set of all graphs; that is $L\langle CG, \geq \rangle$. Graph $g1 \in CG$ subsumes graph $g2 \in CG$, $g1 \geq g2$, if there is a mapping $\pi: g1 \rightarrow g2$, where $\pi g1$ is a subgraph of g2.[6]

If $g1 \geq g2$, g1 is said to be more general that g2 and, conversely, g2 is said to be more specific than g1. The top, $\top$, is the most general graph, i.e. $\top \geq g \ \forall g \in CG$. Similarly, the bottom, $\bot$, is the most specific graph, i.e. $\forall g \in CG \ g \geq \bot$.

An ancestor of graph g is any graph that is more general than g. Thus, $\top$ is an ancestor of all graphs. A descendent of graph g is any graph that is more specific than g. Thus, $\bot$ is a descendent of all graphs.

A parent of $g2 \in CG$ is any ancestor $g1 \in CG$ which is an immediate predecessor of g2, i.e. if $g1 = $ parent(g2), then there doesn't exist a $g \in CG$ where $g \neq g1$ and

---

[6] As used here "subgraph" includes types being subtypes and the possibility of folding, i.e. two or more concepts in the more general graph mapping to the same concept in the specialization graph and relations being simplified. Also, π is not necessarily unique, $[\top] \geq g$ could map to any of g's concepts. Note that for any graph $g \in CG$, $g \geq g$.

$g{\neq}g2$ such that $g1{\geq}g{\geq}g2$. The set of immediate predecessors, IP, of $g2$ is the set of all parents of $g2$, i.e. $IP(g2) = parents(g2)$, and $g1{\geq}g2 \; \forall g1{\in}IP$. A child is defined similarly. The set of immediate successors, IS, of $g2$ is the set of all children of $g2$, i.e. $IS(g2) = children(g2)$, and $g2{\geq}g3 \; \forall g3{\in}IS$.

4 Adding Contexts and Complex Referents

In the black box view of a context [7] it is thought of as a special kind of concept, able to participate in CGs in the same way that any concept can. In the white box view of a context, it is thought of as a special kind of referent. This duality leads to thinking of contexts and concepts as abstraction duals [8, 9]. The idea of substitutability of the defining graph for an individual marker leads to the observation that a concept with a graph or graphs as its referent, called a context, is nothing more than a concept for which an individual has not yet been resolved, else the substitution could be made to simplify the context to a concept. In any case, it is necessary for the CG database to be able to represent contexts; that is, concepts where the referent contains graphs. More generally, it is necessary to represent concepts whose referents contain any of the many other possible combinations of names, variables, literals, graphs, contexts, and nested contexts.

The projection operator π implicitly defines the partial ordering relation $\geq$ that establishes the complete CG lattice. To add contexts to the CG lattice it is necessary to extend π to cover contexts. Extending π starts by adding concepts with various forms of referents. Possible referents are icon, index and symbol. The Concept Restrict Formation Rule maps a generic concept to one with some form of referent. For $g1{\geq}g2$ the projection operator π currently has the property that, for each concept c in $g1$, πc is a concept in $g2$ where $TYPE(c){\geq}TYPE(\pi c)$ and, if c is an individual, then $REFERENT(c)=REFERENT(\pi c)$.

The last part of this needs to be extended for all the different kinds of referents. The extension must be general enough to handle all referent variations. The new definition is: $REFERENT(c){\geq}REFERENT(\pi c)$.

The question becomes, which referents are more general that others. A generic concept of some type T is more general than one of the same type with a referent, so: $[T]{\geq}[T:name]$, $[T]{\geq}[T:{*}x]$, $[T]{\geq}[T:\#1]$, $[T]{\geq}[T:graph]$, and $[T]{\geq}[T:graph]$.

A key idea is that combinations of icons, indexes, or graphs, are more specific that any of them individually, so: $[T:name]{\geq}[T:name{*}x]$, $[T:{*}x]{\geq}[T:name{*}x]$, $[T:{*}x]{\geq}[T:{*}x\#1]$, $[T:\#1]{\geq}[T:{*}x\#1]$, $[T:name1]{\geq}[T:name1 \; name2]$, $[T:name2]{\geq}[T:name1 \; name2]$, $[T:graph1]{\geq}[T: graph1 \; graph2]$, $[T:graph2]{\geq}[T: graph1 \; graph2]$, $[graph1]{\geq}[graph1 \; graph2]$, and $[graph2]{\geq}[graph1 \; graph2]$.

To summarize, every generic concept/context has, potentially, under it in the generalization lattice a sublattice consisting of all possible combinations in any order of icons, indexes, and graphs with the bottom being, implicitly, all of them together.[7] Similarly, every graph has, potentially, under it something like the cross product [6] of these individual sublattices with a common bottom where each concept of the graph has all possible referent values.

---

[7] These concept bottoms are also context bottoms and can be use, in a bit encoded lattice, to limit the search to the context.

A <u>context</u> is a concept that has one or more graphs as its referent. Each of them is an immediate predecessor of the context. Thus, IP([T: g1 g2]) = {[T] g1 g2}. To show how contexts fit into and extend the CG lattice, we add to an example commonly used [4, 14]. It involves the 7 graphs given below. The lattice that results from inserting these 7 graphs into a CG database utilizing the original Method III algorithm is shown in Figure 1.

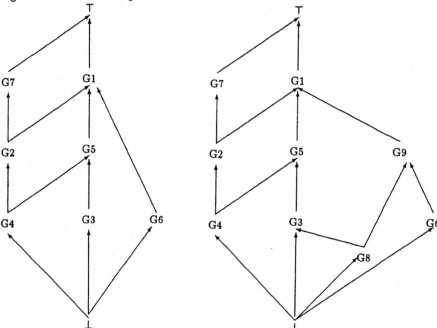

Fig. 1. Lattice Of Initial Graphs.   **Fig. 2.** Lattice With Context Graph Inserted.

G1: [PERSON]←(AGNT)←[EAT]
G2: [GIRL]←(AGNT)←[EAT] →(MANR)→[QUICKLY]
G3: [:Sue]←(AGNT)←[EAT] →(OBJ)→[PIE] →(CONT)→[APPLES]
G4: [:Sue]←(AGNT)←[EAT]-
    →(OBJ)→[PIE]
    →(MANR)→[QUICKLY]
G5: [:Sue]←(AGNT)←[EAT]
G6: [:Dan]←(AGNT)←[EAT]→(OBJ)→[PIE]-
    →(CONT)→[APPLES]
    →(POSS)→[SUE]
G7: [EAT]→(MANR)→[QUICKLY]
We add to this set a graph containing a context. G8: [:Daniel]→(BELIEVE)→[
    [:Sue]←(AGNT)←[EAT] →(OBJ)→[PIE] →(CONT)→[APPLES]
    [:Dan]←(AGNT)←[EAT] →(OBJ)→[PIE] →(CONT)→[APPLES]]
When G8 is added to the previous CG database with the new algorithm, it results in the creation of G9, for Dan eating apple pie, and the lattice shown in Figure 2. Another key idea is how to handle <u>coreference</u>. John Sowa [21] handled the problem of coreference by propagating aliases down each line of identity from all dominant

concepts. For example, [T1:*x]...[T2:*y] becomes [T1:*x *y] [T2:*y *x].[8]

This causes all dominated concepts to acquire multiple indexes, which places them lower in the generalization lattice than the corresponding concept without the coreferent link. That is, T1:*x]≥[T1:*x]...[T2:*y] and [T2:*y]≥[T:*x]...[T:*y].

The impact on the query/insert algorithm, of handling complex referents, is minimal because all these complex referent checks are part of the isomorphism test. In Method V, once bindings are found, they are propagated. [18]

5 Adding Negation

The best way to think about how CGs are stored, when negative contexts are included, is as a lattice over the syntactic structure of graphs, not as all possible semantic implications. The immediate predecessor,IP, and immediate successor, IS, links remain as described in the Lattice Terminology section. Since a node's IS links are part of its children's IP links, it is only necessary to consider either the IP or IS links. Below we concentrate on extending the IP links to cover negative contexts.

5.1 Representing Negative Assertions

To implement negation two bits, PA and NA, are added to each node to indicate whether the graph represented by the node has been asserted positively, negatively, both, or not at all. The interpretation of PA and NA is 00 - No Assertion, 10 - Positive Assertion, 01 - Negative Assertion, and 11 - Both Asserted (a logical inconsistency).[9] Graphically, these are shown as nothing, +, -, & +- in Figure 3.

It is natural in this system to represent arcs from A to NOT B. To extend the IP links to cover negative contexts, an IP arc from a node to a parent is labeled + when the positive version of the parent projects into the node, and labeled - when the negative version of the parent projects into the node. To represent the assertion ¬A, where A is some graph, graph A is added to the hierarchy and its negative assertion bit set. To represent a term ¬A, that is part of some other graph B, the IP link from B to A is labeled negatively. In Figure 3, nodes G10, G13 & G14 are negative assertions and nodes G3, G4, G11 & G12 have links as negative terms.

The full Boolean logic lattice has nodes for all possible disjunctions and conjunctions. Since the lattice is symmetric, each negative conjunct has a corresponding complementary disjunct. Thus, if each conjunctive node can be optionally negated, only half of the lattice is needed to represent all possible logical expressions.

Our proposal is to represent each graph in conjunctive normal form (CNF); that is, as conjuncts of the graph's IPs which may or may not be negated in the conjunction. Disjunctions are represented by allowing negative graphs.

Since one graph's IP link is another graph's IS link, we can think of labeled, bi-directional links between a parent and child; that is, between a graph (the child) and a part of its conjunct (the parent).

---

[8] The original syntax used = signs as in [T:*x=*y].

[9] In some logics a contradiction does not undermine the integrity of the entire database and there are other more "avante gard" solutions, such as not letting the contradiction be used in proofs in which it is not relevant. [10]

If a link is labeled +, the positive version of the parent is an IP of the child. This implies that the child has, as part of its conjunction, the positive version of the parent into which the parent projects to satisfy the subsumption relation.

If a link is labeled - , the negative version of the parent is an IP of the child. This implies that the child has, as part of its conjunction, the negative version of the parent into which the negative version of the parent projects to satisfy the subsumption relation.

5.2 Negation Examples

The following example of disjunction involves negated contexts containing graphs. We use DeMorgan's Law to convert AvB to ¬(¬A&¬B). For example, the disjunction G10: [:DISJ{Sue,Dan}]←(AGNT)← [EAT]→(OBJ)→[PIE] becomes
¬[¬[G11] ¬[G12]] where G11: [:Sue]←(AGNT)← [EAT]→(OBJ)→[PIE] and
G12: [:Dan]←(AGNT)← [EAT]→(OBJ)→[PIE].

When G10 is added, it requires the insertion of G11 and G12 before G10 can be inserted as shown in the Figure 3. Note the - signs beside the links from G10 to G11 & G12 and the assertion indicators before each graph's label.

Implication also involves negation because the graph A⇒B is represented as ¬[A ¬[B]]. For example, the graphs G13: ¬[G11 ¬[G4]], for "If Sue eats pie, Sue eats pie quickly.", and G14: ¬[G5 ¬[G3]], for "If Sue eats, Sue eats apple pie." are also shown in Figure 3.

5.3 Distinguishing Syntactic and Semantic Lattices

At this point we would like to make a distinction between "syntactic" and "semantic" lattices over a set of CGs. The syntactic lattice is the structure implied due to the graphs, formation rules, and Peirce inference rules alone. It is formed from those links that can be inferred directly from two graphs, the formation rules, and Peirce's inference rules without regard to any other graphs in the lattice.

The syntactic lattice is true (and may be assumed such) in all domains because the links do not actually depend on domain knowledge. Which portion of the lattice is actually stored will depend on efficiency considerations.[10]

The semantic lattice adds inferences that require three or more graphs to establish and are not implied by the transitivity of the syntactic lattice. For example, given A, B and A⇒B as graphs, the relationship between A and B is only known given the third graph, A⇒B. In other words, suppose neither A nor B have been asserted but A⇒B has been asserted. Then both A and B will have assertion bits 00, neither positive or negative assertion. On asserting A later, one can think of B as a ramification and mark it as being inferred without adding any new links.

The syntactic lattice identifies links based on structure, formation, and Peirce rules.[11] These links are true regardless of the domain. They depend only on graph

---

[10] An assumption is that ¬T = ⊥, but this is not shown. Also, by representing both nodes, one can have different encodings for them to improve computational efficiency.

[11] A subsequent section defines Peirce's inference rules in terms of the formation rules which define the projection operator and, consequently, the partial ordering relation. Thus, Peirce's inference rules are contained within the definition of projection and fit into the syntactic lattice.

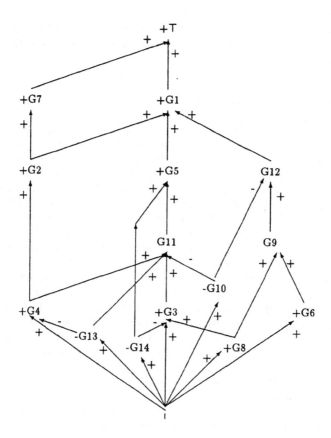

Fig. 2. Lattice With Negative Graphs Inserted.

structures and not on what has been asserted and not asserted.[12] In the syntactic lattice a link means "this link holds in all databases".

Where the formation rules are foundational for the syntactic lattice, the inference rules are foundational for the semantic lattice because they define the relationship that establishes the semantic lattice. The semantic lattice models implications among graphs stored in the database. These implications are identified by using the truth values that have been asserted for the graphs, the implications directly stated (e.g. in the graph for $A \Rightarrow B$), and Peirce's inference rules. Here an inference means "given what has been asserted (axioms), this graph holds (is a theorem)". The most efficient computation of the semantic lattice is an important research question to be addressed in the future.

The two lattices may be stored together by adding two inference bits, PI & NI, to each node indicating whether the graph represented by the node has been inferred

---

[12] They also depend on the equivalence classes since only one representative from each equivalence class need be stored.

positively, negatively, both, or not at all. The interpretation of PI and NI is 00 - Not Inferred, 10 - Positive Inference, 01 - Negative Inference, and 11 - Both Inferred (a logical inconsistency).[13] [14]

6 Augmented Algorithms

In this section we describe how existing algorithms for operating on the generalization hierarchy may be augmented to handle contexts, negation, and assertion.

The algorithms described here search the CG lattice for where a new graph Q, called the query graph, fits into the lattice. The algorithm is divided into five phases. Phase 0 deals with Q's negation and assertion. Phase I finds the immediate predecessors of Q, IP(Q). Phase II finds the immediate successor of Q, IS(Q). Phase III inserts Q in the CG lattice. And Phase IV propagates implications. The consequence of finding IP(Q) and IS(Q) is that each IS(Q) $\Rightarrow$ Q $\Rightarrow$ each IP(Q).

The algorithm considers graphs in topological sort order [4] by using a combination of a FIFO queue called C and the addition of a depth to each graph node. The depth is one greater than the largest depth of its immediate predecessors, IPs. The function pop(C) returns the next graph from C. If IS(X) is the set of immediate successors of graph X, then IS'(X) returns the members of IS(X) which have depth one greater than X. The procedure push(C, SET) pushes set SET onto the FIFO queue C. The augmentation is to the original Method III algorithm [4, 14] to add Phase 0, the greatest lower bound (GLB) test, negation checking, and Phase IV.

Phase 0: Handle negation and assertions.

1. Set Q = Q minus pairs of leading ¬'s.
2. IF Q is being asserted AND Q = ¬Q' THEN Q = Q' AND complement the assertion.
3. IF Q asserted THEN PA = 1 OR IF ¬Q asserted THEN NA = 1.

Phase I: Find IP(Q), the immediate predecessors of Q.

4. IF Q = $\perp$, THEN RETURN IP($\perp$).
5. INITIALIZE C := $\top$ AND S := {}.
6. WHILE not empty(C) DO X := pop(C) AND
   IF All parents of X are predecessors of Q AND
     GLB(S $\cup$ {X}) $\neq$ $\perp$ AND X is a predecessor of Q (isomorphism test)
   THEN mark X as a predecessor of Q;[15] S := S $\cup$ {X} - IP(X);[16] push(C,IS'(X)).
7. RETURN S with negative links marked.

Phase II: Find IS(Q).

8. INITIALIZE S := {} AND C := {} AND Y := some element of IP(Q).
9. I := intersection of the successor sets of each element of IP(Q) except Y.
10. FOR each successor X of Y in topological order DO
    IF X is in I and X is a successor of Q (isomorphism test) THEN S := S $\cup$ {X}.
    Eliminate successors of X from the rest of the FOR loop.
11. RETURN S with negative links marked.

---

[13] It is now possible to have other inconsistencies, like Positive Assertion/Negative Inference
[14] A coding scheme that combines the Assertion and Inference bits is possible.
[15] Based on the isomorphism test, this may be a negative link mark.
[16] Notationally, - stands for set subtraction.

Phase III: Update IP & IS Nodes (possibly inferring Q).[17]

12. For each x in IP(Q) DO IS(x) := IS(x) ∪ {Q} - IS(Q) AND
    IF x negatively asserted or inferred, THEN negatively infer Q.
13. For each x in IS(Q) DO IP(x) := IP(x) ∪ {Q} - IP(Q) AND
    IF x asserted or inferred, THEN infer Q.

Phase IV: Semantic Update.

14. IF Q asserted THEN infer all the IPs of Q[18] ANDIF
    Q is A⇒B[19] AND A is true OR Q⇒B is true THEN infer B.
15. IF ¬Q asserted THEN negatively infer all the ISs of Q ANDIF
    B⇒Q is true THEN negatively infer B.
16. FOR EACH newly inferred node DO Phase IV as if the node had just been asserted.

The algorithm above uses only the syntactic links stored in the lattice. After insertion is completed and the links fixed, semantic update, Phase IV, ensues. It is based on both the new syntactic links formed and the implications of the graph inserted. Phase IV will change inferred bits to be on for graphs that can now be inferred which were not known to be inferred previously. It should be pointed out that Phase IV does not absolutely complete the semantic lattice, but represents the transitive closure of the implications that have been identified. Many inferences remain due to Peirce's inference rules.

The algorithm also needs to be modified slightly for nested contexts. For them the graphs are inserted "inside out" with the deepest nested graphs inserted first, and concept variables obtained, with outer levels referring to inner levels via these variables. When comparing two contexts to each other, the concept variables are compared identically to the way type labels are compared in the type hierarchy.

Just as bitcodes may be given to members of the type hierarchy to allow comparison to take place in constant time, bitcodes may be assigned to members of the graph hierarchy itself. If a graph structure is repeated multiple times in a nested context graph, and this correspondence is known (possibly due to the interface), the query should be organized to ask this graph only once.

7 Peirce Inference Rules

So far we have extended the CG generalization hierarchy/lattice and algorithms to include contexts, other complex referents, and negation. It is now possible to use the CG lattice to perform inference. Peirce defined a graphical system of logic and inference rules upon which CG theory is based. Peirce's inference rules provide a basis for defining a corresponding simple set of inference rules for the CG lattice.

The formation rules defined in Section 2 applied to graphs in the same context. These are augmented by cross context formation rules which consist of join and detach across context domination boundaries.[20]

---

[17] A graph to be inserted, that is not known to be true, will be true if it is more general than a graph that is known to be true or, if it can be derived from a sequence of inferences based on one or more asserted graphs.

[18] Because we are storing CGs in CNF, if a CG is true, then all of its IPs must be true.

[19] By "Q is A⇒B" we mean Q is an implication and, if true, can be used to do inferences.

[20] "Domination" is a key word here since contexts are strictly nested and lines of identity must follow the same nesting.

Whether these cross context formation rules are generalizations or not depends on the even or oddness of the dominant concept(s). For example, it's not enough to detach from even context to even context. That's because it's the even or oddness of the context containing the dominant concept of that line of identity that counts. The basic idea of the Peirce rules is to allow equivalence changes in any context, generalization of even contexts, and specialization of odd contexts. The problem is that, depending on the even or oddness of the contexts involved, the generalization formation rules are not always generalizations. In other words, it is not possible for each formation rule to always fall into only one category, equivalence, generalization, or specialization, when multiply nested, possibly negated contexts are involved.

Every structure in the lattice is of even or odd parity, with its negation taken as the opposite parity. As one traverses to deeper levels of nesting, the parity (and hence which inferences are possible) changes with each negation step. Higher level nested structures may be able to exploit inference links previously formed between lower structures without having to rediscover these inferences.

With the above in mind, here's a restatement/merge of the Peirce rules in John Sowa's 1984 book [21] and in the draft of his new book [23] using our formation rules.

Definition: Erasure. An evenly enclosed context may be generalized by applying the following formation rules to one or more of the graphs contained in the context: erasure, unrestrict, or detach from evenly enclosed dominant concepts.

Definition: Insertion. An oddly enclosed context may be specialized by applying the following formation rules to one or more of the graphs contained in the context: insertion, restrict, or join to oddly enclosed dominant concepts.

Definition: Iteration. A copy of all or part of any graph may be inserted into the same or a dominated context. Note that this is a coreferent copy.[21]

Definition: Deiteration. Any graph or part of a graph which could have been the result of an iteration, may be simplified.

Definition: Double Negation. Two nested negated contexts with nothing between them may be drawn around or removed from around any graph, set of graphs, or context.

8 Conclusions and Summary

The methods and implementation model described in this paper provide the ground work for a full first-order logic theorem prover based on conceptual graphs. The ground work laid includes the handling of negation, nested contexts, complex referents, and consistency checking. While we have clearly defined a distinction between syntactic and semantic implication, with the former corresponding to Sowa's conception of a generalization hierarchy, the most efficient algorithm for discovery of the semantic links remains an important topic for future work. Along similar lines, while we have a mechanism for discovering when syntactically-derived contradictions have

---

[21] The following is a different wording that does not have quite the same meaning and, consequently, needs to be evaluated: "The result of a copy formation rule may go into the same or a dominated context." (Note that this may not be true since the copy formation rule is not strictly a copy.

occurred, we are not yet guaranteed of finding any inconsistencies that also may require semantic links other then a brute force generation of all implied graphs of a given size. A definition of Peirce's inference rules based directly on the implemented data structure also remains to be completed.

The syntactic lattice provides a mechanism for compilation of powerful macros and proof structures for use in all domains so that a theorem prover can improve with experience. In Lisp-based EG theorem provers developed for a class project, we negate queries and attempt to validate them by deriving an empty context on the sheet. This process is accelerated over time by storing (remembering) graphs from which we have already derived the empty context.

In future work, we also hope to improve (as discussed above) the ability to update the semantic links and, along similar lines, to exploit better the abstractions imposed by the semantic equivalence of two graphs. The ability of the retrieval algorithms to find all syntactic implications of graphs at any contextual level provides exactly the information needed to employ higher-order inference rules. Our other work on heuristic search and machine learning [12] may also be able to improve the efficiency of the inference system.

It is our hope that the implementation model in this paper will extend the scope of discussion in the CG community to include the hard issues that we have ignored up until present. It is also hoped that the beauty and elegance of logical inference based on graphs, first envisioned by Peirce, will be more deeply appreciated and complete implementations of his theory, such as incorporating our approach into the Peirce Conceptual Graphs Workbench [5], will soon come to pass.

References

1. H. Boley. Pattern associativity and the retrieval of semantic networks. *Computers and Mathematics with Applications*, 23(6-9):601–638, 1992. Part 2 of Special Issue on Semantic Networks in Artificial Intelligence, Fritz Lehmann, editor.
2. M. Chein and M.L. Mugnier. Conceptual Graphs: Fundamental Notions In *Revue d'Intelligence Artificielle*, n14, 1992, pp365-406.
3. G. Ellis. Compiled hierarchical retrieval. In E. Way, editor, *Proceedings of Sixth Annual Workshop on Conceptual Structures*, pages 187–208, SUNY-Binghamton, 1991.
4. G. Ellis. Efficient retrieval from hierarchies of objects using lattice operations. In G. Mineau and B. Moulin, editors, *Proceedings of First International Conference on Conceptual Structures (ICCS-93)*, Montreal, 1993.
5. G. Ellis and R. Levinson. Proceedings of the Fourth International Workshop on Peirce: A Conceptual Graph Workbench. In association with the *Second International Conference on Conceptual Structures*, ICCS'94, College Park, Maryland, USA, August 19, 1994.
6. G. Ellis and F. Lehmann. Exploiting the Induced Order on Type-labeled Graphs for Fast Knowledge Retrieval. In W.M. Tepfenhart, J.P. Dick and J.F. Sowa editors *Conceptual Structures: Current Practices*, the proceedings of the Second International Conference on Conceptual Structures, ICCS'94, College Park, Maryland, USA, Springer-Verlag, August, 1994.
7. J.W. Esch. Contexts as White Box Concepts. In *Supplemental Proceedings of First International Conference on Conceptual Structures: Theory and Application*, Quebec City, Canada, August 4-7, 1993, pages 17-29.

8. J.W. Esch. Contexts and Concepts: Abstraction Duals. In em Proceedings of Second International Conference on Conceptual Structures, ICCS'94, College Park, Maryland, USA, August 1994, pages 175-184.

9. J.W. Esch. Contexts, Canons and Coreferent Types. In em Proceedings of Second International Conference on Conceptual Structures, ICCS'94, College Park, Maryland, USA, August 1994, pages 185-195.

10. B. Goertzel. Chaotic logic : language, thought, and reality from the perspective of complex systems science In *IFSR International Series on Systems Science and Engineering* v.9., Plenum Press, New York, 1994.

11. J.A. Goguen and F.J. Varela. Systems and Distinctions: Duality and Complementarity. In *International Journal of Systems*, 5(1):31-43, 1979.

12. J. Gould and R. Levinson. Experience-based adaptive search. In *Machine Learning:A Multi-Strategy Approach*, volume 4. Morgan Kauffman, 1992.

13. R. Levinson. A self-organizing retrieval system for graphs. In *AAAI-84*, pages 203–206. Morgan Kaufman, 1984.

14. R. Levinson. Pattern associativity and the retrieval of semantic networks. *Computers and Mathematics with Applications*, 23(6-9):573–600, 1992. Part 2 of Special Issue on Semantic Networks in Artificial Intelligence, Fritz Lehmann, editor. Also reprinted on pages 573–600 of the book, Semantic Networks in Artificial Intelligence, Fritz Lehmann, editor, Pergammon Press, 1992.

15. R. Levinson and G. Ellis. Multilevel hierarchical retrieval. *Knowledge-Based Systems*, 1992.

16. R. Levinson and Karplus K. Graph-isomorphism and experience-based planning. In D. Subramaniam, editor, *Proceedings of Workshop on Knowledge Compilation and Speed-Up Learning*, Amherst, MA., June 1993.

17. R.A. Levinson. Exploiting the physics of state-space search. In *Proceedings of AAAI Symposium on Games:Planning and Learning*, pages 157–165. AAAI Press, 1993.

18. R. Levinson. UDS: A Universal Data Structure. In *Proceedings of Second International Conference on Conceptual Structures*, ICCS'94, College Park, Maryland, USA, August 1994, pages 230-250.

19. D.D. Roberts. The existential graphs. In *Semantic Networks in Artificial Intelligence*, pages 639–664. Roberts, 1992.

20. E. Sciore. A complete axiomatization for join dependencies. *JACM*, 29(2):373–393, April 1982.

21. J.F. Sowa. *Conceptual Structures*. Addison-Wesley, 1984.

22. J.F. Sowa. Conceptual Graphs Summary. In *Conceptual Structures: Current Research and Practice*, T. Nagle et. al. Ed., Ellis Horwood, 1992.

23. J.F. Sowa. Knowledge Representation: Logical, Philosophical, and Computational Foundations. Book in preparation. To be published by PWS Publishing Company, Boston, Massachusetts.

24. M. Willems. Generalization of Conceptual Graphs. In *Proceedings of Sixth Annual Workshop on Conceptual Graphs*, Binghamton, New York, State Univ. of New York, 1991.

A Linear Descriptor for Conceptual Graphs and a Class for Polynomial Isomorphism Test

O. Cogis   O. Guinaldo

L.I.R.M.M. *(U.M.R. 9928 Université Montpellier II / C.N.R.S.)*
161 rue Ada, 34392 Montpellier cedex 5 - France
tel.: (33) 67 41 85 79   fax: (33) 67 41 85 00
e-mail: ocogis@lirmm.fr guinaldo@lirmm.fr

Abstract. The isomorphism problem is not known to be NP-complete nor polynomial. Yet it is crucial when maintaining large conceptual graphs databases. Taking advantage of conceptual graphs specificities, whenever, by means of structural functions, a linear order of the conceptual nodes of a conceptual graph G can be computed as invariant under automorphism, a descriptor is assigned to G in such a way that any other conceptual graph isomorphic to G has the same descriptor and conversely. The class of conceptual graphs for which the linear ordering of the conceptual nodes succeeds is compared to other relevant classes, namely those of locally injective, C-rigid and irredundant conceptual graphs. Locally injective conceptual graphs are proved to be irredundant, thus linearly ordered by the specialization relation.

Keywords: Isomorphism problem, structural functions, descriptors, classes of conceptual graphs.

1   Introduction

Conceptual graphs have been subject to a large amount of research since they have been introduced by John Sowa in 1984 [Sow84]. See [Way94] for a survey.

These researches address a wide variety of topics, ranging from the suitability of conceptual graphs for knowledge representation in some·domains to the power, as well as efficiency, of their treatment. Our work deals with the later.

In order to structure any set of conceptual graphs, the specialization hierarchy is fundamental and is based on the notion of projection [Sow84], [CM92], to which isomorphism is a special case.

Applications that use conceptual graphs as a knowledge representation formalism are then faced with a crucial problem: projections and isomorphisms are the basic tools used to query and maintain knowledge bases expressed in terms of conceptual graphs but the "projection problem" is known to be NP-complete while the "isomorphism problem" is presently unclassified.

Taking this into account, one can, possibly in a cooperative perspective:

1. use backtraking techniques [Gre82];
2. take advantage of specificities of the application domain, with the hope that not too many too regular graphs will be encountered (it is the case when labels on nodes are reasonably different one from another) [MLL91];
3. restrict the problem to conceptual graphs classes for which the problem turns to be polynomial [MC92], [CM92], [LB94];
4. use efficient pre-processing on conceptual graphs [LE91], [Lev92], [Lev94], [EL94];
5. organise the database according to the specialization hierarchy [Ell91], [Lev92], [Ell93], [Lev94], and
   (a) traverse the hierarchy so that predictable comparisons are avoided [LE91], [Ell91], [Lev92], [Ell93], [Lev94], (see [Woo91], [BFH$^+$92] for a general discussion);
   (b) restricting the problem to some finite set of graphs, for each graph define a unique code significant to the hierarchy and use a hash function [EL94].

In [EL94], a pioneering algebraic approach on treating altogether the graph structure and the labels set structure, each graph is given a "name", that is a code which is unique and can respond for any specialization test. This is done when first given a finite set of graphs and a finite structured set of labels.

Focussing our attention on the "isomorphism problem", conceptual graphs are essentially graphs with labelled nodes.

We link our work to several of the points just mentioned by associating to each conceptual graph a code depending on the graph structure and on the nodes labelling (not seen as structured). We rather call it a descriptor, for the uniqueness property of the code depends entirely on the finite set of conceptual graphs we are dealing with (thus it is heavily linked to the application which produced this set).

This descriptor, which is polynomialy computed, takes its place among the various tools appeared on the "isomorphism problem" field long ago [Ung64] (see [CK80], [RC77] for overviews): define from a graph some item invariant under isomorphism; this item should be the most specific you can to this graph, while remaining polynomialy computed. Presently, no such polynomialy computable item is known on the class of graphs (otherwise the "isomorphism problem" would be known to be tractable).

Basically, most methods start using neighbourhoods, degrees and refinement procedures, as in [Lev92], [EL94]. We also do so, but we decompose the problem:

1. we define a standard total preorder (a reflexive and transitive binary relation such that, given any two elements, the relation holds at least in one way) on the concept nodes of a conceptual graph G, which we call $\Omega(G)$;
2. if $\Omega(G)$ turns out to be a linear order (an antisymetric total preorder), it is used to define a descriptor for the conceptual graph G in such a way that any other graph isomorphic to G has the same descriptor and conversely.

Though this is a heuristic on the class of conceptual graphs, it is worth pointing out that:

1. as for many other methods testing the conceptual graphs isomorphism, there are domains for which, when restricted to the set of their associated conceptual graphs, our heuristic is a polynomial algorithm;

2. provided that it stays invariant under isomorphism, Ω can be change into any other total preordering function on the concept nodes, with our result still holding: the descriptor is unique up to isomorphism;

3. in case $\Omega(G)$ is not antisymetric, it can be used as the result of a first step in some other heuristic for isomorphism testing (it holds valuable information about the automorphism partition).

Function Ω introduces the class of so-called Ω-sorted conceptual graphs, a new class with a tractable "isomorphism problem".

Calling C-rigid the conceptual graphs G such that any automorphism of G restricted to the concept nodes is the identity mapping, it is not surprising that Ω-sorted conceptual graphs must be C-rigid.

On the other hand, Ω-sorted conceptual graphs can be, but need not be, irredundant keeping in mind that irredundant conceptual graphs, introduced in [CM92], play a central role in the specialization hierarchy (see the section 2).

This situation is converse to the one of another class of conceptual graphs with a polynomial isomorphism algorithm, namely the class of locally injective conceptual graphs introduced in [LB94] (conceptual graphs such that no two edges from different relation nodes with the same label, say l, bearing themselves the same label, say l', can meet on the same concept node). It is easy to produce a locally injective conceptual graph not C-rigid, while we prove that locally injective conceptual graphs must be irredundant. As a by-product, locally injective conceptual graphs turn out to be ordered by the specialization relation.

In section 2, we recall formal definitions useful to the paper. Section 3 is devoted to defining functional preorderings and intrinsic sortings on which are founded the heuristic and the class of Ω-sorted conceptual graphs. Comparisons, w.r.t. to inclusion, between different classes of conceptual graphs are exposed in section 4. Finally, section 5 briefly addresses some implementation points.

2 Conceptual Graphs, Projections, Isomorphisms and Irredondance

We restrict ourselves to simple conceptual graphs in the sense of [CM92] from which we only import what is needed here:

Definition 1 *A conceptual graph is a connected bipartite labelled graph $G = (R, C, E, Lab)$. R and C denote the set of R-vertices (relation nodes) and C-vertices (conceptual nodes) respectively.*

Each vertex $x \in R \cup C$ has a label $Lab(x)$. The labels of C-vertices are ordered by a lattice $(T_C, \leq_{T_C})$. The labels of R-vertices impose the degree on the R-vertex they label.

E is the set of edges. The set of the edges adjacent to each R-vertex r is numbered from 1 to $degree(r)$, and $G_i(r)$ denotes the ith C-vertex adjacent to r.

A sub-conceptual-graph *of a conceptual graph G is an induced subgraph of G which is itself a conceptual graph.*

Definition 2 *[CM92] A projection from a conceptual graph $G = (R, C, E, Lab)$ to a conceptual graph $G' = (R', C', E', Lab')$ is an ordered pair $h = (f, g)$ of mappings, f from R to R', g from C to C', such that:*

(i)   *for all r in R and for all i, $1 \leq i \leq degree(r)$, $g(G_i(r)) = G'_i(f(r))$*
(ii) *$Lab'(f(r)) = Lab(r)$ for all r in R*
(iii) *$Lab'(g(c)) \leq_{T_C} Lab(c)$ for all c in C.*

By extension, we write h in place of f and g when needed and we write $G' = h(G)$.

Definition 3 *[CM92] An isomorphism from a conceptual graph G to a conceptual graph G' is a projection $h = (f, g)$ where f and g are bijections, with point (iii) of projection definition being specialized as:*

(iii)' *$Lab'(g(c)) = Lab(c)$ for all c in C.*

We then write $G \cong G'$ or $G \stackrel{h}{\cong} G'$. When $G' = G$, h is called an automorphism.

In some works, projections are meant to be injective, even when not mentioned. But mappings f and g in the projection definition need not be injective.

This implies, as it has been shown in [CM92], that the reflexo-transitive closure of the relation "can be projected into", which is equivalent to the specialization relation, is a preorder, but not an order, on the class of conceptual graphs. Each equivalence class for this preorder possesses a conceptual graph, unique up to isomorphism, with the least possible number of vertices among the conceptual graphs of the class. It is the irredundant conceptual graph of the equivalence class and can be defined as follows:

Definition 4 *[CM92] A conceptual graph G is* irredundant *when there exists no projection from G to one of its strict sub-conceptual-graph. Otherwise, it is called* redundant.

While it is proved that to decide whether or not G is irredundant is a NP-complete problem [CM92], projections from G to its irredundant equivalent is still of interest and deal with projections of special kind defined in section 4.

3   Functional Preorderings and Intrinsic Sortings of C-vertices Inducing a Conceptual Graph Descriptor

3.1   A Linear Descriptor for Conceptual Graphs

Given a conceptual graph $G=(R, C, E, Lab)$, we assume that $|R|=p$ and $|C|=q$.

Let L_R and L_C be linear orders defined on R and C respectively, and let denote by num_R and num_C the numberings induced by these linear orders on R and C respectively (that is the least element of R according to L_R, say r, has $num_R(r) = 1$, and so on).

Definition 5 *Given the conceptual graph* $G = (R, C, E, Lab)$ *and linear orders* L_C *and* L_R, *let the descriptor of* G *according to* L_C *and* L_R *be the sequence:*

$$\Delta(G, L_C, L_R) = a_1, ... a_q, b_1, ..., b_p$$

where : $a_i = Lab(num_C^{-1}(i))$
$\qquad b_i = e_i n_1^i ... n_{d_i}^i$ $\qquad$ *with* $\qquad e_i = Lab(num_R^{-1}(i))$
$\qquad\qquad\qquad\qquad\qquad\qquad\qquad\qquad\qquad n_j^i = num_C(G_j(num_R^{-1}(i)))$

In other words, the sequence $\Delta(G, L_C, L_R)$ is made from the labels of the C-vertices of G, sorted according to L_C, followed by the sequence of descriptors for the R-vertices, sorted according to L_R. Each of the R-vertices descriptors is made of the label of the R-vertex itself, followed by the numberings, according to num_C, of the neighbours of the R-vertex, the first neighbour coming first and so forth. See the example in Figure 1.

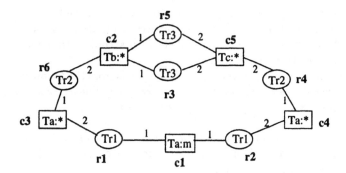

$L_C : c1 < c2 < c3 < c4 < c5$ $\quad$ and $\quad$ $L_R : r1 < r2 < r3 < r4 < r5 < r6$

$\Delta(G, L_C, L_R)$ = ([Ta:m][Tb:*][Ta:*][Ta:*][Tc:*]
$\qquad\qquad\qquad$ (Tr1,1,3)(Tr1,1,4)(Tr3,2,5)(Tr2,4,5)(Tr3,2,5)(Tr2,3,2))

Fig. 1. Descriptor of G

Given L_C and L_R, the conceptual graph descriptor can obviously be computed in polynomial time (we will be more precise in section 3.5), and keeps track of the conceptual graph structure in the following sense:

Theorem 1 *Given two conceptual graphs* G *and* G', *and given linear orders* $L_R, L_C, L_{R'}$ *and* $L_{C'}$: *if* $\Delta(G, L_C, L_R) = \Delta(G', L_{C'}, L_{R'})$, *then* $G \cong G'$.

proof : let $f(r) = num_{R'}^{-1}(num_R(r))$ and $g(c) = num_{C'}^{-1}(num_C(c))$ for all $r \in R$ and $c \in C$.
1) clearly f and g are bijections from R to R' and C to C' respectively;
2) $a_i' = Lab(num_{C'}^{-1}(i)) = Lab(num_{C'}^{-1}(num_C(num_C^{-1}(i)))) = Lab(g(num_C^{-1}(i)))$
while $a_i = Lab(num_C^{-1}(i))$; from $a_i = a_i'$ for all i, $1 \le i \le q$, we deduce that $Lab(g(c)) = Lab(c)$ for all $c \in C$;

3) in the same way, from $e_i = e'_i$ for all i, $1 \leq i \leq p$, we deduce that $Lab(f(r)) = Lab(r)$ for all $r \in R$;

4) $n''^i_j = num_{C'}(G'_j(num_{R'}^{-1}(i))) = num_{C'}(G'_j(num_{R'}^{-1}(num_R(num_R^{-1}(i))))) = num_{C'}(G'_j(f(num_R^{-1}(i))))$ while $n^i_j = num_C(G_j(num_R^{-1}(i))) = num_{C'}(num_{C'}^{-1}$ $(num_C(G_j(num_R^{-1}(i))))) = num_{C'}(g(G_j(num_R^{-1}(i))))$; from $n^i_j = n''^i_j$ for all i, $1 \leq i \leq p$ and all i, $1 \leq j \leq d_i$, we deduce that $G'_j(f(r)) = g(G_j(r))$. $\square$

3.2 Relational Preorders Induced by Conceptual Linear Orders

Definition 6 *Given a conceptual graph G and a linear order L_C on C, let $Q_R(G, L_C)$ denote the total preorder induced on R by the lexicographic order on the R-vertices descriptors $s(r, G, L_C)$ defined as:*

$$s(r, G, L_C) = Lab(r)m_1^r, ...m_{d(r)}^r \qquad where \qquad m_j^r = num_C(G_j(r)).$$

In other words, r and r' being two R-vertices of G, r and r' are equivalent w.r.t. $Q_R(G, L_C)$ when they are twin vertices, and, otherwise, are ordered w.r.t. $Q_R(G, L_C)$ first according to their labels, then according to the numbering (induced by L_C) of their neighbourhoods.

Our interest in the total preorder $Q_R(G, L_C)$ yields from the following proposition, where we call *reduction* of a total preorder Q any linear order L induced by Q in the sense: $x <_L y$ implies $x <_Q y$ or $x \equiv_Q y$

Proposition 1 *Given a conceptual graph G and a linear order L_C on its conceptual vertices, if L_R and L'_R are any two reductions of $Q_R(G, L_C)$, then :*

$$\Delta(G, L_C, L_R) = \Delta(G, L_C, L'_R)$$

proof: 1) because the two descriptors of G use the same L_C, it suffices to compare b_i to b'_i for $1 \leq i \leq p$;

2) $b_i = s(r, G, L_C)$ with $num_R(r) = i$, while $b'_i = s(r', G, L_C)$ with $num'_R(r') = i$; because they are numbered the same by two reductions of $Q_R(G, L_C)$, r and r' must be equivalent w.r.t. $Q_R(G, L_C)$, and therefore must have the same R-vertex descriptor; hence $b_i = b'_i$. $\square$

Definition 7 *Given a conceptual graph G and a linear order L_C on its conceptual vertices, let L_R be any reduction of $Q_R(G, L_C)$. The descriptor $\Delta(G, L_C, L_R)$, uniquely defined according to proposition 1, is called the normal descriptor of G induced by L_C and is denoted $\Delta_Q(G, L_C)$.*

3.3 Structural Functions associated to Conceptual Graphs

Putting $h = (f, g)$, $G \stackrel{h}{\cong} G'$ denotes that (f, g) is an isomorphism from G to G' and we write $G' = h(G)$. We extend the use of h to denote the correspondence, through the isomorphism (f, g), between similar items attached to G and G'. For instance:

- if $rc \in U$ for some $r \in R$ and $c \in C$, then we write $h(rc) \in h(E)$, which stands for $(f(r), g(c)) \in U'$;
- if L_R is a linear order on R, $h(L_R)$ is the linear order on $R'=h(R)$ defined by: r'_1 precedes r'_2 in $h(L_R)$ if and only if $r_1=f^{-1}(r'_1)$ precedes $r_2=f^{-1}(r'_2)$ in L_R.

In other words, based on f and g, h carries every structure which can be expressed from R and C by means of repeated power-sets and Cartesian products. More formally:

Definition 8 *Let (f, g) be an isomorphism from G to G'. We define the function h as the least defined function such that:*

(i) *if $x = r \in R$, then $h(x) = f(r)$*
(ii) *if $x = c \in C$, then $h(x) = g(c)$*
(iii) *if $x = (y, z)$ and $h(y)$ and $h(z)$ are defined, then $h(x) = (h(y), h(z))$*
(iv) *if $x = \{y_i | i \in I\}$ and $h(y_i)$ is defined $\forall\ i \in I$, then $h(x) = \{h(y_i) | i \in I\}$*
(v) *otherwise $h(x)$ is not defined.*

As there is no risk of ambiguity, we still write $h = (f, g)$ when needed.

Definition 9 *A structural C-preordering (resp. R-preordering) function on conceptual graphs is a function Θ, not necessarily a total function, such that:*

(i) *when defined for a conceptual graph G, $\Theta(G)$ is a total preorder on C (resp. a total preorder on R),*

(ii) *for any conceptual graph isomorphism h such that $G \stackrel{h}{\cong} G'$, if $\Theta(G)$ is defined, then $\Theta(G')$ is defined too and $\Theta(G')=h(\Theta(G))$, that is $\Theta(h(G))=h(\Theta(G))$.*

The definition of h allows the definition of structural functions other than C-preorderings or R-preordering, as exposed in [Cog76]. A complete description being largely beyond our aim, we restrict ourselves to the above definition.

Proposition 2 *If Φ is a structural C-preordering function, then the function Ψ defined on those conceptual graphs for which $\Phi(G)$ is a linear order on C by $\Psi(G) = Q_R(G, \Phi(G))$, is a structural R-preordering function.*

proof : let G be a conceptual graph for which $\Psi(G)$ is defined, let h be some G to G' conceptual graph isomorphism, $L_R = \Psi(G)$ and $L_{R'} = \Psi(h(G))$; we now prove that $L_{R'} = h(L_R)$;
1) because h is the $R \times R$ to $R' \times R'$ extension of the R to R bijection f, and because L_R and $L_{R'}$ are total preorders, it suffices to prove that given any two R-vertices r_1 and r_2 of G, if $r_1 <_{L_R} r_2$ then $f(r_1) <_{L_{R'}} f(r_2)$; so assume that the two R-vertices descriptors are such that $s(r_1) < s(r_2)$ and let's verify that $s(f(r_1)) < s(f(r_2))$;
2) if $Lab(r_1) < Lab(r_2)$ then, labels being invariant through h, $s(f(r_1)) < s(f(r_2))$;

3) if $Lab(r_1)=Lab(r_2)$, then $m_j^{r_1} < m_j^{r_2}$ for some j; therefore, for some i_1 and i_2, $num_C(G_j(num_R^{-1}(i_1))) < num_C(G_j(num_R^{-1}(i_2)))$,
$num_{C'}(g(G_j(num_R^{-1}(i_1)))) < num_{C'}(g(G_j(num_R^{-1}(i_2))))$,
$num_{C'}(G_j'(f(num_R^{-1}(i_1)))) < num_{C'}(G_j'(f(num_R^{-1}(i_2))))$,
$num_{C'}(G_j'(num_{R'}^{-1}(i_1))) < num_{C'}(G_j'(num_{R'}^{-1}(i_2)))$,
$num_{C'}(G_j'(f(r_1))) < num_{C'}(G_j'(f(r_2)))$; hence $s(f(r_1)) < s(f(r_2))$; $\square$

3.4 Intrinsically C-sorted Conceptual Graphs

Definition 10 *A conceptual graph G is said to be* intrinsically C-sorted *by a structural C-preordering function Φ when $\Phi(G)$ is a linear order on C. Then the descriptor $\Delta(G, \Phi(G), Q_R(G, \Phi(G))) = \Delta_Q(G, \Phi(G))$ is called the* normal descriptor *of G induced by Φ, and is denoted by $\Delta_\Phi(G)$.*

And our interest in $\Delta_\Phi(G)$ lies in the following converse result to theorem 1 when restricted to intrinsically C-sorted Conceptual Graphs.

Theorem 2 *Let G be a conceptual graph intrinsically C-sorted by a structural C-preordering function Φ. Then, for any conceptual graph G':*
if $G \cong G'$, then G' is also intrinsically C-sorted by Φ and $\Delta_\Phi(G) = \Delta_\Phi(G')$.

proof : Because $\Phi(G)$ is defined and Φ is structural, $\Phi(G')$ is defined too and G' is intrinsically C-sorted by Φ. Then we have:
$$\Delta_\Phi(G) = a_1, ...a_q, b_1, ...b_p \text{ with } b_i = e_i n_1^i ... n_{d_i}^i$$
$$\Delta_\Phi(G') = a_1', ...a_q', b_1', ...b_p' \text{ with } b_i' = e_i' n_1'^i ... n_{d_i'}'^i$$
Let h be such that $G \overset{h}{\cong} G'$. Then $\Phi(G') = h(\Phi(G))$, $num_{R'}^{-1}(num_R(r)) = f(r)$ for all $r \in R$ and $num_{C'}^{-1}(num_C(c)) = g(c)$ for all $c \in C$;
1) $a_i = Lab(num_C^{-1}(i)) = Lab(g(num_C^{-1}(i))) = Lab(num_{C'}^{-1}(num_C(num_C^{-1}(i))))$
$= Lab(num_{C'}^{-1}(i)) = a_i'$
2) similarly, $e_i = e_i'$
3) $n_j^i = num_C(G_j(num_R^{-1}(i)) = num_{C'}(g(G_j(num_R^{-1}(i)))$
$= num_{C'}(G_j'(f(num_R^{-1}(i)))) = num_{C'}(G_j'(num_{R'}^{-1}(i))) = n_j'^i$ $\square$

The preceding results can now be summarised into:

Theorem 3 *Given two conceptual graph G and G' and a structural C-preordering function Φ:*

1. *if $G \cong G'$, then G is intrinsically C-sorted by Φ if and only if G' is intrinsically C-sorted by Φ;*
2. *if G is intrinsically C-sorted by Φ, then $G \cong G'$ if and only if G' is intrinsically C-sorted by Φ and $\Delta_\Phi(G) = \Delta_\Phi(G')$.*

proof: It is a straightforward deduction from definition 10, theorem 1 and 2. $\square$

Given any structural C-preordering function Φ, theorem 3 yields a heuristic for conceptual graphs isomorphism testing:

isomorphic? (G, G') ;

{given some structural C-preordering function Φ}

   compute $\Phi(G)$ and $\Phi(G')$;

   <u>if</u> $\Phi(G)$ and $\Phi(G')$ are linear orders

       <u>then</u> $G \cong G'$ if and only if $\Delta_\Phi(G) = \Delta_\Phi(G')$

   <u>else</u> <u>if</u> neither of $\Phi(G)$ and $\Phi(G')$ is a linear order

       <u>then</u> the heuristic fails

          {a different conceptual graphs isomorphism test must be used}

   <u>else</u> G and G' are not isomorphic

Remark. The heuristic is based on the question whether G and G' are intrinsically C-sorted by Φ or not. It fails in the sole case when the answer is negative for both of them. But then, $\Phi(G)$ and $\Phi(G')$ have not been computed in vain: if $G \overset{h}{\cong} G'$, for any isomorphism h, then $\Phi(G') = h(\Phi(G))$; namely, if equivalence classes of C w.r.t. $\Phi(G)$ (resp. of C' w.r.t. $\Phi(G')$) have been numbered $C_1, ..., C_k$, so that $C_i <_{\Phi(G)} C_j$ if and only if $i < j$ (resp. $C'_1, ..., C'_{k'}$, $C'_i <_{\Phi(G')} C'_j$), then $h(C_i) = C'_i$. That is to say that any candidate to match any $c \in C_i$ in any isomorphism h from G to G' must belong to C'_i. This is a first step in many of graph isomorphism testing techniques [CG70], [Sir71], [Sau71], [RC77], [CK80], ..., (in other words, the automorphism partition of C is finer than the class partition of C induced by $\Phi(G)$).

3.5 A Polynomial Structural C-Preordering Function for Conceptual Graphs: the Search-C-Preordering Ω

Theorem 3, together with the induced heuristic for conceptual graphs isomorphism testing, justifies the interest for structural C-preordering functions Φ such that, on the one hand, as many conceptual graphs as possible are intrinsically C-sorted by Φ, but, on the other hand, the normal descriptor $\Delta_\Phi(G)$ for these conceptual graphs G are as "easily" computed as possible.

We propose here a structural C-preordering function Ω derived from a width first search of conceptual graphs and defined for all conceptual graphs.

A width first search, starting from vertex x, of a bipartite graph G with vertex set X, computes a partition $\{X_0, X_1, ..., X_{k(x)}\}$ such that:

1. $X_0 = \{x\}$;
2. $X_i = \{y \in X / i$ is the least possible number of edges in a path connecting x to $y\}$; therefore, $k(x)$ is the greatest possible i;
3. if there is an edge in G from $y \in X_i$ to $z \in X_j$, then $j \in \{i - 1, i + 1\}$.

Therefore the edge set E can be partition into $\{E_1, ..., E_{k(x)}\}$, E_i being viewed as the set of edges of G at distance i from x.

Definition 11 *Let G be a conceptual graph. To any C-vertex c, computing a width first search starting from vertex c, we associate the descriptor:*

$$\omega(c) = Lab(c) m_1 m_2 ... m_{k(c)}$$

where each m_i is computed as follows:

1. *to each edge $e \in E_i$ we assign a descriptor (v, u, w) where $v = Lab(x)$, x being the end point of e in X_{i-1}, $u = Lab(e)$, and $w = Lab(y)$, y being the end point of e in X_i ;*
2. *m_i is the list, sorted in lexicographic order, of the descriptors of all the edges of E_i.*

The total preorder $\Omega(G)$ on C is the total preorder induced by the lexicographic order on the descriptors $\omega(c)$. We call Ω the search-C-preordering.

Proposition 3 *Ω is a structural C-preordering function.*

proof : Let $G \overset{h}{\cong} G'$, and let $c \in C$ and $c' = g(c) \in C'$; let $\{X_0, X_1, ..., X_{k(c)}\}$ (resp. $\{X_0', X_1', ..., X_{k(c')}'\}$) be the partition of X (resp. X'), $\{E_1, ..., E_{k(c)}\}$ (resp. $\{E_1', ..., E_{k(c')}'\}$) be the partition of E (resp. E'), computed by a width first search starting from vertex c (resp. c');
clearly $X_i' = h(X_i)$ and $E_i' = h(E_i)$, and because Lab is invariant under conceptual graphs isomorphism, $\Omega(c) = \Omega(c')$; therefore $c_1 <_{\Omega(c)} c_2$ if and only if $h(c_1) <_{\Omega(h(c))} h(c_2)$ for any $c_1, c_2 \in C$, that is $\Omega(h(G)) = h(\Omega(G))$. □

Proposition 4 *For any conceptual graph G, the computation of $\Omega(G)$ is in $O(|C| \times |E| \times (|C| + \log|E|))$, and so is the computing of $\Delta_\Omega(G)$ when G is intrinsically C-sorted by Ω.*

proof : Briefly speaking, given $c \in C$, we first compute $\omega(c)$; using a width first search is in $O(m)$ (because G is connected), where $n = p + q = |R| + |C|$ and $m = |E|$, and ordering every E_i is, on the whole, in $O(m \log m)$.
Therefore, the computation of all the $\omega(c)$ for $c \in C$, is in $O(q \times m \times \log m)$.
Then to compare $\omega(c)$ to $\omega(c')$ is in $O(m)$, so comparing all pairs is in $O(q^2 \times m)$.
If G is intrinsically C-sorted by Ω, computing $\Delta_\Omega(G)$ from $\Omega(G)$ is in $O(q \times m \times (q + \log m))$. □

We summarise now how compute the descriptor $\Delta_\Omega(G)$ of the graph in fig.1.
$lab(c3) =_L lab(c4) <_L lab(c1) <_L lab(c2) <_L lab(c5)$ implies $\omega(c4) <_L \omega(c1) <_L \omega(c2) <_L \omega(c5)$ and $\omega(c3) <_L \omega(c1)$. So, $c3 <_{\Omega(G)} c1$, $c4 <_{\Omega(G)} c1$ and $c1 <_{\Omega(G)} c2 <_{\Omega(G)} c5$.
$\omega(c3) = $ Ta:*, ((Ta:*,1,Tr2), (Ta:*,2,Tr1)), ((Tr1,1,Ta:m),(Tr2,2,Tb:*)), and so on for m_3, m_4, m_5 and m_6.
$\omega(c4) = $ Ta:*, ((Ta:*,1,Tr2), (Ta:*,2,Tr1)), ((Tr1,1,Ta:m),(Tr2,2,Tc:*)), and so on for m_3, m_4, m_5 and m_6.
$\omega(c3) <_L \omega(c4)$, so, $c3 <_{\Omega(G)} c4$. Thus $\Omega(G)$ is a linear order on C, that is G is intrinsically C-sorted by Ω.
$lab(r1) = lab(r2)$, $G_1(r1) = G_1(r2) = c1$, $num_C(G_2(r1)) < num_C(G_2(r2))$ so $s(r1, G, \Omega(G)) <_L s(r2, G, \Omega(G))$ and $r1 <_{Q_R(G,\Omega(G))} r2$. In the same way, we prove that $r2 <_{Q_R(G,\Omega(G))} r6 <_{Q_R(G,\Omega(G))} r4 <_{Q_R(G,\Omega(G))} r5$. r5 and r3 are twin nodes, so they are equivalent w.r.t. $Q_R(G, \Omega(G))$. So,

$$\Delta_\Omega(G) = ([\text{Ta:*}][\text{Ta:*}][\text{Ta:m}][\text{Tb:*}][\text{Tc:*}]$$
$$(\text{Tr1,3,1})(\text{Tr1,3,2})(\text{Tr2,1,4})(\text{Tr2,2,5})(\text{Tr3,4,5})(\text{Tr3,4,5}))$$

4 Ω-sorted Conceptual Graphs and other Classes of Conceptual Graphs

Definition 12 *We call* Irredundant-CG *the class of irredundant conceptual graphs,* Ω-Sorted-CG *the class of conceptual graphs for which* $\Omega(G)$ *is a linear order on* C *and* C-Rigid-CG *the class of conceptual graphs for which every automorphism keeps each of the* C-vertices invariant.

Proposition 5 Ω-Sorted-CG $\subset$ C-Rigid-CG.

proof : If $G \notin$ C-Rigid-CG, then there exist $c, c' \in C$ with $c \neq c'$ and $c' = g(c)$ for some automorphism $h = (f, g)$ of G; then $\omega(c) = \omega(c')$, and $\Omega(G)$ is not a linear order on C. $\Box$

From [LB94] we derive:

Definition 13 *We call* Locally-Injective-CG *the class of conceptual graphs satisfying: for any* C-vertex *and any two different* R-vertices *r and r', if* $c = G_i(r) = G_i(r')$ *for some* i, *then* $Lab(r) \neq Lab(r')$.

Conceptual graphs in Locally-Injective-CG can be tested for isomorphism in polynomial time, as shown in [LB94]. As expressed in the following results, they also have properties of interest with regards to redundancy.

Theorem 4 G *is irredundant if and only if every projection from* G *to itself is an automorphism.*

proof: Assume G is irredundant and let $h=(f, g)$ be a projection from G to itself.
1) f and g must be surjective, hence bijective.
2) Any bijection $\sigma : X \to X$ induces a partition of X into orbits which are sequences of the form $(x, \sigma(x), \sigma^2(x), ..., \sigma^{k-1}(x))$ where k is the least positive integer n such that $\sigma^n(x) = x$ (for any x, provided that σ is a bijection, such a k does exist).
Therefore, $Lab(c) \geq Lab(g(c)) \geq Lab(g^2(c)) \geq ... \geq Lab(g^k(c)) = Lab(c)$ for each C-vertex c, for some k. Hence $Lab(c) = Lab(g(c))$ for each C-vertex c, which means that h is an automorphism of G.
Conversely, it is clear that, if G is redundant, any projection to one of its irredundant sub-conceptual-graph is not an automorphism. $\Box$

Definition 14 *Let* $h = (f, g)$ *be a projection of* G *into itself and let* $G' = h(G)$. *We call the projection* h *a* folding *when every vertex in* G' *is kept invariant through* h, *that is:* $\forall x \in R' \cup C'\ h(x) = x$.

Proposition 6 *Let* G *be a redundant conceptual graph and* G' *one of its equivalent irredundant sub-conceptual-graph (there may exist several of them, all isomorphic to each other). Then there exists a folding from* G *onto* G'.

proof : Let $h=(f,g)$ be a projection of G to G'. The restriction, say h', of h to G' is itself a projection of G' to G', then necessarily an automorphism (theorem 4). Thus h'', defined by $h''(G)=h'^{-1}(h(G))$, is also a projection from G to G', therefore a folding. Note that G' being irredundant is essential[1] to prop. 6. $\square$

We now can state:

Proposition 7 *Locally-Injective-CG $\subset$ Irredundant-CG.*

proof: Assume $G \in$ Locally-Injective-CG while it is redundant; then, from proposition 6, there exists a projection $h = (f,g)$ from G to one of its strict sub-S-graph G' which is a folding of G onto G'.

Let $r \in R\text{-}R'$ and adjacent to some $c \in C'$, with $c = G_i(r)$ for some i (such a R-vertex must exist: let x be a vertex of $G\text{-}G'$ and y a vertex of G'; because G is connected, there exists a chain joining x to y, and, along this chain, at least an edge joining x' to y', with $x' \in G\text{-}G'$ and $y' \in G'$; if y' is a R-vertex, its degree in G' does not agree with its label; thus y' is some $c \in C$ and x' is some $r \in R\text{-}R'$). Let $r' = f(r)$ (thus $r' \neq r$). Because h is projection of G onto G', $Lab(r) = Lab(r')$, $g(c) = g(G_i(r)) = G_i(f(r)) = G_i(r')$, and because h is a folding of G onto G' with $c \in C'$, $c = g(c)$; therefore $G_i(r) = G_i(r')$, which contradicts the local injectivity of G on vertex c. $\square$

Note that the problem of irredundancy is NP-complete, while the problem of the local injectivity is a polynomial one. Also, as a straightforward corollary, we answer a question[2] about locally injective conceptual graphs:

Theorem 5 *Restricted to Locally-Injective-CG, the specialization preorder is an order relation.*

We summarise the inclusion relationship between the classes of conceptual graphs we mentioned above:

Theorem 6 *Among conceptual graphs, the inclusion relationship between Ω-Sorted-CG, C-Rigid-CG, Irredundant-CG and Locally-Injective-CG, is the one shown in the diagram of fig. 2.*

proof : From prop. 5 and 7 and making use of graphs given in fig. 3, 4 and 5.

5 About the Actual Use of the Structural Preordering Function Ω and the Descriptor Δ_Ω

CORALI is a platform designed and developed at the LIRMM. It is meant to allow knowledge acquisition and representation based on conceptual graphs, together with a system for ontology building, knowledge base handling and questioning. The knowledge base is made of conceptual graphs which are organised in such a way that it does not store two isomorphic conceptual graphs. To achieve this goal, the computation of $\Omega(G)$ is taken as an entry point to the base:

---

[1] Marie-Laure Mugnier, private communication.
[2] asked by Michel Liquière, private communication.

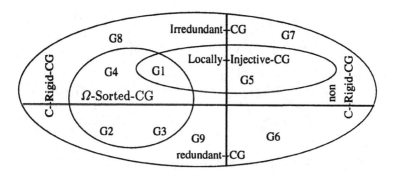

Fig. 2. Classes of Conceptual Graphs

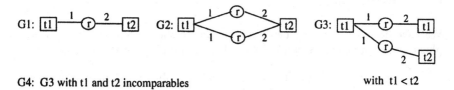

G4: G3 with t1 and t2 incomparables   with t1 < t2

Fig. 3. Ω-sorted conceptuals graphs

- if G is Ω-Sorted, the use of a hash function applied to $\Delta_\Omega(G)$ solves the problem;
- otherwise, more computation is carried on, taking into account the partition of C induced by $\Omega(G)$.

This has been tested on two different applications dealing with natural languages, for which every conceptual graph turned out to be Ω-Sorted, and one related to chemistry, where, when conceptual graphs were not Ω-sorted, which happened quite frequently, $\Omega(G)$ was actually the automorphism partition of C.

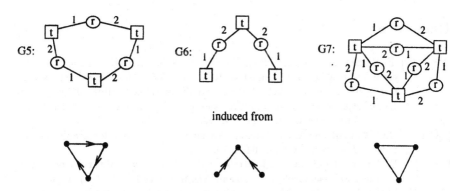

induced from

Fig. 4. non C-rigid conceptuals graphs

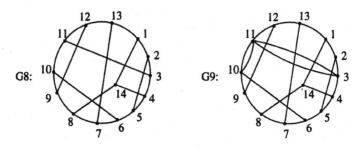

Fig. 5. C-rigid and non Ω-sorted conceptuals graphs

6 Conclusion

We have designed a simple heuristic relevant to the conceptual graphs isomorphism problem which we can use to code conceptual graphs and such that:

- on a practical ground and depending on the application domain, the heuristic may turn out to be quite efficient, not surprisingly when used in applications with domains such as natural language;
- from a theoretical point of view, "bad" conceptual graphs for the heuristic are those induced by simple graphs known for long as crucial for the isomorphism problem; on the other hand, being abstract up to the point of using any structural preordering function, the heuristic is designed as to provide flexibility to the choice of actual functions; in particular, many of the structural functions used for simple graph isomorphism testing can be adapted into structural preordering functions.

We also have introduced a new class of conceptual graphs for which the isomorphism problem is tractable, and we have compared it, w.r.t. inclusion, with other relevant classes of conceptual graphs. We think this kind of classification is worth studying and we would welcome new classes to fit in.

References

[BFH+92] F. Baader, E. Franconi, B. Hollunder, B. Nebel, and H.J. Profitlich. An empirical analysis of optimization techniques for terminological representation systems. In B. Nebel, C. Rich, and W. Swartout, editors, *Principles of Knowledge Representation and Reasoning - Proceedings of the 3rd International Conference*, Cambridge, MA, 1992.

[CG70] D.G. Corneil and C.C. Gotlieb. An efficient algorithm for graph isomorphism. *J.A.C.M.*, 17(1):51–64, 1970.

[CK80] D.G. Corneil and D.G. Kirkpatrick. A theorical analysis of various heuristics for the graph isomorphism problem. *SIAM J. COMPUT.*, 9(2), 1980.

[CM92] M. Chein and M.L. Mugnier. Conceptual graphs : fundamental notions. *Revue d'Intelligence Artificielle*, 6(4):365–406, 1992.

[Cog76]   O. Cogis. A formalism relevant to a classical strategy for the graph isomorphism testing. In *Proceedings of the 7th S-E Conf. Combinatorics, Graph Theory and Computing*, pages 229–238, Bâton Rouge, 1976.

[EL94]   G. Ellis and F. Lehmann. Exploiting the induced order on type-labeled graphs for fast knowledge retrieval. In *Proceedings of the 2nd Int. Conf. on Conceptual Structures*, Maryland, USA, August 1994. Springer-Verlag. Published as No. 835 of Lecture Notes in Artificial Intelligence.

[Ell91]   G. Ellis. Compiled hierarchical retrieval. *Proceedings of the 6th Annual Workshop on Conceptual Graphs*, pages 187–207, Binghamton, 1991.

[Ell93]   G. Ellis. Efficient retrieval from hierarchies of objects using lattice operations. In *Proceedings of the 1st Int. Conf. on Conceptual Structures*, Montreal, Canada, August 1993. Springer-Verlag. Published as No. 699 of Lecture Notes in Artificial Intelligence.

[Gre82]   J.J. Mc Gregor. Backtrack search algorithms and the maximal common subgraph problem. *Software-Practice and Experience*, 12:23–34, 1982.

[LB94]   M. Liquière and O. Brissac. A class of conceptual graphs with polynomial iso-projection. In *Supplement Proceedings of the 2nd International Conference on Conceptual Structures*, College Park, Maryland, USA, 1994.

[LE91]   R. Levinson and G. Ellis. Multi-level hierarchical retrieval. In *Proceedings of the 6th Annual Workshop on Conceptual Graphs*, pages 67–81, 1991.

[Lev92]   R. Levinson. Pattern associativity and the retrieval of semantic networks. In F. Lehmann, editor, *Semantic Networks in Artificial Intelligence*, pages 573–600, Pergamon Press, Oxford, 1992.

[Lev94]   R. Levinson. Uds: A universal data structure. In *Proceedings of the 2nd Int. Conf. on Conceptual Structures*, Maryland, USA, August 1994. Springer Verlag. Lecture Note in Artificial intelligence #835.

[MC92]   M.L. Mugnier and M. Chein. Polynomial algorithms for projection and matching. In Heather D. Pfeiffer, editor, *Proceedings of the 7th Annual Workshop on Conceptual Graphs*, pages 49–58, New Mexico University, 1992.

[MLL91]   S.H. Myaeng and A. Lopez-Lopez. A flexive matching algorithm for matching conceptual graphs. In *Proceedings of the 6th Annual Workshop on Conceptual Graphs*, pages 135–151, 1991.

[RC77]   R.C. Read and D.G. Corneil. The graph isomorphism disease. *Journal of Graph Theory*, 1:339–363, 1977.

[Sau71]   G. Saucier. Un algorithme efficace recherchant l'isomorphisme de deux graphes. *R.I.R.O.*, 5ème année:39–51, 1971.

[Sir71]   F. Sirovich. Isomorfi fra grafi : un algoritmo efficiente per trovare tutti gli isomorfismi. *Calcolo 8*, pages 301–337, 1971.

[Sow84]   J.F. Sowa. *Conceptual Structures - Information Processing in Mind and Machine*. Addison-Wesley, Reading, Massachusetts, 1984.

[Ung64]   S.H. Unger. GIT - a heuristic program for testing pairs of directed line graphs for isomorphism. *C.A.C.M*, 7(1):26–34, 1964.

[Way94]   E. Way. Conceptual graphs - past, present and future. In *Proceedings of the 2nd Int. Conf. on Conceptual Structures*, Maryland, USA, August 1994. Springer Verlag. Lecture Note in Artificial intelligence #835.

[Woo91]   W.A. Woods. Understanding subsumption and taxonomy: A framework for progress. In J.F. Sowa, editor, *Principles of Semantic Networks, Explorations in the Representation of Knowledge*, pages 45–94, Morgan Kaufmann, San Mateo (USA), 1991.

Projection and Unification for Conceptual Graphs

M. Willems
Faculty of Mathematics and Computer Science
Vrije Universiteit Amsterdam
De Boelelaan 1081a
1081 HV Amsterdam
The Netherlands
e-mail: willems@cs.vu.nl
tel: +31-20-5485520

Abstract. In this paper we will investigate subsumption and unification for structured descripitons by considering conceptual graphs with their projection and (maximal) join. The importance of projection for conceptual graphs is well-known as it essentially defines a partial order (subsumption hierarchy) on the graphs, that allows one to speed up search considerably. We investigate the complexity of projection by introducing a weaker notion of structural similarity, polyprojection. We prove that a polyprojection implies a projection for so-called non-repeating conceptual graphs. Furthermore, we show that a polyprojection can be determined by a polynomial algorithm. Indeed, the algorithm presented generalizes well-known algorithms for subtree isomorphism, and subsumption between feature term graphs.

A maximal join is defined as the join on a maximally extended compatible projection. The operation is closely related to unification in feature logics and logic programming, but it is allows more flexibility of representation. In essence, a maximal join corresponds to the greatest lower bound of two conceptual graphs when the partial order due to projection is a lattice. Finally, unification of structured descriptions as maximal join is shown to be polynomially related to projection.

1 Introduction

Research in knowledge representation has led to a distinction between terminological logics and assertional logics. Assertional logics contain languages to assert what is the case at some time in some world, whereas terminological logics describe the concepts by which an assertion can be made and associations between them. However, the formalisms proposed to capture these levels are very similar because assertions connect different individuals by relations, and term descriptions combine different (primitive) concepts by roles. Independently, research in computational linguistics has developed so-called feature logics that aim at representing the information conveyed by natural languages. Here too, a relational notion (features) is essential for the description of complex linguistic categories.

Conceptual graphs [Sowa, 1984] take such a relational view as their basis. In this paper we will investigate subsumption and unification for structured descripitons by considering conceptual graphs with their projection and (maximal) join. The importance of projection for conceptual graphs is well-known as it essentially defines a partial order (subsumption hierarchy) on the graphs, that allows one to speed up search considerably. We investigate the complexity of projection by introducing a weaker notion of structural similarity, polyprojection, and we give a polynomial algorithm that unlike other approaches does not start at one particular concept in the graph.

Maximal join is defined as the join on a maximally extended compatible projection. The operation is closely related to unification in feature logics and logic programming, but it is allows more flexibility of representation. In essence, a maximal join corresponds to the greatest lower bound of two conceptual graphs when the partial order due to projection is a lattice. Unification of structured descriptions as maximal join is shown to be polynomially related to projection.

2 Conceptual Graphs

First we shall repeat some of the important notions for a conceptual graph calculus, see [Sowa, 1984], in a slightly different formulation which is based on our own preferences. A conceptual graph can be said to describe a set of concepts that are connected by conceptual relations, or more logically, a set of typed variables that constrain each other through predicates. The components from which one builds a conceptual graph are given in a *canon* that consists of a type hierarchy T with a subtype relation $\leq$, and a set of individuals I along with a conformity relation ::.

Definition 1. A canon is a tuple $(T, I, \leq, ::)$ where

- T is the set of types. We will further assume that T contains two disjuntive subsets T_C and T_R containing types for concepts and relations.
- I is the set of individuals.
- $\leq \subseteq T \times T$ is the subtype relation. It is assumed to be a lattice (so there are types $\top$ and $\bot$ and operations $\cup$ and $\cap$).
- $:: \subseteq I \times T$ is the conformity relation. It is assumed to obey certain properties.

With this canon one can define a conceptual graph. The nodes are divided into concepts and relations, that are typed by a type function. The concepts are related to individuals by a referent function. The structure of the graph is indicated by a set of argument functions that indicate what are the argument concepts of a relation.

Definition 2. Let $(T, I, \leq, ::)$ be a canon. A *conceptual graph* with respect to a canon is a tuple $G = (C, R, type, referent, arg_1, ..., arg_m)$ where

C is the set of concepts, $type : C \rightarrow T$ indicates the type of a concept, and $referent : C \rightarrow I$ indicates the referent marker of a concept.

R is the set of conceptual relations, $type : R \rightarrow T$ indicates the type of a relation, and

each $arg_i : R \rightarrow C$ is a partial function where $arg_i(r)$ indicates the i-th argument of the relation r. The argument functions are partial as they are undefined for arguments higher than the relation's arity.

We often write $c \in G$ instead of $c \in C$ when it is clear that c is a concept (similarily for relations $r \in G$).

When translated to first-order logic the individual concepts are constants, and the generic concepts are existentially quantified variables. Concept types are unary predicates on these variables and constants and straightfordwardly the n-ary relations are translated to n-ary predicates.

Following the ideas of C.S. Peirce, negation is captured by means of *contexts*. A context is a concept that in turn contains a conceptual graph. A special kind of context is the *negative context*; it can be interpreted as saying that the graph it contains is not the case. Together with *coreference* between concepts in different contexts this gives conceptual graphs the power of first-order logic.

When we want a terminological logic, it is easy to switch to descriptions: instead of asserting a graph one uses it to describe a type. To do this one encapsulates a conceptual graph inside a lambda-expression that defines a type. The conceptual graph is the body of the lambda-expression and it asserts the necessary and sufficient conditions that describe a concept of that type. A particularity of a lambda-expression is that several concepts are mark as *parameters*. For a type-definition only one concept of the body graph is a parameter, i.e. the *head* of the description. Below we will use a coreference between a concept in a defined graph and a concept in the defining graph to show the parameters.

```
type [FATHER:*x] is
     [MAN:*x]-(CHILD)->[PERSON].
```

Fig. 1. A father is a man that has a child.

The type-definition above corresponds to FATHER $=$ MAN$\sqcap\exists$CHILD.PERSON in a more traditional notation of terminological logics. Basically the language we use here allows for conjunction $\sqcap$, existential roles $\exists R$, and existential role-value-maps $P = Q$. By adding negation and coreference we can easily enhance the expressive power.

3 Projection

Before we proceed we will define some graph-theoretic notions (walk, path, cycle, tree) for a conceptual graph, that are important for later proofs.

Definition 3. Let $G = (C, R, type, referent, arg_1, \ldots, arg_m)$ be an arbitrary conceptual graph. A *walk* is an alternating sequence of concepts and relations $W = (c_1, r_1, c_2, r_2, \ldots, r_{n-1}, c_n)$ such that each r_i for $i = 1, \ldots, n\text{-}1$ there are an ith and a jth argument-function such that $arg_j(r_i) = c_i$ and $arg_k(r_i) = c_{i+1}$. If $c_1 = c_n$ the walk is *closed*. A *path* is a walk W such that no concept occurs more than once in w; except for $c_1 = c_n$. If this exception occurs the path is closed and it is called a *cycle*. A conceptual graph without cycles is called a *tree*.

In the definitions below we will sometimes use the infix notion for functions; we will write either $y = f(x)$ or xfy. The latter notation is useful when we generalize the notion of projection to polyprojection later on. Functions can be viewed as a special kind of subset of a Cartesian product. The "special" property of functions is their single valuedness: every element of the domain has at most one image: if xfy and xfz, then $y = z$. For subsets of Cartesian products one can easily define composition as the set of pairs $x\,(f \circ g)\,y$ such that $\exists y: xfy$ and ygz, and identity as the identity of the sets of pairs.

We define projection directly as a mapping between two graphs instead of using a sequence of canonical derivation rules (see [Sowa, 1984]) that define specialization. The notions of projection and specialization are completely identical: not only does every specialization imply a projection, but inversely every projection implies a specialization [Chein & Mugnier, to appear].

Definition 4. Consider two graphs $G = (C, R, type, referent, arg_1, \ldots, arg_m)$ and $G' = (C', R', type', referent', arg'_1, \ldots, arg'_m)$ with respect to the same canon. A *projection* from G to G' is a pair of functions $\pi_C : C \to C'$ and $\pi_R : R \to R'$ that are

1. Type preserving: for all concepts $c \in C$ and $c' \in C'$, $c\,\pi_C\,c'$ only if $type(c) \geq type'(c')$, and $referent(c) = *$ or $referent(c) = referent'(c')$,
2. Type preserving: for all relations $r \in R$ and $r' \in R'$, $r\,\pi_R\,r'$ only if $type(r) \geq type'(r')$, [1]
3. Structure preserving: for all relations $r \in R$ there holds that $arg'_i(\pi_R(r)) = \pi_C(arg_i(r))$ or equivalently in infix-notation $\pi_R \circ arg'_i = arg_i \circ \pi_C$.
4. Non Empty: for all concepts $c \in C$ there is a concept $c' \in C'$ such that $c\,\pi_C\,c'$.

We say that G *projects into* G' if there is a projection $\pi : G \to G'$ and we will write $G \to G'$. Identity and composition of projections are straightforwardly defined, and for the latter we will write $G \to G' \to G''$.

Note that non-empty condition holds for relations too, because any relation will have an argument that has an image due to non-emptiness of concepts, and by structure preservation the relation must have an image.

---

[1] Note the $\geq$ that allows for a type hierarchy of relation types

4 Structural Similarity by Polyprojection

In order to investigate projection we will first consider a slightly weaker kind of structural similarity by using arbitrary subsets of the Cartesian products instead of functions only, see [Willems, 1991b, Willems, 1991a, Willems, 1993]. Indeed, a *polyprojection* will be defined as a pair of non-empty sets of element pairs (not necessarily functions) between two simple conceptual graphs that preserve the types of the concepts as well as the structure of the graph. If a relation in the first graph is paired with a relation in the second graph, then its i-th argument concept is paired with the i-th argument concept of the second relation.

Definition 5. Consider two graphs $G = (C, R, type, referent, arg_1, \ldots, arg_m)$ and $G' = (C', R', type', referent', arg'_1, \ldots, arg'_m)$. A *polyprojection* μ from G to G' is a pair of Cartesian product subsets $\mu_C \subseteq C \times C'$ and $\mu_R \subseteq R \times R'$ that are

1. Type preserving: for all concepts $c \in C$ and $c' \in C'$, $c \, \mu_C \, c'$ only if $type(c) \geq type'(c')$, and $referent(c) = *$ or $referent(c) = referent'(c')$,
2. Type preserving: for all relations $r \in R$ and $r' \in R'$, $r \, \mu_R \, r'$ only if $type(r) \geq type'(r')$,
3. Structure preserving: $\mu_R \circ arg'_i = arg_i \circ \mu_C$.
4. Non Empty: for all concepts $c \in C$ there is a concept $c' \in C'$ such that $c \, \mu_C \, c'$.

We say that G' *is structurally similar to* G, if there is a polyprojection μ between G' and G, and we will write $G'\mu G$.

Compare this definition to Definition 4 of projections. Note that the infix notation $c\mu c'$ is necessary now because for any c there need not be only one single concept $c' = \mu(c)$. All other conditions remain, the type preserving conditions, the structure preserving conditions, and the non-emptiness that ensures that every concept in G has an image in G'. Thus the following theorem that any projection is a polyprojection too, can be given without proof.

Theorem 6. *If $\pi : G \to G'$ is a projection, then $G\pi G'$ is a polyprojection.*

The reverse implication, that any polyprojection is a projection does not always hold. The important difference is that of single-valuedness: a subset of the Cartesian product is not a function if some concept has more than one 'image'. However, if G is structurally similar to G' then each concept or relation in G has *at least one* image in G'. Thus, if a polyprojection exists, every walk in G is paralleled by a walk in G'.

Lemma 7. *Let G and G' be two conceptual graphs such that a polyprojection $G\mu G'$ exists. If c is the start concept of some walk $W = (c_1, r_1, c_2, \ldots, r_{n-1}, c_n)$ in G and the pair (c, c') is in μ_C, then c' is the start concept of a walk $W' = (c', r'_1, c'_2, \ldots, r'_{n-1}, c'_n)$ in G' such that $(c_i, c'_i) \in \mu_C$ and $(r_i, r'_i) \in \mu_R$ for every $i = 1, \ldots, n$.*

Proof. We will first show that any relation r that has concept c as its i-th argument, has for any concept c' that is an image of c, an image relation r' that has c' as its i-th argument; formally if $(c, c') \in \mu_C$, then for any relation $r \in G$ with $arg_i(r) = c$ there is a relation $r' \in G'$ with $arg'_i(r') = c'$ such that $(r, r') \in \mu_R$. From $arg_i(r) = c$ and $(c, c') \in \mu_C$ yielding $(r, c') \in arg_i \circ \mu_C$, and from the structure preserving condition of Definition 5 it follows that $(r, c') \in \mu_R \circ arg'_i$. So there must be an relation $r' \in G'$ such that $(r, r') \in \mu_R$ and $arg'_i(r') = c'$.

Analogously any concept c that is the i-th argument of some relation r, has for any relation r' that is an image of r, an image concept c' that is the i-th argument of relation r'. Indeed if $(r, r') \in \mu_R$ then for $c' = arg'_i(r')$ it holds that $(r, c') \in \mu_R \circ arg'_i$. Hence $(r, c') \in arg_i \circ \mu_C$ and thus $c = arg_i(r)$ implies the pair $(c, c') \in \mu_C$.

From the above two facts it follows that whenever concept c_i has an image c'_i and a successor relation r_i in W, then there is an image r'_i that is the successor of c'_i. And if relation r_i has an image r'_i and a successor concept c_{i+1} in W, then there is an image c'_{i+1} that is the successor of r'_i. Together this gives us the walk W' that parallels W.

Corollary 8. *If $G\mu G'$ is a polyprojection, then any path P in G projects onto a walk W' in G'.*

Proof. By the non-empty condition of Definition 5 every concept c_1 that is the start concept of some path has an image c'_1 with $(c_1, c'_1) \in \mu_C$. The image walk then follows from Lemma 7.

5 Algorithm

The notion of polyprojection has an important, desirable property: its existence can be determined by a polynomial algorithm. For two arbitrary graphs, the algorithm starts with a pair of relations that are type-preserving i.e. the labels of pairs are subtypes.

Algorithm 1 *Given two finite conceptual graphs G and G', determine a pair of sets $(\mathbf{M}_C, \mathbf{M}_R)$ with $\mathbf{M}_C \subseteq C \times C'$ and $\mathbf{M}_R \subseteq R \times R'$ that are type preserving. Let the result be* **Type-preserving(G,G')**.

- **Type-preserving(G,G')** *contains all pairs that obey the first two conditions of a polyprojection; all pairs $(c, c') \in C \times C'$ such that* type$(c) \geq$ type$'(c')$, *and* referent$(c) = *$ *or* referent$(c) =$ referent$'(c')$, *and all pairs $(r, r') \in G_R \times G'_R$ such that* type$(r) \geq$ type$'(r')$.

The next step towards a polyprojection is to iteratively determine subrelations by deleting those pairs that do not obey the structure-preservation condition. When this process ends only the third condition needs to be checked to get a polyprojection.

Algorithm 2 *Given two finite conceptual graphs G and G', and a type-preserving pair of sets $M^0 \subseteq \textbf{Type-preserving(G,G')}$. Determine a pair of sets $M \subseteq M^0$ that is structure-preserving. Let the result be* **Structure-preserving(M)**.

- *Construct M^{n+1} from M^n (for $n \geq 0$) by deleting all pairs $(c, c') \in M^n$ and $(r, r') \in M^n$, that do not obey condition of structure·preservation of a polyprojection; for instance, remove the pair (c, c') if there is an relation $r \in R$ such that $(r, c') \in (\arg_i \circ M_C^n)$, but not (r, c') $in(M_R^n \circ \arg_i')$; analogously, remove the pair (r, r') if there is a concept $c' \in C'$ such that $(r, c') \in (M_R^n \circ \arg_i')$, but not $(r, c') \in (\arg \circ M_C^n)$.*
- *Stop with $M = M^n$, when $M^{n+1} = M^n$ (for $n \geq 0$).*

Note that the algorithm always ends, because there are only a finite number of pairs to delete. It can be executed in time polynomial in the sizes of the graphs, because in the worst case it deletes all pairs of M one by one, whose number is at most the square of the number of concepts, the concepts and relations. Below it is proven that the algorithm is sound, because the result is a polyprojection or it does not obey the non-empty condition. Also the algorithm is complete because any possible polyprojection is contained in the result of the algorithm.

Theorem 9. *Let G and G' be two conceptual graphs. A polyprojection $G\mu G'$ exists if and only if the result of Algorithm 2 obeys the non-empty condition of Definition 5.*

Proof. The algorithm is sound, the reverse implication holds, because any result obeys the first two conditions for a polyprojection, and the third condition is checked in the statement of the theorem.

To prove completeness, the forward implication. Let us assume that a polyprojection $G\mu G$ exists. By induction we will show that all M^n of the algorithm contain μ. First, $\mu \subseteq M^0$ because all pairs in μ have type-preserving labels. Next assume that for an arbitrary n, $\mu \subseteq M^n$. Then because all pairs of μ obey the structural-conditions, no pairs of μ are deleted from M^n, and so $\mu \subseteq M^{n+1}$. We conclude that if μ exists, all relations occurring in the course of the algorithm contain μ. In particular, the result M of the algorithm will contain μ.

6 Graphs with Known Polynomial Projection

The algorithm described above is remeniscent to the one given in [Reyner, 1977]. There the subgraph isomorphism problem is shown to be solvable between trees in polynomial time. In [Mugnier & Chein, 1992] another algorithm similar to ours is given for projection from trees. It also involves all possible images for all nodes in the graph. Actually when the first graph is a tree the results are the same.

Indeed, when the first graph is a tree we can prove that a polyprojection exists if and only if a projection exists. We prove this by induction, the usual proof-technique for trees.

Theorem 10. *Let T and G' be two conceptual graphs where T is a tree. Then a polyprojection $T\mu G'$ exists if and only if $T \to G'$.*

Proof. The reverse implication is obvious by Theorem 6, so it remains to be shown that $T\mu G'$ implies $T \to G'$. By induction to the number of concepts in T we will prove that any polyprojection $T\mu G'$ contains a projection $\pi : T \to G'$, i.e. $\pi \subseteq \mu$.

For this we consider the two graphs and the polyprojection $T\mu G'$. If T has only one concept c, there is a pair $(c, c') \in \mu$ by the non-empty condition. This pair by itself constitutes a projection.

Next if T has n concepts for some $n > 1$, we can assume that any polyprojection from a tree with n-1 concepts contains a projection. By deleting a leaf c and the (only) relation r with $arg_i(r) = c$ from T we construct a graph T^* that has n-1 concepts. A polyprojection $T^*\mu^*G$ can be found by leaving out all pairs with deleted elements from μ. By induction there is a projection $\pi^* : T^* \to G$ with $\pi^* \subseteq \mu^*$. This projection can be extended to a projection $\pi : T \to G'$ by adding those pairs (c, c') and (r, r') that obey $(arg_i \circ \pi^*)(r) = arg'_i(r')$.

In this way a polyprojection $T\mu G'$ inductively implies a graph projection $\pi : T \to G'$ with $\pi \subseteq \mu$.

The previous theorem proves when the first graph is a tree that Algorithm 2 results in a polyprojection if and only if a projection exists. Thus we have a polynomial algorithm for determining the existence of a projection between trees (even though we do not find the projection explicitly).

The reverse of this theorem (although this seems to be the case) is not true; a polyprojection can exist from a random graph to a tree even when no projection exists. Still the class of graphs for which Algorithm 2 works is larger than for trees only.

A nice example of a class of graphs with polynomial subsumption are the ψ-terms introduced in [Aït-Kaci, 1986]. They extend trees because closed walks are allowed, but there are only binary relations. Basically these ψ-terms or feature terms can be seen as conceptual graphs that are restrained by the following conditions: all concepts are on at least on directed path that starts at a special head concept, and each relation is thought of as a function. See Figure 2 for a feature term in the conceptual graph notation using functional relations.

```
[MAN]-(NAME)=>[WORD:*x],
     -(CHILD)=>[PERSON]-(NAME)=>[WORD:*x].
```

Fig. 2. A feature term graph of 'a man who has the same name as his (only) child'.

Definition 11. A conceptual graph G is a *feature term graph* if it obeys the following conditions:

- the relations are all *binary*, for any relation r only $arg_1(r)$ and $arg_2(r)$ are defined.
- the relations are *functional*, for any relations r and $r' \in A$, $arg_1(r) = arg_1(r')$ and $type(r) = type(r')$ implies that $r = r'$, and
- there is a *head* concept $h \in C$ such that for all $c \in C$ there is a path $(c_1, r_1, \ldots, r_{n-1}, c_n)$ with $arg_1(r_i) = c_i$ and $arg_2(r_i) = c_{i+1}$ such that $c_1 = h$ and $c_n = c$. Note that when $n = 1$ this includes the case $c = h$.

The class of conceptual graphs is essentially larger than the class of feature term graphs. Allowing n-ary relations is not too difficult an adjustment. The second condition, implying that no two identically labeled relations can leave the same concept, would then need to be adjusted. But this is the property that essentially distinguishes feature term graphs from general conceptual graphs. Below is a conceptual graph that is not a feature term graph.

```
[MAN]-(PART)->[HAND],
     -(PART)->[FOOT].
```

Fig. 3. A conceptual graph of 'a man with two body-parts' is not a feature term graph.

In [Liquière & Brissac, 1994] a variant to this functionality condition is introduced. Local label injectivity states that any two neighbors can be distinguished by their types, i.e. the typing function is injective on the neighbor-sets. However, this property is more restrictive than functionality of feature terms because any two relations that 'point to' the same concept should be different too, whereas two identical functions can point to the same node.

Feature terms graphs have polynomial projection, if one adds the restriction that the heads should be mapped onto each other. Indeed, feature term graphs are a class of graphs for which a polyprojection mapping the heads on each other exists if and only if a projection mapping the heads on each other exists. It follows that we can use the polynomial Algorithm 2 to check for projection, that is similar to subsumption.

Theorem 12. *Let G and G' be two feature term graphs with heads h and h' respectively. There exists a polyprojection $G\mu G'$ with only one pair containing h, $h\mu h'$, if and only if $\pi : G \to G'$ with $\pi(h) = h'$.*

Proof. Again the reverse implication is obvious by Theorem 6, so it remains to be shown that $G\mu G'$ implies $G \to G'$. To prove this let us assume that $G\mu G'$, but that there is no projection $G \to G'$. The only possible reason why μ is not a projection is that the relations do not obey single-valuedness. Thus there is a concept $c \in G$ and concepts $c', c'' \in G'$ such that (c, c') and (c, c'') in μ.

The concept c cannot be the head h of G, because the occurrence of a second pair in μ is excluded. So c is not the head of G and there is a directed path P

from head h to c. By Corollary 8 all paths in G have an image in G', and in particular this holds for the directed path P. This path has two image paths because c has two images, say P' from h' to c' and P'' from h' to c''. Now c' and c'' are different but the paths must be the same, or else the second condition of Definition 11 is contradicted. This proves the theorem.

7 Graphs with Polynomial Projection

If we continue our search for graphs that have the property that a polyprojection implies a projection, we find a problem with cyclic structures. Consider the two graphs in Figure 4 describing a man who helps himself (an egoist?) and a man who helps an (other) man who helps him.

According to Definition 5, the graph G is structurally similar to G'; the polyprojection includes the pairs (x,y) and (x,z). However, there is no projection $G \to G'$ (a man who helps a man is not necessarily an egoist).

The characteristic of a circuit that spoils the polyprojection as a projection is its internal 'repetition'. The circuit G', for instance, contains a path repeating the

```
[MAN:*x]-(HELPS)->[MAN:*x].
```

```
[MAN:*y]-(HELPS)->[MAN:*z]-(HELPS)->[MAN:*y].
```

Fig. 4. No downwards projection exists.

relation $-(\texttt{R})->$ twice. Cycles with such repetition are not necessarily directed, because we can have an undirected cycle that repeats an undirected sequence like $<-(\texttt{R})-, -(\texttt{S})->$ twice. We will define a conceptual graph to be *non-repeating* if it does not contain such *repeating closed walks*.

Definition 13. A closed walk $C = (c_1, r_1, c_2, \ldots, r_{n-1}, c_n)$, $c_1 = c_n$, is a *repeating closed walk* if it consists of a number of identical subwalks: a projection maps a path $P' = (c'_1, r'_1, \ldots, c'_{p-1}, r'_p)$ to each of the subwalks $W_j = (c_{jp+1}, \ldots, c_{jp+p})$ for all $j = 0, \ldots, \frac{n-p}{p}$ of C. The fact that $P \to W_j$ are projections ensures that the types of concepts and relations are repeated in all subwalks of C. A repeating closed walk is called *trivial* if W_0 is closed.

A conceptual graph G is *repeating* if it contains a non-trivial repeating closed walk C, otherwise the graph is *non-repeating*.

Non-repeatingness is an important condition for graph projections, because it defines a class of graphs for which a polyprojection implies a graph projection. The following theorem shows this.

Theorem 14. *Let G and G' be conceptual graphs that are non-repeating. Then a polyprojection $G\mu G'$ exists if and only if $G \to G'$.*

Proof. Only a proof of the forward implication is necessary, so let us assume that $G\mu G'$ but that there is no projection $G \to G'$. Again the only possible reason why μ is not a projection is that the relations do not obey single-valuedness. Thus there is a concept $c \in G$ and concepts $c', c'' \in G'$ with both (c, c') and (c, c'') in μ.

By Theorem 10 we know that every tree in G projects onto G', so G must contain a cycle $W = (c_1, r_1, c_2, \ldots, r_{n-1}, c_n)$ that does not occur in G'. By Corollary 8 we know that W has an image walk $W' = (c'_1, r'_1, c'_2, \ldots, r'_{n-1}, c'_n)$ such that $(c_i, c'_i) \in \mu_C$ and $(r_i, r'_i) \in \mu_R$. But as there is no projection to G' the walk W', unlike W, is not closed. Single-valuedness of projections is violated by the start concept c_1 and its images c'_1 and c'_n that are different.

Because $c_1 = c_n$ we have $(c_1, c'_n) \in \mu$, and so c'_n must be the start concept of another walk W'' that is an image of W. Continuing in this way the endpoint of W'', c''_n must be the start concept of a walk W''' that is an image of W. At some point the walk consisting of all these walks $C' = W', W'', W''', \ldots$ must close as otherwise G' would not be finite. The way in which this closed walk C' is achieved makes sure that it contains a number of copies of W, and thus it is a repeating closed walk. This closed walk C' is not trivial because the walk W' closed because c'_1 is not equal to c'_n.

The non-trivial closed walk that is constructed in this way contradicts the assumption that G' is non-repeating, and the theorem is proven.

This theorem shows that non-repeating conceptual graphs are a class of graphs that have a polynomial algorithm to check for projections. We can apply Algorithm 2 and if the result is a polyprojection, then Theorem 14 guarantees us a projection.

A similar algorithm was presented for subgraph isomorphism in [Ullmann, 1976]. Basicallly the trick there was to refine an adjacency matrix as much as possible before performing a brute-force search. The adjacency matrix is almost identical to the ultimate relation **M** in Algorithm 2. The result in [Ullmann, 1976] shows that our algorithm can be applied to graphs in general, but further processing is needed for repeating graphs.

A condition that is often implemented in other algorithms is that the heads (if such a notion exists) be identified. Similarly when a node disobeys single-valuedness, mapping onto more nodes, one can non-deterministically choose one of the mappings. However, this will lead to backtracking, that is explicitly avoided in our algorithm. In another paper we will show a variant that (partly) checks single-valuedness.

A natural question that comes to mind is the practical relevance of non-repeating graphs. More specifically, whether their number in practice is small or large. From the example in Figure 4 one might be tempted to conclude that repeating graphs (the lower graph) will occur quite often. Against this one should remark that the algorithm presented only depends on non-repeatingness in the worst case. For instance, consider a tree and a repeating graph; a polyprojection exists only if a projection exists see theorem 10 but theorem 14 does not guarantee it.

In addition, graphs in practice are not as symmetric as in Figure ??. This assymmetry helps to determine projections through polyprojections. In fact, the definition of the repeating-property might be formulated less strict to include these graphs.

8 Unification of Conceptual Graphs

We will use a notion of *pushout* from category theory [Herrlich & Strecker, 1973] to describe the combination of two graphs. This is an idea taken from work on graph grammars [Ehrig, 1987]. There is a strong parallel between the pushout and *unification* [Knight, 1989] in knowledge representation and computational linguistics. [Sowa, 1984] uses the term *join* to describe this. Basically it consists of finding identical subgraphs of two graphs and gluing the graphs together by unifying the nodes and arcs of the subgraphs.

Definition 15. Let G_1 and G_2 be two conceptual graphs. A *unifier* for G_1 and G_2 is a graph U such that

1. *projection:* projections $U \rightarrow G_1$ and $U \rightarrow G_2$ exist, and
2. *compatiblity:* for any concept $u \in U$ the images $u_1 \in U_1$ and $u_2 \in U_2$ must be compatible, i.e. $type(u_1) \cap type(u_2) \neq \perp$.

The *join* of G_1 and G_2 on a unifier U is the graph G such that

1. *commutativity:* there exist morphisms $G_1 \rightarrow G$ and $G_2 \rightarrow G$ such that $U \rightarrow G_1 \rightarrow G = U \rightarrow G_2 \rightarrow G$, and
2. *universality:* for any other graph G' satisfying the previous condition there exists a morphism $G \rightarrow G'$ such that $G_1 \rightarrow G' = G_1 \rightarrow G \rightarrow G'$ and $G_2 \rightarrow G' = G_2 \rightarrow G \rightarrow G'$.

The notation for the join of G_1 and G_2 on U is $G_1 +_U G_2$.

The graph G is uniquely determined up to isomorphism. The commutativity property ensures that the images of U are identified in G. The universal property ensures that no more nodes are identified, and that no nodes are added except those from G_1 and G_2. The compatibility property is added to keep the result sensible. Only concepts that can in principle describe the same individual may be joined.

Another name for a join is the *gluing of G_1 and G_2 along U*, because it can be constructed by taking the disjoint union of G_1 and G_2 and then *gluing* together the corresponding images of U under the given projections. Figure 5 may clarify this by example.

In actual applications the aim is to find the unifier U. The idea underlying logic programming is to find an instantiation of variables that make the query true. This can be performed by repeatedly matching rules with a query. In languages like PROLOG matching amounts to unification along a single predicate, i.e. a concept or a relation. This is extended in a language like LIFE [Aït-Kaci,

```
U:  [PERSON:*y]-(NAME)=>[WORD:*x].


G1: [MAN]-(NAME)=>[WORD:*x].
         -(CHILD)->[GIRL:*y]-(NAME)=>[WORD:*x].


G2: [PERSON:*y]-(NAME)=>[WORD:*x 'Smith'].


G:  [MAN]-(NAME)=>[WORD:*x].
         -(CHILD)->[GIRL:*y]-(NAME)=>[WORD:*x 'Smith'].
```

Fig. 5. Given unifier U and projections $U \rightarrow G_1$ and $U \rightarrow G_2$ the unification G is found.

1986] where ψ-terms are matched. An important advantage of this is that more "predicates" and variables can be unified in one step, instead of having several repeated steps.

Unification of feature term graphs makes use of the fact that the the relations are functions. Joining two concepts then necessarily propagates along the functional relations, because the same object can not be related by the same function to two different objects. For conceptual graphs the unifier is enlarged maximally in the definition of maximal join [Sowa, 1984].

Definition 16. A projection $U' \rightarrow G$ *extends* a projection $U \rightarrow G$ if there is a projection $U \rightarrow U'$ such that $U \rightarrow U' \rightarrow G = U \rightarrow G$. We also say that U' extends U in G.

A *maximal join* is a join $G +_U H$ that is *maximally extended*, i.e. there is no unifier U' that extends U in both G and H.

Obviously it is convenient if one can find these unifiers easily. In the following theorem we show that the number of maximally extended unifiers is polynomial in the size of the graphs. Thus it follows that maximal join is polynomially related to projection, that is used to find a unifier, and therefore for the class of non-repeating graphs finding all maximal joins is polynomial.

Theorem 17. *Consider two conceptual graphs G and H with n and m concepts respectively. The number of maximally extended unifiers of G and H is $\mathcal{O}(nm)$.*

Proof. We will show that the number of possible maximal unifiers is bounded by the number of pairs of compatible projections from a unifier containing only one concept. First, consider U as a graph containing only one concept. Now we must keep in mind that each projection of U into graph G can be compatible with any projection of U into graph H. Thus, because the number of projections of U into G is $\mathcal{O}(n)$ and the number of projections of U into H is $\mathcal{O}(m)$, we know that the number of compatible projections of U into G and H is $\mathcal{O}(nm)$.

Next, let us consider the case where U is a graph containing an arbitrary (non-zero) number of concepts (and possibly some relations) with compatible projections $U \to G$ and $U \to H$. Now, for any injective projection $U' \to U$ there are also compatible projections $U' \to G = U' \to U \to G$ and $U' \to H = U' \to U \to H$ which are extended by the original compatible projections.

As $U' \to U$ is injective, we know that the number of concepts in U' is smaller (or equal) to the number of concepts in U. Also note that there is at least one such injective projection $U' \to U$ namely for $U' = U$. From these observations it is possible to deduce that the number of compatible projections from U' into G and H that are extended by U is at least the number of compatible projections from U into G and H (equality only occurs if U contains only one concept).

From these observations it is possible to deduce that for each pair of compatible projections (note that this is not only a unifier but that it also takes the projections into the graphs into account) with U two concepts there exists a larger number of smaller pairs of projections from unifiers U' extended by it with one concept. In general one can prove that for each size of unifier n there exists a larger number of smaller compatible projections from unifiers with size n-1. Therefore the number of compatible projections from a graph U' that are extended by a graph U is at least the number of compatible projections from the larger graph U.

In short, the number of unifiers (projections from it) of a certain size becomes smaller if the unifier is larger. Moreover, all unifiers that are extended by a maximal unifier are themselves not maximal. Hence the number of maximally extended unifiers is smaller than the number of unifiers with one concept, and the theorem is proven.

9   Conclusions

The polynomial algorithm presented in this paper determines a projection between any two non-repeating conceptual graphs. The class of non-repeating graphs contains graphs containing cycles are non-repeating and have a polynomial projection with our algorithm.

Unlike many other algorithms ours does not start at one particular concept in the graph, like a root or a head. Thus in a sense the constraints imposed on a projection propagate uniformly from all concepts. Thus it profits from the fact that no backtracking occurs.

Further it was sketched that the number of compatible projections from maximally extended unifiers between two graphs is smaller than the number of pairs of compatible projections from a unifier with one concept. It follows that the number of maximally extended unifications between two graphs is polynomial in the number of concepts. This result will certainly contribute to a polynomial algorithm for maximal join.

Acknowledgements

I would like to thank Marie-Laure Mugnier her careful scrutinizing of my ideas. However, any mistakes that still remain are my own.

References

[Aït-Kaci, 1986] H. Aït-Kaci. An algebraic semantics approach to the effective resolution of type equations. *Theor. Comp. Sc.*, 45:293–351, 1986.

[Chein & Mugnier, to appear] M. Chein and M.-L. Mugnier. Conceptual graphs: fundamental notions. *Revue d'Intelligence Artificielle*, to appear.

[Ehrig, 1987] H. Ehrig. *Tutorial Introduction to the Algebraic Approach to Graph Grammars*, volume 291 of *Lecture Notes in Computer Science*. Springer-Verlag, 1987.

[Herrlich & Strecker, 1973] H. Herrlich and G.E. Strecker. *Category Theory*. Allyn and Bacon Inc., Boston, 1973.

[Knight, 1989] K. Knight. Unification: a multidisciplinary survey. *ACM Computing Surveys*, 21(1), 1989.

[Liquière & Brissac, 1994] M. Liquière and O. Brissac. A class of conceptual graphs with polynomial iso-projection. In W.M. Tepfenhart, J.P. Dick, and J.F. Sowa, editors, *Proceedings Supplement of the 2nd International Conference on Conceptual Structures*. University of Maryland, 1994.

[Mugnier & Chein, 1992] M.-L. Mugnier and M. Chein. Polynomial algorithms for projection and matching. In H. Pfeiffer, editor, *Proceedings of the 7th Annual Workshop on Conceptual Graphs*. State University of New Mexico, 1992.

[Reyner, 1977] S.W. Reyner. An analysis of a good algorithm for the subtree problem. *SIAM J. Computer*, 6(4):730–732, 1977.

[Sowa, 1984] J. F. Sowa. *Conceptual Structures: Information Processing in Mind and Machine*. Addison–Wesley, Reading, 1984.

[Ullmann, 1976] J.R. Ullmann. An algorithm for subgraph isomorphism. *Journal of the ACM*, 23(1):31–42, 1976.

[Willems, 1991a] M. Willems. Generalization of conceptual graphs. In E.C. Way, editor, *Proceedings of the 6th Annual Workshop on Conceptual Graphs*, Binghamton, 1991. State University of New York.

[Willems, 1991b] M. Willems. Subsumption of knowledge graphs. In H. Boley and M.M. Richter, editors, *Processing Declarative Knowledge*, volume 567 of *Lecture Notes in Artificial Intelligence*. Springer-Verlag, 1991.

[Willems, 1993] M. Willems. *Chemistry of Language: a graph-theoretical study of linguistic semantics*. PhD thesis, University of Twente, Enschede, 1993.

A Novel Algorithm for Matching Conceptual and Related Graphs

Jonathan Poole* and J. A. Campbell

Department of Computer Science
University College London
Gower Street
London WC1E 6BT
United Kingdom

Abstract. This paper presents a new similarity metric and algorithm for situations represented as graphs. The metric is based on the concept of shared information, and there is discussion of how this would apply for different forms of similarity—including surface, structural and thematic similarity. An algorithm is presented which will determine the similarity of two conceptual graphs for any given measure of information content, which can, as a result, be used for any similarity measure that is based on the concept of shared information. It therefore allows the very flexible use of domain and application specific factors. While the algorithm is not polynomial time, it is argued that for real examples of a useful size it can give an answer in a reasonable time.

1 Introduction

This paper addresses the question of measuring the similarity of situations represented as attributed graphs, which includes the use of conceptual graphs. Similarity measurement is an important component in many types of systems, including those for Case Based Reasoning, Conceptual Retrieval and Machine Learning.

Previous research on algorithms for graph comparison has not tended to look at the types of similarity that are being measured. The psychological literature on similarity tends to distinguish several forms (see e.g. [12]) but graph comparison measures tend to take a very narrow view of similarity, assuming it is entirely dependent on the 'shapes' of the graphs.

This paper defines similarity in terms of shared information. It requires the definition of an 'interest' function on graphs that returns a numeric value indicating the amount of (useful) information contained in an attributed (conceptual) graph. Given this the algorithm presented here will find, for any two graphs, the common generalization that maximizes the interest measure. From this the similarity of the two graphs can be calculated. The particular information measure will depend on the domain and the purpose of the match: the algorithm presented here allows the substitution of any interest function, allowing the definition of virtually any practical similarity metric.

---

* email: J.Poole@cs.ucl.ac.uk

The project from which this work comes has been evaluating the use of graph-based representations of situations in Case-Based Reasoning (CBR)[16]. The project uses situations from law cases as a testbed for ideas of similarity, though legal reasoning and analysis is not a particular goal. Section 1.3 discusses the representation of legal situations as graphs.

1.1 Previous Work on Graph Comparison

Methods of comparing graph representations can be divided into two groups: those that measure common structure and those making use of transformation distance.

Common-structure approaches essentially look for maximal common subgraphs between the two input graphs. The similarity of the graphs increases as the size of this common subgraph increases. Examples of such an approach include Leishman's work on analogy [9], and the algorithms of McGregor [11] and Myaeng and Lopez-Lopez [14]. Leishman's work is unusual in that she considers evaluation functions that chose between competing minimal common generalizations.

The problem with most algorithms reported is that they tend not to consider exactly what form of similarity they are trying to measure, but instead implicitly measure only structural similarity.

Transformation-distance approaches require the definition of a set of legal graph transformations, each of which has an associated cost. The similarity of two graphs is then inversely related to the least-cost set of transformations that can transform one graph into the other [2]. This is a more general model than that of common structure. In practice, however, it is difficult to define a set of general but consistent transformations that are much more than the operations used to find generalizations in conceptual graph theory.

The algorithm in this paper is based on finding common structure, but the issue of defining 'maximality' in the context of attributed graphs is also tackled. The flexibility of transformation approaches can be useful, but defining a general set of transformations for a usefully large domain appears to be beyond the current state of knowledge. The most similar previous work is Leishman's work on analogy transfer on conceptual graphs ([9],[10]), where there is consideration not only of finding the least common generalizations, but also to how the significance of these can be analyzed.

1.2 Essentials of Description Graphs

Conceptual Graphs were intended as a general theory of representation and logic and have been used by different authors in many different ways. The chief intention of introducing the new term 'description graph' is to emphasize that the algorithm presented here and the related work on the project is interested merely in the use of conceptual graph theory and related work as it relates to the representation (description) of situations.

We have introduced some minor changes to the form of description graphs compared to conceptual graphs as described in [18]. These changes have little relevance to the paper in hand, and are thus just described briefly:

- description graphs are oriented graphs, so there is a maximum of one arc between any two nodes, and this arc has a direction. This single arc is seen as a descriptor of all the connections between the objects represented by the nodes, and can be arbitrarily detailed.
- nodes represent just objects and events: attributes are not represented as separate nodes but rather as features in the nodes that represent the objects that the attributes refer to.

The metric and algorithm described apply almost unchanged to standard conceptual graphs. This paper does not consider co-reference links or hierarchical graphs, where individual objects and events are themselves described using graphs, as the implications of these in general descriptions are not clear at this point. Extensions to the algorithm to incorporate these are, however, quite straightforward.

1.3 Legal Situations as Description Graphs

In this project, description graphs are being used for the representation of the *concrete facts* of law cases. This domain has been chosen as it contains easily accessible and arbitrarily complicated examples of situations. In addition, there is a certain amount of previous research on representation of law cases that has used complex representations (see for example [3],[1],[19]). Finally, the task of finding most similar past law cases given a description of a new situation is clearly useful as much of legal decision-making is precedent-based.

Our analysis of law cases is naive. The aim has been to try to capture the 'common-sense' similarity of cases that is visible to non-specialists. For this purpose cases from a small and relatively well-demarcated area of U.K. law have been used, namely that of 'Nervous Shock'.

The description graph representations used are of the concrete objects and concrete events of the case: from a legal point of view the most interesting components of the case are perhaps the duties involved. At present it is not clear how best to integrate these higher-order relations into description graphs, so the representations are relatively low level. Despite this it is argued that the representations are useful both as examples of complex causal situations, and in the longer term as a low-level framework on which to attach higher-order relations.

2 A Similarity Metric Based on Shared Information

The starting point is that we consider the similarity of two representations to be a function of the amount of information they share. We call the information measure on representations *interest*, to distinguish it from the more formal use of the term information in information theory.

If we have two representations $\mathcal{A}$ and $\mathcal{B}$ (be they graphs or whatever), we first find the largest set of information common to them which we call $\mathcal{G}$. The similarity function is then the proportion of the information in the original graphs that is in $\mathcal{G}$ and therefore shared:

$$similarity[\mathcal{A}, \mathcal{B}] = \frac{interest[\mathcal{G}]}{max[interest[\mathcal{A}], interest[\mathcal{B}]]}$$

This gives a value from 0, for least similar representations, to 1 for identical representations.

If all possible representations form a generalization hierarchy, then a common generalization represents a set of shared information. All that is required then is to find the common generalization that maximizes the interest function.

We first consider non-graph representations, such as concepts from a concept lattice as in conceptual graph theory. Each element of the lattice can be assigned an interest value, allowing the similarity of all possible pairs of concepts to be calculated in terms of the interest of their least upper bound. This method is used in Case Based Reasoning ([7]). The method of assigning the interest value might be as naive as having it the depth of the concept in the hierarchy. An alternative is for each concept in the hierarchy to be manually assigned a specific interest value, being how much more interesting that concept is than the sum of its super-types. Basic-level terms [8] and terms with particular importance for the domain would tend to add the most interest. The interest of a concept is then recursively defined as the sum of the interest of the concept and that of its super-types.

This method can then be generalized to graph representations, where the common generalization that maximizes the interest function is found, and then similarity can be calculated. Note that this is *not* the same as finding the maximal common subgraph and then evaluating this, as is implicitly assumed in some graph comparison algorithms. While the most interesting common generalization must be a least common generalization, it is not necessarily the one with the most arcs.

With graphs it will not be possible to manually associate with all graphs an interest value, so this will have to be calculated as a function of the interest of the components and the interest of the interactions between the components. Here we consider just three possible ways we might measure interest, representing three different aspects of similarity.

The example is adapted from [1], and concerns a hypothetical system for legal information retrieval. Given a case base of previous situations represented as description graphs, we want to retrieve the case most similar to a new situation, also represented as a description graph. We shall see how the different similarity metrics lead to different cases being judged most similar.

The probe case is as follows, and the graph of it is shown in Fig. 1[2].

During a hockey game, John intentionally hit a hockey stick held by Bill.

---

[2] Note that real case will tend to be several times as large as this

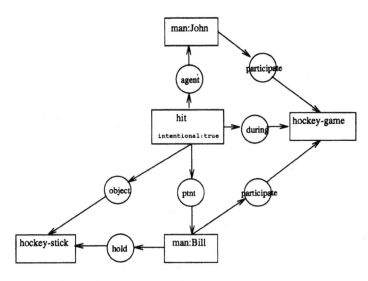

Fig. 1. The Description Graph of the Probe Case

The three relevant stored cases that are to be compared are as below. Due to lack of space we do not present the graphs for these.

1. Michael and Robert stole a hockey stick from the changing rooms during the hockey game
2. Arthur shoved an umbrella held by Mary during an argument over some items on sale at a department store
3. Fred punched Roger on the nose during a boxing match

Below we look at how the probe case could be found to be most similar to any one of these three stored cases, depending on the type of similarity function used. We describe the form the interest function would take in each case to implement the appropriate form of similarity.

surface similarity: this is similarity based on the matching of particular objects and attributes rather than on relations or patterns. This can be modeled by having an interest function on description graphs that just sums up the individual interests of the nodes in the graph. With such a measure the most similar case might be judged to be stored case 1 above, giving the least common generalization shown in Fig. 2

structural similarity: here the interest measure would be designed to give most weight to the number of arcs, or perhaps to factors such as general connectivity or path lengths. Case 2 is judged to be most similar as it matches the most arcs, as shown in Fig. 3. Note that the ultimate structural similarity measure would be if the interest function returns just the number of arcs, at which point the algorithm becomes Maximal Common Subgraph (as in [11]).

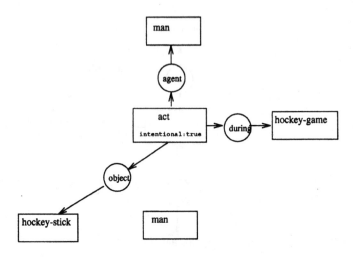

Fig. 2. Generalization with Case 1 using a Surface Interest Function

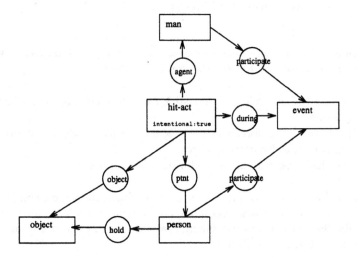

Fig. 3. Generalization with Case 2 using a Structural Interest Function

thematic similarity: here interest would not be a function of either number of nodes or number of arcs, but rather would depend on the presence of particular patterns of concepts and relations. These patterns might be domain specific patterns of interest or arise from connections in script-like knowledge [17]. Case 3 is chosen, in this case, and the least common generalization is shown in Fig. 4. In this case the pattern of interest is that of physical contact happening during a contact sport in which both people were participating. Clearly for this common generalization to be judged by the system to be interesting it needs to be aware that contact sports involve physical contact.

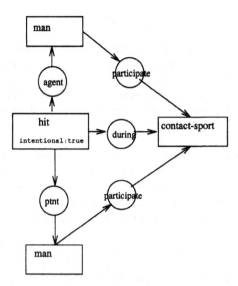

Fig. 4. Generalization with Case 3 using Thematic Interest Function

The space of possible interest measures is very large. It is expected that real interest measures for a particular domain would be a combination of the three factors discussed above. This paper does not consider particular measures, but presents an algorithm that can take an arbitrary interest measure, and determine similarity of graphs based on this.

2.1 Minimum Requirements of an Interest Function

For the algorithm to give correct results there are certain properties the interest function is required to have, as follows:

1. empty nodes and arcs (i.e. those with descriptors containing just 'T') do not add interest to a graph
2. removing an arc or node cannot increase interest
3. removing detail from a node or arc (i.e. generalizing a concept or relation) cannot add interest

Simply stated these mean that for a graph $\mathcal{G}$, a generalization of $\mathcal{G}$ cannot be more interesting than $\mathcal{G}$. These restrictions on the interest measure should not usually be limiting, but they do mean that interest cannot be increased by missing information[3].

---

[3] For example, the representation of 'An animal was reading a book' cannot be more interesting than that of 'A person was reading a book'

2.2 Assumptions regarding Generalization

The algorithm presented here looks for the most interesting set of shared information. The possible sets of shared information are all common generalizations of the two input graphs: that is they can all be formed by taking an input graph and removing arcs or nodes, and generalizing the concept and relation labels.

For the algorithm presented to be correct it is therefore necessary that two assumptions hold:

1. all generalizations of a graph must contain a subset of the information in the original graph (i.e. all generalizations formed by the syntactic operations of removing nodes and arcs and generalizing concepts and relations must represent actual generalizations of the situation)
2. all possible subsets of information must be represented by a generalization

In general, for realistically complex domains, these assumptions will not hold completely. For them to hold it would be necessary that all implicit information about a situation would be found using generalization. We can help this by adding as much implicit information as possible before doing the matches, using rules such as in [13]. However, in general it will not be possible to make all the implicit information explicit in advance. A better solution would be to dynamically apply transformations as the search proceeds.

This issue is not pursued further here, as a solution appears extremely difficult with the current state of knowledge. It is an issue that applies to all symbolic representations when we try to judge similarity of objects from the similarity of representations of the objects.

3 Finding the Most Interesting Common Generalization

This section presents an algorithm that finds the most interesting common generalization of two description graphs, as defined in the previous section. In the form in which we state it here the algorithm is not especially efficient, but it is an *extremely* flexible framework into which it is easy to insert heuristics. Later sections look at how the basic algorithm can be modified to make it more efficient.

3.1 Overview

Given two description graphs A and B and a function *interest* that assigns interest values to graphs, the goal is to return a graph G that is a common generalization of A and B that maximizes the interest function.

The algorithm works in two stages. First a **product graph**[4] P is formed. This product graph has the property that all the maximal common generalizations of A and B are themselves subgraphs of P.

---

[4] the term 'product graph' is sometimes given a slightly different definition in graph theory

The second stage is to use an A* search [6] to find the subgraph of $\mathcal{P}$ that maximizes the interest function. The algorithm makes use of the observation that the interest of $\mathcal{P}$ itself forms an upper bound on the interest of any common generalization.

3.2 Forming a Product Graph

The first step is to create a product graph. It is equivalent to the association graph formed as the first step in the algorithm presented in [14]. Essentially the cartesian product of the nodes and arcs is formed, where the pairs are combined using the join function. Matched node pairs or arc pairs with empty common generalizations (i.e. $\top$) are removed.

We assume that the labels on nodes (concepts) and arcs (relations) can be treated as if they formed lattices, so that the join operation on a pair of concepts or a pair of relations returns an unambiguous label, being the least upper bound in the lattice. In the following algorithms we make use of a procedure $\text{JOIN}(n, m)$ that takes a pair of nodes n and m and returns a *product-node* p, being a node containing the least upper bound of the concepts from n and m, and with slots $\text{NODE-A}(p)$ which returns a reference to n and $\text{NODE-B}(p)$ which returns a reference to m. An analogous procedure $\text{JOIN}(a,b)$ for arcs a and b returns a *product-arc*.

Given two description graphs $\mathcal{A}$ and $\mathcal{B}$, the product graph $\mathcal{P}$ is formed by $\text{PRODUCT}(\mathcal{A},\mathcal{B})$, as defined below:

```
    PRODUCT(Description Graph A, B)
1       P ← MAKE-PRODUCT-GRAPH()
2       Included ← MAKE-ASSOCIATIVE-ARRAY()
3       for each n_A ∈ NODES(A)
4           do for each n_B ∈ NODES(B)
5               do n_P ← JOIN(n_A, n_B)
6                   if n_P ≠ ⊤
7                   then INSERT-NODE(P, n_P)
8                       ASSOCIATE(Included, < n_A, n_B >, n_P)
9       for each a_A ∈ ARCS(A)
10          do for each a_B ∈ ARCS(B)
11              do a_P ← JOIN(a_A, a_B)
12                  if a_P ≠ ⊤
13                  then source ← MAP(Included, <SOURCE(a_A), SOURCE(a_B)>)
14                      target ← MAP(Included, <TARGET(a_A), TARGET(a_B) >)
15                      if source ≠ Nil and target ≠ Nil
16                      then INSERT-ARC(P, a_P, source, target)
17      return P
```

The algorithm uses an association list called *Included* to keep track of pairs of nodes with a non-empty join. Line 8 adds an association between pairs of nodes joined and the corresponding node in the product graph. The function MAP() is then used to find the association for a particular pair of nodes, and returns *Nil* if there is no association. The procedure INSERT-ARC(*DG,descriptor,source,target*) takes a graph, a relational descriptor and source and target nodes, and inserts an arc between source and target with that descriptor.

The important point about the product graph $\mathcal{P}$ is that all maximal common generalizations of $\mathcal{A}$ and $\mathcal{B}$ are subgraphs of $\mathcal{P}$ [5]. Most of the subgraphs of $\mathcal{P}$ are not meaningful, as they include inconsistent nodes—that is pairs of product nodes that arise from matching the same nodes in $\mathcal{A}$ or $\mathcal{B}$. If we think of each node in $\mathcal{P}$ as representing a match between a node in $\mathcal{A}$ and a node in $\mathcal{B}$, then a consistent subgraph of $\mathcal{P}$ is one where no node from $\mathcal{A}$ or $\mathcal{B}$ is required to match more than one node from the other graph.[6]

The task of finding the most interesting common subgraph is now equivalent to finding the most interesting consistent *subgraph* of $\mathcal{P}$. This is relatively easy to implement as A* (A-star) search.

We assume the definition of a function INCONSISTENT() that takes two product nodes and returns true iff they are have a common source, that is if either the NODE-As are equal or the NODE-Bs are equal. We also assume a function CONSISTENT() on product graphs that returns true iff there is no pair of product nodes that are mutually inconsistent.

3.3 The search itself

The aim is to search within the product graph $\mathcal{P}$ for a consistent subgraph that maximizes the function *interest*. We use an A*-based search [6][7] to do this.

The start state of the search is the original product graph $\mathcal{P}$. Each subsequent state is a subgraph of this, created by copying it and deleting particular nodes from it. These graphs are kept in a priority queue according to their *interest*.

A move involves taking the most promising state graph $\mathcal{S}$ from the queue and checking if it is consistent. If $\mathcal{S}$ is consistent it is guaranteed to be the maximal consistent common generalization. In that case it is returned and the algorithm terminates. Otherwise $\mathcal{S}$ is split to form two new state graphs. The split involves first copying $\mathcal{S}$, then choosing a node n that is inconsistent with at least one other node in $\mathcal{S}$. In one of the copies of $\mathcal{S}$ the node n is *asserted*, which involves deleting all nodes that are inconsistent with n, and in the other copy n is deleted. Note that copying graphs can be computationally expensive. However, because all state graphs are subgraphs of $\mathcal{P}$ we can represent them as what we term *virtual graphs*. These are a type derived from description graphs, and contain only a handle to a 'parent' graph, and a bit vector implementing a mapping from NODES($\mathcal{P}$) to {**true, false**} that indicates which nodes are included in the virtual graph. Virtual graphs can only represent induced subgraphs on their parent graph, but copying them is extremely efficient.

---

[5] proofs of this and other such statements will appear in forthcoming thesis by J. Poole
[6] in [15] there is discussion of algorithms that allow so-called 'folding': here it is not allowed.
[7] A* search is also described in most introductory AI texts

In the original formulation of the A* algorithm the search proceeded so as to minimize a cost function [6], but here the purpose is to *maximize* the interest function. Each internal node S is itself a description graph, and all leaves below S are subgraphs of S, so the interest of S forms an upper bound on the interest of the subtree rooted at S. In the simplest case, therefore, we just define f', which estimates the potential value of a (partially completed) search path, to be the interest of the description graph S representing the incomplete search state. As this can never underestimate the interest of a leaf node below S in the search tree, our algorithm is *guaranteed* to find the optimal solution. In Sect. 4 we examine how it might be possible to improve the accuracy of this estimation.

The algorithm JOIN() below takes two description graphs and returns the common generalization that maximizes the interest function.

```
JOIN(Description Graphs A,B, function interest)
1      Open ← MAKE-PRIORITY-QUEUE(Eval-Function ← interest)
2      C ← PRODUCT(A,B)
3      while not COMPLETE(C)
4          do best-node ← CHOOSE-NODE(C)
5              C' ← COPY(C)
6              ASSERT(C, best-node)
7              INSERT(Open, C)
8              DELETE(C', best-node)
9              INSERT(Open, C')
10             C ← EXTRACT-MAX(Open)
11     return C
```

```
ASSERT(Description Graph G, Node n)
1      for each m ∈ NODES(G)
2          if   inconsistent[m,n]
3          then   DELETE(G,m)
```

We set up the procedure CHOOSE-NODE() to just choose the node that is inconsistent with the greatest number of other nodes. The pseudo-code for the procedure is trivial but tedious and is not presented here. This policy may be tuned according to the particular interest function that one selects; the optimal policy will depend on the particular interest function used.

4   Estimating Incomplete Search Paths

This section examines in more detail an issue passed over quickly in the description of the algorithm above: that of estimating how good a partially completed search path is. The potential interest of a partial match is an estimate of the best possible interest of a full match completing the partial match. The quality of this estimation, more than any other single factor, determines the efficiency of the algorithm. Note that if we want to *guarantee* the best match, then the potential interest estimate must *never* underestimate.

A naive way of evaluating the potential interest of a partial path is just to measure the interest of the state graph representing the partial path. This will always provide an upper bound on the interest, but it will usually be a wildly over-generous estimate. Depending on the interest function, however, it is possible to put a much tighter upper bound on the potential interest.

To examine possible improvements to the estimation we consider the overall interest function to be able to be broken down into three separate parts:

interest of the nodes: this is the 'context-free' contribution to the interest of the whole graph made by the member nodes (and in particular their attached concepts), independent of the particular arcs to which they are connected. This might be seen as the 'surface' interest.

interest of the arcs: this is the interest added by the arcs. The interest given by a particular arc can depend on the interest of its source and target. This can be seen as part of the structural interest.

thematic interest : this is the interest due not to local nodes, arcs or combinations, but due to larger patterns. Such interest can therefore only be determined reliably by looking at the whole graph.

The total interest of a graph is the sum of these three components. The potential thematic interest of a partially completed search path can only reliably be determined by looking at the thematic interest of the whole state graph. The potential interest of the nodes and arcs can be much more accurately estimated however, by forming the arcs and nodes into groups that are mutually inconsistent. Only one member of each such group can be included in the final graph, so we need only consider the greatest interest value of each such group towards the potential interest estimate.

Clearly this will make a huge difference to the estimates of potential interest. It is worth devoting attention to these estimation heuristics, as even slight improvements in the estimates can lead to great reduction in the number of moves made in finding the solution.

In addition to these methods which guarantee an overestimate, one of the attractions of the A* algorithm is that it exhibits graceful degradation: if the estimation heuristic only rarely underestimates (and by a small amount) the potential interest of a search state, then the final result will only rarely be suboptimal (and then only by a small amount). This allows efficient and flexible means of making fast estimates of best matches. Investigations so far suggest that even relatively poor estimates (up to 50% underestimate) still tend to result in optimal or near optimal solutions.

5   Evaluating the Algorithm

It is expected that the algorithm is non-polynomial as the known *NP*-complete problem Maximum Subgraph Matching [5] is a special case of the matching problem (where the interest function is defined to return the number of arcs in the graphs).

However, we argue that the worst-case instances are not actually interesting in practice (see [4] p. 286, and [20]). Formalizing the notion of 'interesting' or 'average' cases is difficult, but we can report an interim empirical examination of the algorithm which gives some indication of its behaviour.

The algorithm has been implemented in Common Lisp. In practice it returns solutions quickly for real description graphs: for a small set of description graphs, with around 15 nodes each, the algorithm finds the optimal matches in 40–50 moves. About 3 moves per second are made on the current implementation[8], which is not particularly efficiently coded.

Figure 5 shows the number of moves made by the algorithm for pairs of pseudo-random graphs with numbers of nodes from 4 to 15, in each case with twice the number of arcs. The data shown in each case are for an average of 25 comparisons. Bear in mind that for a graph with 14 nodes an exhaustive search would make over 25 million moves.[9]

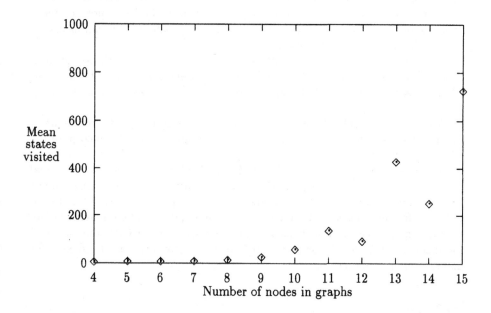

Fig. 5. A* graph match for pseudo-random graphs of various sizes

---

[8] running on a 486-based Unix box

[9] based on the assumption that half the nodes represent objects and half events, so each node is a potential match for half the nodes in the other graph. Thus the number of possible matches for two graphs with N nodes each is at least $((N/2)!)^2$.

Further investigation, both formal and experimental, will be carried out to determine more specific information about the efficiency of the algorithm under various conditions. As with all *NP*-complete problems, there will not be an efficient general purpose algorithm, but for particular types of graphs and interest functions, it appears that an efficient similarity algorithm is possible.

6 Conclusions and Further Work

The metric and algorithm presented are both extremely flexible and also potentially efficient for practical general use.

The intention now is to examine the behaviour of the algorithm for different interest measures and for different forms of source graph. In addition we are intending to test out the forms of interest functions appropriate for different domains and applications.

For these purposes it would be helpful to use a domain where there is a well-defined set of examples, a well defined, and objective, similarity measure, and where the representation of situations is possible with current knowledge. The legal domain does not provide these properties to quite the necessary extent. Currently we are investigating other domains that might have more of these desirable qualities and would welcome suggestions of material that can be shown to have the needed characteristics.

At present the algorithm is implemented just for simple description graphs. However, it appears that only straightforward extensions need to be made to allow hierarchical graphs, and graphs with co-reference links, to be matched by the algorithm.

Another goal would be to find a more general solution to the problems of implicit information hinted at in Sect. 2.2, for example a method of incorporating dynamic transformations into the search process.

For most tasks envisaged, such as CBR and information retrieval, the goal will not be to match just a pair of graphs but rather to match a single input graph against a case base of stored graphs. For this purpose it would be useful to have a fast filtering algorithm that assesses an upper bound on matches before attempting an accurate match such as in this algorithm. In CBR this is called the indexing problem. We are interested to investigate methods of indexing for different interest measures.

References

1. L. Karl Branting. *Integrating Rules and Precedents for Classification and Explanation: Automating Legal Analysis*. PhD thesis, The University of Austin at Texas, 1991.
2. H. Bunke and B. T. Messmer. Similarity measures for structured representations. In *First European Workshop on Case-Based Reasoning, Posters and Presentations, Volume 1*, pages 26–31, 1993.

3. Judith P. Dick. *A conceptual, case-relation representation of text for intelligent retrieval.* PhD thesis, University of Toronto, 1991.

4. Jon Doyle and Ramesh S. Patil. Two theses of knowledge representation: language restrictions, taxonomic classification, and the utility of representation services. *Artificial Intelligence*, 48:261–297, 1991.

5. Michael R. Garey and David S. Johnson. *Computers and intractability : a guide to the theory of NP-completeness.* San Francisco : W. H. Freeman, 1979.

6. P. E. Hart, N.J. Nilsson, and B. Raphael. A formal basis for the heuristic determination of minimum cost paths. *IEEE Transactions on Systems Science and Cybernetics*, 4:100–107, 1968.

7. Janet Kolodner. *Case-Based Reasoning.* Morgan Kaufmann Publishers, Inc., 1993.

8. George Lakoff. *Women, Fire, and Dangerous Things.* The University of Chicago Press, 1987.

9. Debbie Leishman. Analogy as a constrained partial correspondence over conceptual graphs. In *Proceedings of the First International Conference on the Principles of Knowledge Representation and Reasoning*, pages 223–234, 1989. Toronto.

10. Debbie Leishman. An analogical tool: Based in evaluations of partial correspondences over conceptual graphs. In Timothy E.Nagle, Janice A. Nagle, Laurie L. Gerholz, and Peter W. Eklund, editors, *Conceptual Structures: Current Research and Practice.* Ellis Horwood Workshops, 1992.

11. James J. McGregor. Backtrack search algorithms and the maximal common subgraph problem. *Software—Practice and Experience*, 12:23–34, 1982.

12. Douglas L. Medin, Robert L. Goldstone, and Dedre Gentner. Respects for similarity. *Psychological Bulletin*, 100(2):254–278, 1993.

13. Guy W. Mineau. Normalizing conceptual graphs. In Timothy E Nagle, Janice A. Nagle, Laurie L. Gerholz, and Peter W. Eklund, editors, *Conceptual Structures, Current Research and Practice.* Ellis Horwood (Series in Workshops), 1992.

14. Sung H. Myaeng and Aurelio Lopez-Lopez. Conceptual graph matching: a flexible algorithm and experiments. *Journal of Experimental and Theoretical Artificial Intelligence*, 4:107–126, April–June 1992.

15. Timothy E. Nagle and John W. Esch. A notation for conceptual structure graph matchers. In Timothy E.Nagle, Janice A. Nagle, Laurie L. Gerholz, and Peter W. Eklund, editors, *Conceptual Structures: Current Research and Practice.* Ellis Horwood Workshops, 1992.

16. Jonathan Poole. Similarity in legal case based reasoning as degree of matching between conceptual graphs. In *Preprints of the First European Conference on Case Based Reasoning*, pages 54–58, 1993.

17. R. Schank and R. Abelson. *Scripts, Plans, Goals and Understanding.* Lawrence Erlbaum Associates, Hillsdale, N.J., 1977.

18. John F. Sowa. *Conceptual Structures: Information Processing in Mind and Machine.* Addison-Wesley Pub. Co., 1984.

19. Anne v. d. L. Gardner. *An Artificial Intelligence Approach to Legal Reasoning.* The MIT Press, 1987.

20. W. A. Woods. Understanding subsumption and taxonomy: A framework for progress. In John F. Sowa, editor, *Principles of Semantic Networks: Explorations in the Representation of Knowledge.* Morgan Kaufmann Publishers Inc., 1991.

On the ontology of knowledge graphs

C. Hoede

Department of Applied Mathematics
University of Twente
P.O. Box 217, 7500 AE Enschede
The Netherlands

Abstract. Knowledge graphs are a special kind of conceptual graphs. They were developed independently from 1982 on and are special in that a very restricted set of types of relationships is used. In this paper the ontology of knowledge graphs and the philosophy behind its choice are discussed.

1   Introduction

Conceptual graphs are discussed in the book of Sowa [11], and belong to the vast field of semantic networks. This knowledge representation technique plays an important role in artificial intelligence, see the collection of papers in the book edited by Lehmann [8]. A paper in this book, by Sowa, has the title "Conceptual graphs as a universal knowledge representation", indicating the extent of the claim of conceptual graph theory. The foundations of this theory are therein attributed to C.S. Peirce and dated circa 1883.

Knowledge graphs were studied in the Netherlands for the first time by de Vries Robbé [13], who wrote a thesis on formal medicine. Kidney diseases were qualitatively modeled by representing concepts by vertices and relating these concepts by arcs that represented causal relationships. They were labeled directed graphs in the graph theoretical terminology, for which we refer to Bondy and Murty [4]. Stokman and the author started a project called "Knowledge Graphs" that aimed at representing medical or sociological literature by labeled digraphs, where now three types of relationships were distinguished. Next to the CAU-link, describing causal relationships, a PAR-link (part of) and an AKO-link (a kind of) were considered. Three theses, by Bakker [1], de Vries [12] and Smit [10] contain the results of the project in the years 1982–1988. The main aspects are the introduction of "constructs" as interesting subgraphs, the development of a "relational path algebra" that enabled inference of new relationships, the problems in actual text analysis and the study of robustness, consistency and similarity. A computer program named Knowledge Integration and Structuring System (KISS) implemented the theory.

In order to see how much could be represented with only three types of links the results of this theoretical study were tested on several texts, see the survey paper of James [7]. The resulting knowledge graphs represented only parts of the texts and can best be seen as qualitative models of the systems described in the texts. Constructs yielded e.g. subprocesses in knowledge graphs containing CAU-links. For our story the important point is that such subgraphs might be baptized with the name of the subprocess. The level of abstraction of the description could then be raised by representing the subprocess by a single vertex and considering the graph resulting from contracting all arcs in the representing subgraph. This operation was called "framing and naming". If the subprocess is called P then all vertices and arcs in the describing subgraph are related to P by a type of relationship called FPAR (frame part of). Herewith the number of types of relationships was extended to 4.

The consequence of this development was that the focus shifted from the structuring of knowledge, as far as allowed by the restricted set of types of relationships, to the representation of knowledge by graphs. It was decided to study the potential of knowledge graphs for representing natural language and logic. In the period 1989–1993 Willems [14] wrote a thesis on natural language with the suggestive title "The chemistry of language" and van den Berg [2] did so on logic. It was in this period that the members of the project group got notice of conceptual graphs.

Both theses are interesting but only the main results can be mentioned. The important point in Willems' thesis is the use of graphs not only for the semantics but also for the syntax of a sentence. The way of dealing with semantics that was developed in the project group can best be described as being "subjectivistic". It was extensively discussed by van den Berg, who also carried out a substantial part of Peirce's program by giving graph theoretical versions of various systems of modal propositional logic. The relation between subjectivistic and objectivistic, truth conditional, semantics will be discussed in this paper later on.

As a consequence of the research in the "second phase" of the project, the set of types of relationships has grown and changed. When we speak of the ontology of knowledge graphs, or our restricted ontology for conceptual graphs if one prefers, we mean almost exclusively this set of about ten types of links. Its growth in size was accompanied by a loss of importance of the labels of the vertices. In its most extreme form this can be expressed by stating that the only label of importance for the vertices is the label "something". The goal of this paper is to defend the extreme stand that the basics, primitives if one likes, of conceptual graphs should be a restricted set of types of links and one type of vertex. The crux of the defence is to express any word in terms of these primitives. The account that follows aims at making plausible that this is indeed possible. As there are many aspects, discussed at length in centuries of philosophy, the technique of simply stating the ideas will be used.

In Section 2 the proposed types of links are discussed. Section 2.1 discusses types due to the granular structure of the world. Section 2.2 discusses types due to the space-time nature of the world. Section 2.3 discusses types due to the existence of minds in the world. In Section 3 similarity, meaning and the role of truth is discussed. In Section 4 a program and some specific features of knowledge graphs, constructs and relational path algebra, are described shortly.

2 Thinking is linking somethings

The title of this section is chosen for two reasons. Firstly it illustrates the style of stating ideas and secondly it underlines the importance of the subjectivistic nature of the theory.

2.1 Types of links due to the granular structure of the world

A mind is able to distinguish somethings, a word we use for perceptions and other awarenesses. We suppose one, outer, world and many sets of somethings, as many as there are minds (and computers). The reason for difference of some-things, at least of the perceptions of their outer world counterparts, lies in the quantum mechanical nature of the world that enables the existence of distinct particles. The granularity of the world has led both to the awareness of what is called "something" and to the idea of "set", the awareness of a composite something. The idea of a set leads to the introduction of four types of links that reflect four set–theoretical configurations. Let a and b be two sets, then we can distinguish $a = b$, $a \subset b$, $a \cap b \neq \emptyset$ and $a \cap b = \emptyset$. Verbally we describe this as, respectively, a is equal to b, a is a subset of b, a and b have some elements in common and a and b are disparate. Looking on a and b as sets that specify composite somethings, it is natural to introduce the following types of links: EQU, SUB, ALI and DIS. Only a link of type SUB is asymmetric, but the symmetric links can be seen as consisting of two (directed) arcs. The set–theoretical configurations thus lead to the choice of the four types of links described in Figure 1.

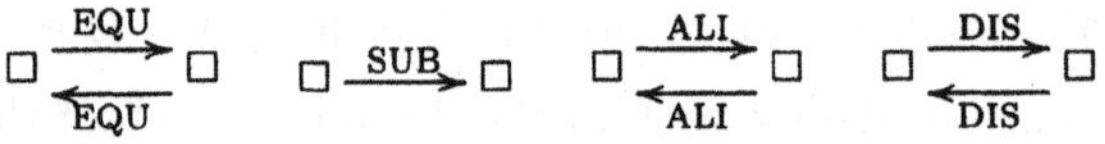

Figure 1

The unlabeled vertices, simple squares in Figure 1, are "tokens", nameless representations of awarenesses. When linking somethings, minds are bound to use these types, amongst others. The most basic type of link is the ALI-link. A mind reflecting on two awarenesses begins its thinking by establishing the existence of common somethings in them. Similarity, a measure for the alikeness of awarenesses, plays an important role in this theory. Two awarenesses can not be completely equal, unless they are the same awareness. The symmetrical EQU-relationship therefore offers itself as the ideal co-reference link. If something common is distinguished in two awarenesses, a substructure of both, it is called a common *property*. This property is itself an awareness and related to both by the SUB-relationship. We can say that the property is a subawareness and that the two awarenesses are similar, to a yet unspecified extent. We will postpone the discussion of the DIS-link to Section 2.3.

We now have to deal with words, used for describing types and names. SOME-THING is the *type* of different awarenesses. We represent this in knowledge graph theory by a token, for the awareness, and a vertex labeled by the word something. Drawing a token expresses that something exists for a mind, that the mind is *aware* of that something without having a word for it yet. The link between these vertices is an arc from the vertex labeled SOMETHING to the token. The type of this arc is called ALI (alike), see Figure 2.

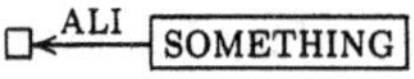

$$\Box \xleftarrow{\text{ALI}} \boxed{\text{SOMETHING}}$$

Figure 2

Less abstract types for a token will be represented in a similar way. Figure 3 expresses that something (a primitive) is of type DOG.

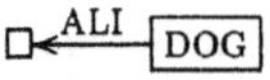

$$\Box \xleftarrow{\text{ALI}} \boxed{\text{DOG}}$$

Figure 3

A mind is considered to contain a *mind graph*, consisting of tokens and links between them. Each substructure of this mind graph is a candidate for attachment of a word. Hence the choice of the direction of the arc from the word to the token. The choice of the type ALI, as in Figures 2 and 3, is due to the view that the mind is trying to "bring a (sub)structure of its mind graph under words" and does so by establishing (high) similarity with structures for which it has words available. Structures for which a word is used are called *concepts*. If a structure A, subgraph of the mind graph, and its substructure B, a subgraph of A, both represent types in the vocabulary of a mind, then the SUB-relationship between B and A leads to the well known *lattice of types* and the statement that an A *is a* B.

Mathematically, two sets a and b are equal, if and only if $a \subseteq b$ and $b \subseteq a$. Note that the SUB–relationship expressed e.g. $a \subset b$, where a is a proper subset of b. Had we interpreted the SUB–link as expressing inclusion or equality, we should have dissolved the symmetric EQU–link into two SUB–links. The asymmetric EQU–link is used in the project to describe values.

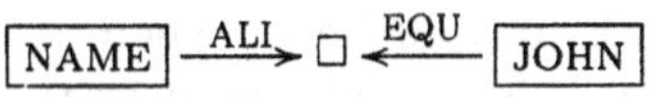

$$\boxed{\text{NAME}} \xrightarrow{\text{ALI}} \Box \xleftarrow{\text{EQU}} \boxed{\text{JOHN}}$$

Figure 4

Figure 4 expresses that something of type NAME has the value JOHN. We shall come back to naming in Section 2.3. There is an essential difference between the symmetric ALI– or EQU–links between two tokens and ALI– and EQU–links between tokens, unlabeled vertices, and labeled vertices. The labels are symbols, words are (composite) symbols, used by the mind to distinguish alikenesses and identify an awareness. In communicating with other minds utterance is given to these labels. "Bij het communiceren met andere denkvermogens wordt uiting gegeven aan deze labels" is the way the author would utter the thoughts behind the former sentence in Dutch. A striking point is that the English word "label" is used in Dutch as well, for lack of another word. However, Flemish speaking people would probably use the word "naamkaart", which is a card with a word on it, which is attached to something and thereby names it. Two other types of link should be mentioned in this section. If $a \cap b = \emptyset$, we often are aware of the existence of a set c for which $a \subset c$ and $b \subset c$. This relates a and b indirectly. As two disparate somethings, that a mind is aware of, can always be seen as part of the full content of awarenesses, this indirect relationship always exists. We may say that two concepts are always associated, and introduce the, symmetric, ASS–link to express this. But if "everything has to do with everything else" this is hardly fruitful. However, considering a general setting for two unrelated concepts is one of the activities of a mind. For what we call the process of "framing and naming" of complex parts of the mind graph, association of concepts is vital. Seeing a certain substructure of the mind graph as of sufficient interest to be able to utter statements about it, the mind is considered to create a word for this substructure. Determining the substructure is called *framing* and creating the word is called *naming*. Let the substructure be baptized s and let it consist of the simple graph $a \xrightarrow{\text{SUB}} b$, then the relationship of the substructure s to the elements of the substructure $(a, b,$ the arc(a, b) with type SUB$)$ is a second type of a part–relationship. Vertices and links of the substructure, that is framed and named s, will be said to be related to s by an asymmetric link of type FPAR (frame par). The very moment a frame[1] concept is established by a mind the relationship between its constituents has changed indirectly and this relationship might be called ASS in case no prior relationship was present. As we said before ASS is a very weak type of relationship. However, it is not without reason that the gentlemen Gold, Gold, Gold and Silver of the firm Moneymaking are called associates.

2.2 Types of links due to the space–time nature of the world

So far we have considered the granular aspect of the world and the types of relationships of mereological nature. The physical aspects space and time give rise to thinking in terms of two other types of relationships. The first type that comes to mind is CAU, describing causal relationship. In fact knowledge graphs were first formulated entirely in terms of CAU–relationships. All processes are essentially directed networks of CAU–links. There is however another type of link that seems more basic. It is the link that expresses ordering, in space or in time, and that is innate in words like before, after, above, behind, etc. The type of this link will be called ORD. CAU–links are especially important as all verbs can be considered to express causations. In Figure 5 the verb HIT is described.

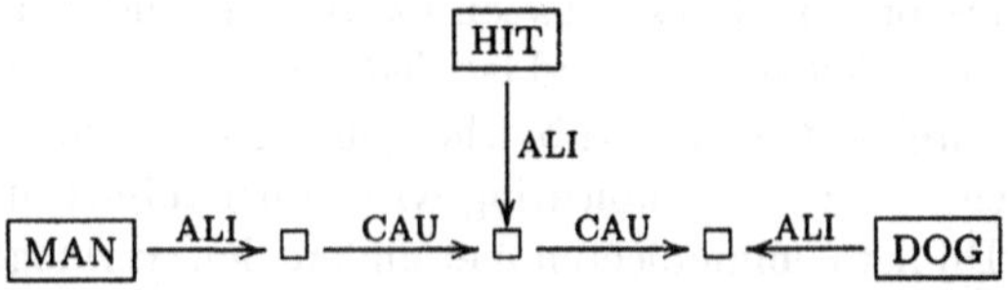

Figure 5

In "man hit(s) dog" the man is causational for the hitting, which is causational

---

[1] The word frame is often used in AI theory. We simply mean that a subgraph of the mind graph has been set apart, putting an imaginary frame around it, very much like a frame is put around an painting.

for the dog. To hit is a transitive verb, hence the two CAU–links. An intransitive verb may be represented by the first CAU–link only. "Man sleep(s)" can be expressed as in Figure 6.

$$\boxed{\text{MAN}} \xrightarrow{\ \text{ALI}\ } \square \xrightarrow{\ \text{CAU}\ } \square \xleftarrow{\ \text{ALI}\ } \boxed{\text{SLEEP}}$$

Figure 6

This way of using the CAU–link to express an intransitive verb is interpreting man as cause for the sleeping process. This is somewhat awkward. Seeing man as some *part* of the sleeping process is another interpretation of "man sleep(s)". Let us look upon the verb as essentially a frame. Then the FPAR–link may be used to relate man to sleep, replacing in Figure 6 the CAU–link by the FPAR–link. The man is then deemed to be a part of his own sleeping. For the transitive verb "hit" both CAU–links should be replaced by FPAR–links towards the token representing the process HIT, indicating that the tokens for man and dog are part of the hitting process. The causality may then be given by an ORD–link as in Figure 7.

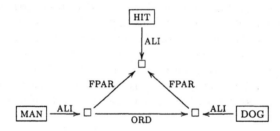

Figure 7

This figure has two weaknesses. First of all it has not been indicated which aspect of ordering has been expressed. One can attribute TIME to both MAN and DOG and consider the ordering as pertaining to the time attributes. The "hitting" comes before the "being hit". This analysis replaces CAU–links by part of and ordering links and resembles the analysis of causality given by Hume, for which we refer to Russell [9]. Secondly there is an aspect of the FPAR–relationship that is not well represented. The ordering itself, the ORD–arc, also belongs to the frame HIT. Quite generally links should be linked to a token representing a frame by a FPAR–link as well. This is a technical difficulty as well as a conceptual difficulty. Links should be seen as concepts too, in the extreme also represented by a token which is typed by the type of link. The technical solution is to consider "total" graphs. A total graph $T(G)$ of a directed graph G has as vertices both the vertices and the arcs of G. The vertices of $T(G)$ are made adjacent by unlabeled (directed) arcs. These arcs connect a vertex representing a vertex of G to a vertex representing an arc of G or vice versa, depending on whether in G the arc is incident *out* (respectively incident *in*) the vertex. We choose the arc $\boxed{a} \xrightarrow{\ \text{CAU}\ } \boxed{b}$ to be represented as $\boxed{a} \longrightarrow \boxed{\text{CAU}} \longrightarrow \boxed{b}$ and use a link of type PAR for attribution, that we will discuss in Section 2.3. With these changes Figure 7 becomes Figure 8.

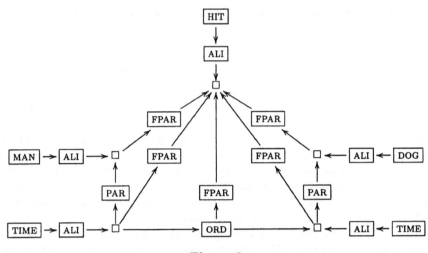

Figure 8

This graph is already quite complicated. In the extreme all vertices representing types of links could be replaced by tokens that are linked to words like PAR and FPAR by means of an ALI–link. For the ALI–link itself this does not work, because the replacement by token and type ALI would be indefinitely repeated. This puts the ALI–link on a pedestal as primes inter pares of the primitives. Thinking has in this view two main features. The tokens (somethings) that are linked and the ALI–links that express the similarities that have been established by the mind.

2.3 Types of links due to the existence of minds in the world

The existence of minds leads to the consideration of reasoning, but also plays a role in our choice of types of links made so far. We consider a mind to have a mind graph reflecting the world as perceived by the mind. Certain substructures are carrying names. Each word in the vocabulary of a mind has a *word graph* corresponding with it. When describing a certain part of its mind graph, for example when describing things happening, the mind is supposed to compare the perceived structure with the structures of the word graphs at its disposal. Those words that come closest, are most "alike", will be chosen. The perceived structures are typed. Naming is slightly different from typing, in that the mind focuses on a certain structure. It is the difference between "a dog bites a man" and "Fido bites a man". In the second sentence the mind expresses an *attribution* it made. This leads to the introduction of yet another type of link expressing how attributes are related to somethings.

The SUB–relationship, exemplified by set inclusion and mathematically transitive, stands central in what is called mereology. This is the theory of the part–whole configuration, in which the various aspects of the "part of"–relationship are considered. As an example we mention a paper by Iris *et al.* [6], in which four types of part of–relationships are considered. Problems of mathematical nature arise when considering element inclusion. We may have $1 \in \{1,2,3\}$ and $\{1,2,3\} \in \{\{1,2,3\},\{3,4\}\}$, so element inclusion is not transitive as $1 \notin \{\{1,2,3\},\{3,4\}\}$. We thus have a reason to introduce at least one other type of link next to the SUB–link and the FPAR–link. This also comes forward when analysing the word combination "yellow ball". Yellow is the colour of the ball. Yellow evaluates the attribute colour. The relationship between colour and ball is not well represented by the SUB–link or the FPAR–link as "ball", does not contain the concept colour for all minds. For a mind that would not call something a ball unless its has colour the FPAR–relationship will be a satisfactory description of the colour–ball situation. If colour is considered to be irrelevant for "ball", an attribute of the mind so to say, that however may be considered in relation to ball, then one needs another part of–link. We will use the PAR–link to describe this relationship. Then Figure 9 describes "yellow ball".

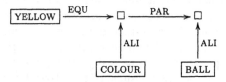

Figure 9

So "yellow" is considered to be the name, value, of a colour, very much like "2 metres" is considered to be the value, name, of a length. Looking upon "element" as an attribute of "set" we may express $1 \in \{1,2,3\}$ as in Figure 10.

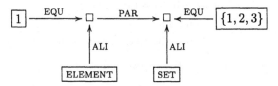

Figure 10

For the time being SUB, FPAR and PAR are the only types necessary, other relationships considered in mereology probably stemming from structural complexification, that we will discuss later in Section 3. For universal quantification Van den Berg and Willems [3] proposed the socalled Skolem–link, an arc of type SKO. The universal quantifier can be represented by the SKO–link. $x \xrightarrow{\text{SKO}} y$ expresses that y is informationally dependent on x. A token that carries a loop of type SKO is something that is informationally dependent only on itself, i.e. may be anything. The DIS–link introduces an aspect of quite another nature, namely logic, in particular negation. However, the DIS–link is not the type of link that is used to deal with negation. In line with our policy of reluctantly extending the number of types, we gave the DIS–link some attention and discussed its potential. It turned out that negation should better be dealt with in another way. In his thesis on knowledge graphs and logic, Van den Berg [2] deals with negation in the way Peirce did it. The proposition expressed by the graph is framed and the contents of the frame are linked to the frame token by a NEG–link. "Fido is not a dog" is expressed, in total graph form, as in Figure 11. "No dog is named Fido" e.g. would involve a SKO–link and a token of type NAME.

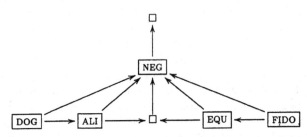

Figure 11

So the NEG–link acts like the FPAR–link. The reader is referred to Van den Berg's thesis for a detailed account of the use of graphs for representing propositional logic, predicate logic and various systems of propositional modal logic. Recently van den Berg *et al.* [3] showed that predicate modal logic can be treated in a similar way. Also possibility and necessity are representable by links similar to the FPAR-link. In Figure 12 a resume of primitives for knowledge graphs is given. It should be clear from this figure that the primitives chosen are domain independent. One may therefore hope for universal applicability. Note that we remarked that some of these primitives may be redundant, like the CAU–link, and that representation of knowledge by knowledge graphs may force us to include yet other primitives.

Vertex primitive (awareness)

□	Token	Something
	Link primitives (granular aspects)	
EQU / EQU	EQU–link	Identity of two somethings
EQU [word]	EQU–link	Valuating of something by word
SUB	SUB–link*	Inclusional part–of
ALI / ALI	ALI–link	Similarity of two somethings
ALI [word]	ALI –link	Typing of something by word
DIS / DIS	DIS–link	Disparateness of two somethings
PAR	PAR–link*	Exterior attributing of something to something
FPAR	FPAR-link*	Interior attributing of something to something (frame)
ASS / ASS	ASS–link	Associatedness of two disparate somethings (being part of the same frame)
		* Mereological links

Link primitives (space–time aspects)		
ORD	ORD–link	Something ordered in comparison with something else
CAU	CAU–link	Something influencing something
Link primitives (logical aspects)		
SKO	SKO–link	Something being informationally dependent on something else
SKO (loop)	SKO–loop	Something being only informationally dependent on itself (universal** quantification)
NEG	NEG–link	Something being part of something (frame) that is negated ** Existential quantification by drawing of the graph

Figure 12

3 Similarity, Meaning and Truth

Our discussion of primitives showed that the ALI–link is very special. The alikeness of somethings is at the basis of typing. Similarity of something with something else leads to an awareness that may be uttered by a word (symbol) that expresses the similarity, i.e. communality, of the two somethings. This communality may be complex. It may consist of a frame in which other types of somethings have been distinguished and have been linked. This process in the mind has been called "framing and naming", although "framing and typing" is more precise in relation to the use of the ALI–link in Figure 1. In the inverse process a word invokes a structure of the knowledge graph of a mind, namely the frame that was named by that word. The meaning of the word, to the mind, is that structure. This is a very important part of the theory. The mind graph is subjectively used to give meaning to a word. Giving meaning to a word, that corresponds to a certain subgraph of the mind graph, is done by embedding that subgraph in the mind graph. As all mind graphs are supposedly different, a word has as many meanings as there are mind graphs. It is again the similarity, now between mind graphs and subgraphs of mind graphs, that enables communication. However, every word has a large number of homonyms and this in two ways: First there is the difference in structure of the parts of mind graphs that minds bring in correspondence with the word. This is *interpersonal homonymity*. Different minds give different meanings to the word ELEPHANT. In objectivistic theories, aiming at unique and generally accepted meaning, the prototype ELEPHANT is discussed, a discussion that has as essential difficulty the fact that minds, in an exchange of information, have different ELEPHANT in mind. Secondly there is *intrapersonal homonymity*. Every mind has the problem that there are not enough words to name all structures that the mind could frame. A well-known example is that of gradually changing forms. Plate — bowl — cup — vase are four words used to describe an infinity of forms. Where a form changes from a plate into a bowl is a highly subjective matter. For our discussion it is important that there is abundant homonymity. The conclusion is that minds subjectively give meaning to words. Theory making is therefore always subjective. Objective theory making is the goal of a community of minds, say those of scientists. What is commonly accepted, read embedded in many mind graphs, achieves the status of objective knowledge. We will return to this discussion in Section 4, when we consider truth and objectivist theories of semantics. Of course the theory making in this paper itself belongs to the attempts to establish "objective" insight. The whole of philosophy (including science) has this goal.

The *meaning of a word* is ultimately the mind graph in which its corresponding frame is embedded. This definition has a major drawback. All concepts that occur in a specific mind graph have the same meaning, which is that mind graph. Say one word and the listening mind links its up with everything known to that mind (we assume that mind graphs are connected, i.e. have only one component). Usually one wants to focus upon part of the mind graph. It is therefore useful to distinguish between foreground knowledge and background knowledge. Let v be a vertex token in a mind graph M and let $N(v)$ be the set of its neighbours, then the induced graph $\langle v \cup N(v) \rangle$ is called the foreground meaning of v and the induced graph $\langle V(M) \backslash \{v\} \rangle$ minus the links in $\langle N(v) \rangle$ is called the background meaning of v. The meaning of v and w may be the same, namely the mind graph M that they are part of, their foreground and background meanings always differ. Objective theory making, in the sense of defining concepts, consists of attaining consensus on foreground meanings for minds. Linguistic problems mainly concern the differences in background meanings for minds. The ideas brought forward here can be illustrated by discussing the second basic concept next to the token (something), which is the concept of link or relationship. Again ultimately, the *meaning of the link* between two tokens can be defined as the component of the mind graph that they both belong too. This has the same drawback as before. If the mind graph is connected all links between pairs of tokens have the same meaning. All pairs can be said to be associated or related at least by the ASS–link, whereas the DIS–link might be said to be present if the tokens would belong to different components of the mind graph. As in the case of single tokens, one may consider pairs of tokens that are part of subgraphs of the mind graph. Any such subgraph establishes a relationship between the two tokens, and gives the pair a foreground meaning, the background meaning being the rest of the mind graph. So two tokens can be related by as many connected subgraphs there are that contain them both. Figure 13 depicts this symbolically.

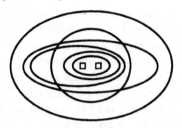

Figure 13

Six relationships between the two tokens have been indicated, ranging from the whole mind graph to a subgraph that may consist of a single arc, possibly even unlabeled. The most basic awareness, next to that of the token, is that two somethings are linked, the linkage being the similarity between a (possibly infinite) number of pairs. The many relationships possible between two tokens lead to abundant homonymity of relationships, both intrapersonal and interpersonal. This homonymity is so enormous that one may doubt whether a mind really uses synonyms. Two words used to indicate the same awareness, of a relationship or of something, may easily have meanings that are extremely similar but yet are not completely identical. As similarity is considered to stand at the basis of the development of types in the mind, any attempt to have a theory that might be used in artificial intelligence, should have measures for similarity. The author [5] considered the similarity of sets to be the starting point for such measures. As we are to compare graphs we compare vertex sets and link sets. Given two graphs G_1 and G_2, the similarity of $V(G_1)$ and $V(G_2)$ may be considered, as a start. The similarity of $E(G_1)$ and $E(G_2)$, as sets, can be taken into account as well. However, for graphs and in particular for graphs with many identical links between tokens, the comparison of the explicit structures should be made. Mappings may be considered, of part of the vertices of G_1 to part of the vertices of G_2, that are governed by identity of types and names. A mapping for which the number of common links is maximum, optimal embedding, may then determine the intersections of the vertex sets and the link sets from which similarity can be calculated.

Given some choice for the similarity measure a computer program can simulate thinking in that similar structures in the mind graph of the computer can be determined. The framing and naming process could easily be implemented in principle. Feeding the computer with "impressions" that it transforms into a mind graph, the computer may start "thinking" about its unstructured mind graph and come up with a set of words for the frames it distinguished as similar substructures.

Truth is something attributed by the mind to a substructure of the mind graph. The substructure does not necessarily correspond to a proposition. The values for the truth may be chosen to be true or false or, if one wants to include this, indeterminate.

The mind graph can be seen as the world model of the mind. Most of it corresponds, and is by perception due to, a presupposed outer world, part is created endogenously and structures the awarenesses about the inner world. Truth attributed to a substructure seen in correspondence with the outer world is usually called *truth*. If attributed to a substructure corresponding to the inner world, truth is usually called *belief*. In our subjectivistic theory there is no essential difference. Also substructures corresponding to perceptions of the outer world are not true per se, but are believed to be true by the mind. Therefore we have no reason, apart from the presupposed origin of the substructure, to make a distinction between truth and belief.

Consistency is another concept that should be mentioned in connection with truth. The contemplating mind attributes the values right or wrong to a substructure. A typical example is a circuit of SUB–links, leading to the conclusion that "a set is a proper subset of itself", see Smit [10]. This is an inconsistency if the mind attributes the truth value false to this statement. If reasoning about the substructure does not lead to a contradiction, the substructure is called *consistent*. It is held to be true by the mind until some contradicting information becomes available. Classical mechanics is a wrong theory, both from the point of view of relativity theory and from the point of view of quantum mechanics. The theory was held to be right for centuries and is now seen as a good theory for small velocities and large masses.

This example is interesting for its analogy with the theory of semantics. In objectivistic theory so-called well–formed formulas are considered. These formulas are interpreted against a model and assigned the truth value true or false. The "meaning" of the formulas consists of a model comparison with a truth check. Hence the name truth–conditional semantics. The truth check is supposed to be performable by any mind (including computer "minds"). The model is also supposed to be given, identical for all minds doing the checking. From our subjectivistic point of view this theory is wrong in a way similar to the way classical mechanics is wrong. The model used for the checking is the substructure of the mind graph corresponding to the model. The truth check is performed by a single mind. Only in the limit of identical substructures of the various mind graphs and of identical truth attributions the objectivistic theory is right. These conditions are usually met, for example, in computers with a certain data base in its memory (the identical substructures) and unfailing checking potential (the identical truth attributions). Therefore formal semantics is useful in computer science. It is basically wrong when dealing with natural language. To make our point clearer, consider a well–formed statement like "It is raining right now". Does this sentence have a meaning? Given a model and a possibility to check its truth against the model it has, according to objectivistic formal theory. In our subjectivistic theory the sentence always has a meaning, the substructure of the mind graph corresponding to the sentence. No truth check is needed. In fact truth is irrelevant for meaning. Meaning is the structure, the model made by the mind, irrespective of the truth value the mind attributes to this structure.

4 Conclusive remarks

We have stressed the philosophical background of knowledge graph theory. This paper does not so much have a conclusion. It rather implies a program. An important point of this program is to make word graphs. A start has been made with a lexicon. Each (English) word in it is described by a graph. The extent and the structure of this graph is ambiguous. It depends on the encoder, very much like entries in normal dictionaries depend on the person that makes the dictionary. The goal is to use the lexicon to see whether the Chemistry of Language [14] works and the proposed ontological representation is precise and non-ambiguous. In principle the theory of conceptual graphs, as described by Sowa [11], holds for knowledge graphs. A knowledge graph can be said to be a deliberately restricted kind of conceptual graph. The main problem with conceptual graphs is that the relationships admitted are too complex. As an example one of the relation types discussed in Appendix B.3 of Sowa's book [11] will be given in knowledge graph form. UNTL, until, has example sentence "The ticket is valid until 1 a.m." Clearly a time interval is partly described, a period of being valid. As intervals usually exist of an infinite number of points–in–time or values on some scale, universal quantification comes into the proper description.

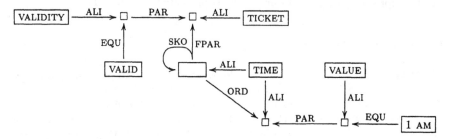

Figure 14

"The knowledge graph of Figure 14 reads" (The) validity of (the) ticket (holds) <u>for all</u> moments (in) time (that are) before (the) moment (in) time of value 1 am.". The conceptual graph representation is

$$[\text{SITUATION} : [\text{TICKET}] \rightarrow (\text{ATTR}) \rightarrow [\text{VALID}]] \rightarrow (\text{UNTIL}) \rightarrow [\text{TIME} : 1\text{AM}]$$

The reader may decide on the subgraph of Figure 14, that is the word graph for "until".

A second important point of the program is to show that linguistic problems involving background information can indeed be solved by knowledge graph theory, in particular by the idea of background meaning.

Special features of knowledge graph theory are constructs and a path algebra. Constructs, introduced by Bakker [1], are subgraphs with the property that a vertex v (with one type of arc), that has an arc *towards* one or more vertices w in the subgraph, has the property that all other vertices in the subgraph can be reached, from some vertex w, by paths completely within the subgraph, and an analogous property for vertices v that have an arc *from* one or more vertices w in the subgraph. In Figure 15 the subgraph induced by the vertex set $\{b, c, d\}$ is one of the constructs.

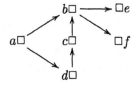

Figure 15

The relational path algebra involves a multiplication of consecutive arcs and an addition of parallel arcs. A CAU–link from x to y followed by a SUB–link from y to z gives a CAU–link from x to z. This is an example of one of the inferences. One of the advantages of a restricted set of types of links is the possibility of a manageable inference system. This is particularly important when knowledge graphs are used for practical applications like decision support systems or expert systems. De Vries Robbé [13] had, in 1978, only the causal link in his theory. By now, using ideas from knowledge graph theory, his group developed MEDES, a prototype of a medical expert system, with an ontology of about 18 types of links. A third important point of the program is to show how other ontologies can be expressed in terms of our ontology, presented in Figure 12.

References

[1] Bakker, R.R., *Knowledge Graphs: Representation and Structuring of scientific knowledge*, Ph.D. Thesis, University of Twente, Enschede. 1987.

[2] Berg, H. van den, *Knowledge Graphs and Logic: One of two kinds.* Ph.D. Thesis, University of Twente, Enschede, 1993.

[3] Berg, H. van den and M. Willems, *Quantification in knowledge graphs*, Memorandum nr. 872, Department of Applied Mathematics, University of Twente, Enschede, 1990.

[4] Bondy, J.A. and U.S.R. Murty, *Graph Theory with Applications*, MacMillan, New York, 1976.

[5] Hoede, C., *Similarity in knowledge graphs*, Memorandum nr. 505, Department of Applied Mathematics, University of Twente, Enschede, 1986.

[6] Iris, M.A., B.E. Litowitz and M.W. Evens, Problems of the part-whole relation. In *Relational models of the lexicon* (M.W. Evens ed.), Cambridge University Press, Cambridge, 261–288, 1988.

[7] James, P., Knowledge graphs. In *Linguistic Instruments in Knowledge Engineering* (R.P. van de Riet and R.A. Meersman, eds.), North-Holland, Amsterdam, 97–117, 1992.

[8] Lehmann, F., *Semantic Networks in Artificial Intelligence*, Pergamon Press, Oxford, 1992.

[9] Russell, B., *History of Western Philosophy*, Unwin University Books, London, 1961.

[10] Smit, H.J., *Consistency and Robustness of Knowledge Graphs*, Ph.D. Thesis, University of Twente, Enschede, 1991.

[11] Sowa, J.F., *Conceptual Structures: Information Processing in Mind and Machine*, Addison-Wesley, Reading, 1984.

[12] Vries, P.H. de, *Representation of Scientific Texts in Knowledge Graphs*, Ph.D. Thesis, University of Groningen, 1989.

[13] Vries Robbé, P.F. de, *Medische Besluitvorming: Een Aanzet tot Formele Geneeskunde*, Ph.D. Thesis, Department of Medicine, University of Groningen, Groningen, 1978.

[14] Willems, M., *Chemistry of Language: A graphtheoretical study of linguistic semantics*, Ph.D. Thesis, University of Twente, Enschede, 1993.

Conceptual Graphs and First-Order Logic

Michel Wermelinger

Departamento de Informática, Universidade Nova de Lisboa
2825 Monte da Caparica, Portugal
E-mail: mw@fct.unl.pt

Abstract. Conceptual Structures (CS) Theory is a logic-based knowledge representation formalism. To show that conceptual graphs have the power of first-order logic, it is necessary to have a mapping between both formalisms. A proof system, i.e. axioms and inference rules, for conceptual graphs is also useful. It must be sound (no false statement is derived from a true one) and complete (all possible tautologies can be derived from the axioms). This paper shows that Sowa's original definition of the mapping is incomplete, incorrect, inconsistent, and unintuitive, and the proof system is incomplete too. To overcome these problems a new translation algorithm is given and a complete proof system is presented. Furthermore, the framework is extended for higher-order types.
Key phrases: logical foundations of Conceptual Structures; ϕ operator; inference rules; logical axioms; higher-order types; meta-level reasoning.

1 Introduction

The logical foundation of CS Theory, as presented in [Sowa, 1984], is based on the definition of the ϕ operator, which translates conceptual graphs into first-order formulas, and on the definition of rules of inference. On page 142 it is claimed that "any formula in first-order logic can be expressed with simply nested contexts and lines of identity", and Theorem 4.4.7 on page 173 states that the inference rules for conceptual graphs are complete. However, as will be shown, the formal definition of ϕ doesn't fulfill the claim and the theorem—which is not proven—is false. Moreover, Sowa has been advocating the use of meta-level graphs. To that end, higher-order types are needed, although they are just as useful to specify finer-grained ontologies.

The purpose of the work to be described in this paper is therefore twofold: on one hand to correct the original definitions, on the other hand to extend them in order to accomodate higher-order types, thus providing a first step towards meta-level reasoning with conceptual graphs. The full reformulation and extension of the (first-order) logical foundations of conceptual graphs is made up of the following steps:

1. Define a first-order language L and an interpretation for it.
2. Define an algorithm to translate conceptual graphs into formulas of L.
3. Define inference rules and logical axioms for conceptual graphs.
4. Prove that they form a sound and complete proof system for first-order logic.

Due to space limitations, steps 1 and 4 have been omitted[1]. They can be found in [Wermelinger, 1995] which also includes the higher-order type framework to be used by ϕ and the inference rules. That framework is a simplified and yet more expressive formulation of the formal proposal for incorporating higher-order types into CS Theory presented in [Wermelinger and Lopes, 1994].

The structure of the paper is straightforward. The next section presents an overview of the adopted higher-order type system, and the other two main sections deal with the ϕ operator and the inference rules, respectively. The reader is expected to have some knowledge of conceptual graphs and logic. Most examples are adapted from [Sowa, 1984; Sowa, 1992].

2 Higher-Order Types

The building units of conceptual graphs are types. There is a concept type hierarchy $\mathcal{T}_C$ and a relation type hierarchy $\mathcal{T}_R$. Both of them are lattices. The top element of $\mathcal{T}_C$ is the universal concept type $\top_c$, and the bottom element is the absurd concept type $\perp_c$. Similarly, the universal relation type $\top_r$ and the absurd relation type $\perp_r$ are the top and bottom elements of $\mathcal{T}_R$. Concept types are classified according to their kind and order, and relation types are classified according to their arity and order.

There are two kinds of concept types: relational and non-relational ones. The former denote relations, the latter do not. The set of all relational concept types is written T^{rc}, and T^{nc} represents all non-relational ones. Concept types can also be classified according to their order: T_i^{rc} is the set of ith-order relational concept types, and the symbol T_i^{nc} stands for the set of all ith-order non-relational concept types. Both $\top_c$ and $\perp_c$ can be of any kind and order. Therefore, they stand apart from the other concept types and aren't included in T^{rc} or T^{nc}. As for relation types, each has an associated arity and order. The set of all n-ary relation types is written as $T_{(n)}^r$ and the set of all ith-order relation types is represented by T_i^r. Again, $\top_r$ and $\perp_r$ do not belong to any of those sets.

Example 1. Following are some types and their classification according to the above scheme:

- CAT, FELINE, ANIMAL, SQUARE, RECTANGLE, RHOMBUS $\in T_1^{nc}$;
- SPECIES, GENUS, SHAPE $\in T_2^{nc}$;
- CATEGORY, CHARACTERISTIC $\in T_3^{nc}$;
- AGNT, OBJ, LOC $\in T_{(2)}^r \cap T_1^r$;
- BETW $\in T_{(3)}^r \cap T_1^r$;
- INVERSE-OF is a relation between two first-order order relations, hence it is an element of $T_{(2)}^r \cap T_2^r$;
- RELATION, BINARY, TRANSITIVE, REFLEXIVE, ANTI-SYM, SYMMETRIC, PARTIAL-ORDER $\in T_2^{rc}$ because there is a second-order relation type.

---

[1] The completeness proof consists mainly in showing how the axioms and inference rules given in [Hamilton, 1988] can be translated to conceptual graphs.

Simply put, a higher-order type denotes a set of lower-order types, and if t_1 is a subtype of t_2 then the denotation of t_1 must be a subset of t_2's denotation. More specifically, relational concept types denote relation types, non-relational concept types denote other non-relational concept types, and relation types denote tuples of concept types[2]. Therefore, if t_1 is a subtype of t_2 then both must be of the same kind. Furthermore, if t_1 and t_2 are relation types they must have the same arity. That way relation nodes can be generalized or specialized without removing or adding concept nodes to the graph.

Example 2. The only subtype relationships (represented by $<$) among the types of the previous example are:

- CAT $<$ FELINE $<$ ANIMAL;
- SQUARE is the maximal common subtype of RECTANGLE and RHOMBUS;
- PARTIAL-ORDER is the maximal common subtype of TRANSITIVE, REFLEXIVE, and ANTI-SYM which in turn are subtypes of BINARY;
- SYMMETRIC $<$ BINARY $<$ RELATION.

A concept $\boxed{t : m}$ indicates that m is an entity of type t. In other words, m is an element of the denotation of t. Therefore, if t is a relational concept type then m must be a relation type. Otherwise, i.e. if t is a non-relational concept type, then so is m. To sum up, relation types and non-relational concept types can be used as markers, too. However, as first-order non-relational concept types denote individuals (and not types), a new set T_0^{nc} of "zero-order types" is needed. The elements of T_0^{nc} are mutually incomparable since they represent individuals. All other markers are organized into (disjoint) lattices since they are types. The marker set $\mathcal{M}$ is therefore a partially ordered set. If we add the generic marker $*$ as a top element and the absurd marker $\maltese$ as a bottom element, then $\mathcal{M}$ becomes a lattice too, like $\mathcal{T_C}$ and $\mathcal{T_R}$.

Example 3. The following concepts show the denotation relationships between the types of Example 1. When a type is used as a marker, it is written in lower case and prefixed with #.

- $\boxed{\text{CAT: #Garfield}}$ where #Garfield $\in T_0^{nc}$;
- $\boxed{\text{SPECIES: #cat}}$ and $\boxed{\text{GENUS: #feline}}$;
- $\boxed{\text{SHAPE: #square}}$ $\boxed{\text{SHAPE: #rectangle}}$ $\boxed{\text{SHAPE: #rhombus}}$;
- $\boxed{\text{CATEGORY: #species}}$ $\boxed{\text{CATEGORY: #genus}}$ $\boxed{\text{CHARACTERISTIC: #shape}}$;
- $\boxed{\text{SYMMETRIC: #inverse-of}}$ which implies $\boxed{\text{BINARY: #inverse-of}}$ and $\boxed{\text{RELATION: #inverse-of}}$;
- $\boxed{\text{BINARY: #agnt}}$ $\boxed{\text{BINARY: #obj}}$ $\boxed{\text{BINARY: #loc}}$.

---

[2] The arguments of a first-order relation are non-relational types, and a higher-order relation has as arguments lower-order relations. But as conceptual graphs are bipartite, those arguments must be represented by concepts. Hence the need for relational concept types.

3 Translation

To show the logical foundations of conceptual graphs, the first step consists in finding a correspondence, i.e. a translation algorithm, between graphs and closed formulas[3] of some first-order language. The latter is implicitly defined by the transformation process. In CS Theory, the translation is given by the ϕ operator. Let us recall its definition as given in [Sowa, 1984].

Assumption 3.3.2. The operator ϕ maps conceptual graphs into formulas in the first-order predicate calculus. If u is any conceptual graph, then ϕu is a formula determined by the following construction:

- If u contains k generic concepts, assign a distinct variable symbol $x_1, \ldots, x_k$ to each one.
- For each concept c of u, let $identifier(c)$ be the variable assigned to c if c is generic or $referent(c)$ if c is individual.
- Represent each concept c as a monadic predicate whose name is the same as $type(c)$ and whose argument is $identifier(c)$.
- Represent each n-adic conceptual relation r of u as an n-adic predicate whose name is the same as $type(r)$. For each i from 1 to n, let the ith argument of the predicate be the identifier of the concept linked to the ith arc of r.
- Then ϕu has a *quantifier prefix* $\exists x_1 \ldots \exists x_k$ and a *body* consisting of the conjunction of all the predicates for the concepts and conceptual relations of u.

Assumption 4.2.3. If p is a proposition asserting the graphs $u_1, \ldots, u_n$, then ϕp is the formula $(\phi u_1 \wedge \ldots \wedge \phi u_n)$. If c is a negative context consisting of (NEG) linked to a proposition p, then ϕc is $\neg \phi p$. All generic concepts that occur in p or any context nested in p must be assigned distinct variable symbols by the formula operator ϕ.

Assumption 4.2.6. If u is a conceptual graph containing one or more lines of identity, compute the formula ϕu by first transforming the graph u according to the following algorithm:

```
assign a unique variable name to every generic concept of u;
for a in the set of dominant concepts of u loop
    x := identifier(a);
    append "=x" to the referent field of every
        concept dominated by a;
end loop;
erase all coreference links in u;
```

The formula ϕu is the result of applying ϕ to the transformed version of u with the following rule for mapping concepts with multiple referents: if b is a concept of u of the form $[t : x_1 = x_2 = \ldots = x_n]$, then ϕb has the form $t(x_1) \wedge x_1 = x_2 \wedge \ldots \wedge x_1 = x_n$.

---

[3] A formula with free variables can be regarded as equivalent to its universal closure.

Although these definitions seem trivial, there are several things to notice about them. In the first place, the universal type $\top$ (using Sowa's notation) and the absurd type $\perp$ aren't handled in any special way. Furthermore, a negation sign can never appear immediately before a predicate; there must be always an existential quantifier between them. Notice also that each context corresponds to a closed formula except for one case. If a concept c_1 in context p_1 is dominated by a generic concept c_2 which is in $p_2 \neq p_1$, then the variable corresponding to c_2 appears free in the translation of p_1, but it is bounded in the formula $\phi(p_2)$.

Having these particularities in mind, consider the formula $\forall x \; P(x)$ where P is any unary predicate. Since ϕ only uses the existential quantifier, the formula must be rewritten as $\neg \exists x \; \neg P(x)$. Because of the negated literal, further transformation is necessary in order to get $\neg \exists x \; \neg \exists y \; P(y) \wedge y = x$ which can be represented by the incomplete graph $\neg [\; \neg [\; [P: \; *y=*x] \;]\;]$ or, in graphical notation, $\neg \boxed{\cdots \neg \boxed{\cdots \boxed{P}}}$. But what concept should be linked to the loose end of the coreference link? Sowa's answer is

Definition 4.2.8. If t is a type label for some concept, the *negated type* $\neg t$ is defined by a type definition of the form **type** $\neg t(x)$ **is** $[\top: \; *x] \; \neg [[t: \; *x]]$.

Therefore, $\neg \exists x \; \neg P(x)$ should be written as $\neg \boxed{\boxed{\neg P}}$ which gets expanded into $\neg \boxed{\boxed{\top} \cdot \neg \boxed{\cdots \boxed{P}}}$. But if we apply the formal definition of ϕ to that graph we obtain $\neg (\exists x \; T(x) \wedge \neg (\exists y \; P(y) \wedge y = x))$ which is only equivalent to the original formula if $T(x)$ is true for any x.

To sum up, the formal definition of the ϕ operator is inconsistent, incomplete, and not intuitive. The fundamental reason is just one: the translation process doesn't reflect the usual interpretation of $\top$ as 'true' and $\perp$ as 'false'. But negated types convey that special meaning of the universal type. Therefore, Assumption 3.3.2, besides not being intuitive, is not consistent with Definition 4.2.8. Furthermore, that interpretation of $\top$ is absolutely necessary for conceptual graphs to be able to represent any closed first-order formula, hence the incompleteness of ϕ.

Adding to the problems mentioned above, the definition of ϕ is not totally correct. On one hand, the empty context gets translated simply into () which is not a well-formed first-order formula because there is no predicate. On the other hand, Assumption 4.2.3 doesn't impose any ordering for the translation of graphs in the same context. That might lead to a formula different from the intended one, if there are coreference links. Consider the graph $\neg \boxed{\boxed{\text{CAT}} \cdots \cdots \boxed{\text{DOG}}}$ which is supposed to state "there is a dog which is not a cat". Applying Assumption 4.2.6, one gets $\neg [\text{CAT}: \; *y=*x] \; [\text{DOG}: *x]$ which can be translated into $\neg (\exists y \; Cat(y) \wedge y = x) \wedge \exists x \; Dog(x)$ or $\exists x \; Dog(x) \wedge \neg (\exists y \; Cat(y) \wedge y = x)$ depending on the chosen order. The formulas are not logically equivalent because x is free in the first formula. Thus, only the second one corresponds to the intuitive meaning of the graph.

The new translation algorithm given by the ten rules of Definition 1 overcomes all these problems and it also handles higher-order types. However, the basic mechanism remains the same as in Sowa's approach: each concept is assigned a unique variable (rule 2) which is existentially quantified (rule 5); those variables are copied from the dominating to the dominated concepts (rules 4 and 6); the formula corresponding to a graph consists of an existential quantifier prefix followed by the conjunction of the predicates generated by the concepts and relations (rule 8); and negative contexts translate into negated formulas (rule 10).

Let us first see how the above mentioned problems are dealt with. In order to be able to represent any closed first-order formula, the universal type must be translated into a true predicate which simultaneously introduces a new variable. The equality predicate is an obvious candidate. For clarity, it will be written as an infix operator. Furthermore, the symbol $\doteq$ was chosen to avoid any confusion with the meta-level equality $=$ used in definitions. Therefore, $\boxed{\top_c}$ will be translated as $x \doteq x$ (rule 6) where x is the variable associated to the concept. Concepts with the absurd type or the absurd marker are always false and correspond thus to the formula $\neg x \doteq x$[4] (rule 6). Similarly for the universal and absurd relation types (rule 7). Handling empty contexts, the second problem, is just as easy. According to [Sowa, 1984, p. 151], "an empty set of graphs makes no assertion whatever. By convention, it is assumed to be true." This means that having no graph is the same as having $\boxed{\top_c}$. Hence, by inserting this concept into each empty context (rule 1), the usual translation process will take care of the rest. Finally, to prevent ϕ from generating incompatible formulas for graphs in the same context, the quantifier prefix of each graph must be moved to the front of the whole formula (rule 9).

Having fixed ϕ's definition, let us extend it to handle higher-order types. In the new framework, types can be used as markers, and therefore there is also a partial order over markers. Furthermore, higher-order graphs are mainly used for meta-level statements. This means that the interpretation of coreference links should be slightly different. Consider the graph

$$g = \boxed{\text{SHAPE: } \#\text{rhombus}} \!\cdots\! \boxed{\text{SHAPE: } \#\text{rectangle}}$$

If the markers represent single individuals, and therefore **SHAPE** is a first-order type, g states that "a rhombus is the same shape as a rectangle". In logic, the equivalent statement is

$$Shape(rhombus) \wedge rhombus \doteq rectangle \wedge Shape(rectangle) \qquad (1)$$

However, if one considers **RHOMBUS** and **RECTANGLE** to be first-order types, and **SHAPE** to be second-order, then the intuitive reading should be "there is a shape which is both a rhombus and a rectangle". The new translation should be

$$\exists x \; Shape(x) \wedge x \sqsubseteq rhombus \wedge x \sqsubseteq rectangle \qquad (2)$$

---

[4] This is the infix form of $\neg \doteq (x, x)$ because only predicates can be negated, not variables.

where $\sqsubseteq$ is a special predicate (written as an infix operator) denoting the partial order among markers. Notice that formula 1 is false, but 2 is true since x can be substituted by SQUARE.

However, the translation generated by ϕ won't be exactly as formula 2. Let us see why. Both relation types and non-relational concept types can appear to the left or to the right of ':' in a concept. In other words, most types can be used as markers too. This means that they can be translated as predicates or constants. For example, $\phi(\boxed{\text{CAT: \#Garfield}}) = Cat(garfield)$ but $\phi(\boxed{\text{SPECIES: \#cat}}) = Species(cat)$ where Cat and cat are different logical symbols. With the purpose of using as few symbols as possible, the form $Holds(cat, garfield)$ will be used instead of $Cat(garfield)$. The "meta-predicate" $Holds$ (similar to the one used in the KIF language [Genesereth and Fikes, 1992]) can also be applied to relations. For example, $Agnt(x, y)$ will be written as $Holds(agnt, x, y)$. Formally, as each predicate must have a fixed arity, there is not a single $Holds$ but a set $\{Holds_i | i > 0\}$ where i is the arity of the relation type that appears as the first argument of the predicate. Concept types can be seen as unary relation types and therefore $\boxed{t : m}$ won't be translated to $T(m)$ anymore but to $Holds_1(t, m)$ instead (rule 6). Similarly the atomic formula $R(x_1, \ldots, x_n)$ will be rewritten as $Holds_n(r, x_1, \ldots, x_n)$ (rule 7). The predicate $Holds_i$ has therefore arity $i + 1$.

The utilization of $Holds_i$ makes the logical vocabulary even smaller, since $\sqsubseteq$ becomes unnecessary. In fact, if m and m' are types, $m \sqsubseteq m'$ can be restated as $\forall x \; Holds(m, x) \rightarrow Holds(m', x)$ which in turn can be written as $\neg \exists x \; Holds_1(m, x) \land \neg Holds_1(m', x)$ (rule 6). Otherwise, i.e. if m and m' are individuals, $m \sqsubseteq m'$ is simply the same as $m \doteq m'$ (rule 6), because neither m nor m' have any subtypes.

Notice that relational concept types can't appear in the referent field of concepts. Therefore, they can't be translated to logical constants and as such can't be quantified over or appear as arguments of some $Holds_i$. This means that $\boxed{t : m}$ will generate $t(m)$ when t is a relational concept type.

Definition 1. The translation of conceptual graphs to first-order logic is done according to the rules that follow. The functions ϕ, ϕ_p, ϕ_b return for each conceptual graph a sequence of logical symbols. The sequence $\phi(g)$ is the *first-order formula* for graph g, and it consists of the *quantifier prefix* $\phi_p(g)$ and the *body* $\phi_b(g)$: For each concept c, the auxiliary functions id, cl, and dom return, respectively, a variable that uniquely identifies c, a boolean that indicates if c is attached to a coreference link, and the set of identifiers of the concepts that dominate c. When necessary, the operator $\odot$ explicity represents the concatenation of symbol sequences.

1. In each empty context of g insert a concept $\boxed{\top_c: *}$.
2. For each concept c let $id(c) = x$, where x is a unique variable.
3. For each concept c let $cl(c) = \text{true}$ if c is attached to some coreference link, otherwise $cl(c) = \text{false}$.
4. For each concept c let $dom(c) = \{id(c') | c' \text{ dominates } c\}$.
5. For each concept c let $\phi_p(c) = \exists id(c)$.

6. For each concept c with $type(c) = t$ and $referent(c) = m$, the formula $\phi_b(c)$ is obtained by the conjunction of all the following sub-formulas that apply:
   - $\neg id(c) \doteq id(c)$ if $t = \perp_c$ or $m = \maltese$;
   - $id(c) \doteq id(c)$ if $t = \top_c$;
   - $t(id(c))$ if $t \in T^{rc}$;
   - $Holds_1(t, id(c))$ otherwise;
   - $\bigwedge\limits_{x \in dom(c)} id(c) \doteq x$;
   - $m \doteq id(c)$ if $m \in T_0^{nc}$ or $cl(c) = \mathtt{false}$ and $m \notin \{\maltese, \maltese\}$;
   - $(\neg \exists x\ Holds_1(id(c), x) \wedge \neg Holds_1(m, x))$, where $x \neq id(c)$, if $m \notin T_0^{nc} \cup \{\maltese, \maltese\}$ and $cl(c) = \mathtt{true}$.

7. Let r be a relation with concepts $c_1, \ldots, c_n$ as arguments. If $type(r) = \top_r$ or $type(r) = \perp_r$ then $\phi_b(r) = id(c_1) \doteq id(c_1)$ or $\phi_b(r) = \neg id(c_1) \doteq id(c_1)$, respectively. Otherwise $\phi_b(r) = Holds_n(type(r), id(c_1), \ldots, id(c_n))$.

8. If g is a conceptual graph without contexts and with concepts C and relations R, then $\phi(g) = \phi_p(g)\phi_b(g)$ where

$$\phi_p(g) = \bigodot_{c \in C} \phi_p(c) \qquad\qquad \phi_b(g) = \bigwedge_{c \in C} \phi_b(c) \wedge \bigwedge_{r \in R} \phi_b(r)$$

9. If p is a proposition containing the set of graphs G then $\phi_p(p)$ is the empty sequence and $\phi(p) = \phi_b(p) = (\bigodot_{g \in G} \phi_p(g) \bigwedge_{g \in G} \phi_b(g))$.

10. If c is a context formed by the negation of proposition p, then $\phi_p(c)$ is the empty sequence and $\phi(c) = \phi_b(c) = \neg\phi(p)$.

Several translation examples follow. They show the difference between the old and the new definition of ϕ, and illustrate how some previously problematic cases are now handled. Table 2 gives further examples.

Example 4. The translation of | CAT: #Garfield |←(AGNT)←| CHASE |→(OBJ)→| DOG | is

$$\exists x \exists y\ Cat(garfield) \wedge Chase(x) \wedge Dog(y) \wedge Agnt(x, garfield) \wedge Obj(x, y)$$

according to Assumption 3.3.2. Applying Definition 1 instead, one has

$$\exists x \exists y \exists z\ Holds_1(cat, z) \wedge z \doteq garfield \wedge Holds_1(chase, x) \wedge Holds_1(dog, y) \wedge$$
$$Holds_2(agnt, x, z) \wedge Holds_2(obj, x, y)$$

Example 5. The formula $\forall x\ P(x)$ states basically that "if x is some entity then $P(x)$ is true". Let P be any relational type[5]. Then

$$\phi(\neg\,\boxed{\top_c}\,\neg\cdots\boxed{P}\,) = \neg(\exists x\ x \doteq x \wedge \neg(\exists y\ P(y) \wedge y \doteq x))$$

---

[5] The result would be similar if P were a non-relational type. Just substitute $Holds_1(p, x)$ for $P(x)$.

This formula is equivalent to $\forall x \; x \doteq x \rightarrow \exists y \; P(y) \wedge y \doteq x$. Due to the properties of equality, $x \doteq x$ is always true and $P(y) \wedge y \doteq x$ corresponds to $P(x)$. Thus one gets $\forall x \; P(x)$ as expected.

Example 6. According to the formulation of rule 9, the graph $\neg\boxed{\boxed{\text{CAT}}\cdots\vdash\cdots\boxed{\text{DOG}}}$ shown before is correctly translated as

$$\exists x \; \neg(\exists y \; Holds_1(cat, y) \wedge y = x) \wedge Holds_1(dog, x)$$

Example 7. The graph $\boxed{\text{SHAPE: #rectangle}}\cdots\boxed{\text{SHAPE: #rhombus}}$ has a coreference link between higher-order concepts. The corresponding formula is therefore

$$\exists x \exists y \; Holds_1(shape, x) \wedge x \doteq y \wedge (\neg\exists z \; Holds_1(x, z) \wedge \neg Holds_1(rectangle, z)) \wedge \\ Holds_1(shape, y) \wedge y \doteq x \wedge (\neg\exists z \; Holds_1(y, z) \wedge \neg Holds_1(rhombus, z))$$

Example 8. The coreference link in $\boxed{\text{PERSON: Rosalie}}\cdots\boxed{\boxed{\text{PERSON: Rosann}}}$ connects two first-order concepts. Applying rule 6 in this case leads to

$$\exists x \, Holds_1(person, x) \wedge x \doteq rosalie \wedge \neg(\exists y \; Holds_1(person, y) \wedge y \doteq rosann \wedge y \doteq x)$$

The ϕ operator just translates a sequence of symbols of some language (Conceptual Graphs) into another sequence of symbols (called formula) of some other language (first-order logic). For this process to have any meaning, the resulting formulas must have an interpretation. Classically, an interpretation of a first-order language L is a pair $\langle D, \delta \rangle$ where the denotation function δ maps constants of L into elements of the domain D and predicates into tuples of elements of D. The new definition of interpretation [Wermelinger, 1995] just adds the constraints presented informally in Section 2.

4 Inference

Theoretically, the translation and interpretation of conceptual graphs is important to show the formalism's expressiveness. But the main goal is to have inference rules that operate directly on conceptual graphs, instead of translating the graphs to formulas, do the proofs with them and then translating back to the graphical form.

The proof system given in [Sowa, 1984] consists of a single axiom, the empty set of graphs, and several first-order rules of inference. These are mainly based on the depth of a graph, i.e. on how many negative contexts one must traverse to reach the graph starting from the outer context. Depending on the depth, the graph is said to be in an evenly enclosed or oddly enclosed context. Even contexts contain true graphs and odd contexts contain false graphs. Therefore, conditions (i.e, graphs and coreference links) can be removed from the former and added to the latter. Moreover, a context c dominating a context c' (that is, $c' = c$ or c' is enclosed in a context dominated by c) corresponds to an implication and therefore the graphs in c (the antecedent) can be copied to c' (the consequent).

As simple and elegant it is, Sowa's system must be changed, even if one considers the corrected version of ϕ and no higher-order types. In fact, there are now several ways of representing truth, and each true graph that can't be derived from others must be an axiom. Otherwise the system won't be complete. Moreover, a new rule must be added: axioms may be inserted and removed from any context. Without these changes the universal instantion rule can't be applied to conceptual graphs. Using the old notation for clarity, consider the example $\forall x\; Cat(x) \vdash Cat(garfield)$. In graphical form the hypothesis is $\neg\boxed{\boxed{T_c}\; \neg \cdots \boxed{CAT}}$. Restricting the referent in the oddly enclosed context one gets $\neg\boxed{\boxed{T_c:\text{\#Garfield}}\; \neg \cdots \boxed{CAT}}$. By the individuation rule, a individual marker can be iterated from a dominating concept to a dominated one[6] and the coreference link may be erased, provided the dominated concept is generic. We thus get $\neg\boxed{\boxed{T_c:\text{\#Garfield}}\; \neg\boxed{CAT:\text{\#Garfield}}}$ but can't proceed any further because graphs can't be removed from odd contexts.

The remaining of this section presents therefore a new formal proof system. For the most part it is similar to Sowa's. The changes that were done (including the above mentioned) are due to the type and marker hierarchies, the new interpretation of universal and absurd types, and the new meaning of coreference links resulting from the use of higher-order types.

In [Sowa, 1984] only concept types formed a hierarchy. Relation types and markers were incomparable. Therefore, the inference rules only enabled one to restrict concept types, i.e. to substitute them by subtypes, and to replace the generic marker by an individual marker, or the other way round. In this framework, relation types and markers may also be (un)restricted but there are some limitations. Let t and t' be any concept or relation types, such that t is a subtype of t'. Therefore, if $Holds_n(t, x_1, \ldots, x_n)$ is true, then $Holds_n(t', x_1, \ldots, x_n)$ is also true, and if the latter is false, so is the former. Thus any type may be unrestricted in evenly enclosed contexts and it may be substituted by a subtype in oddly enclosed contexts. In this respect higher-order types don't change the original inference rules.

However, markers can't be changed at will. Consider Examples 1 and 3: CAT is a subtype of FELINE which is a GENUS while CAT is a SPECIES. If the graph $\boxed{\text{SPECIES: \#cat}}$ is in an evenly enclosed context it is true, but it can't be generalized to the false graph $\boxed{\text{SPECIES: \#feline}}$. Similarly, if the latter is in an oddly enclosed context, it can't be specialized to the former. To sum up, individual markers can't be (un)restricted to other individual markers but they can be transformed into the generic or absurd markers. For example, the true graph $\neg\boxed{\boxed{\text{SPECIES: \#feline}}}$ can be specialized to the equally true $\neg\boxed{\boxed{\text{SPECIES: \#}}}$.

There is however one case where the marker hierarchy can be put to use,

---

[6] A concept c_1 dominates a concept c_2 if there is a coreference link between them and the context of c_1 dominates the context of c_2.

Context of c	Coreference link?	m	Action
even	no	⊼	unrestrict
even	no	$\neq$⊼	unrestrict to ∗
even	yes	any	unrestrict
odd	no	∗	restrict
odd	no	$\neq$∗	restrict to ⊼
odd	yes	any	restrict

Table 1. Conditions for changing referent m of concept c

namely if a coreference link is present. Let m be the marker of some dominating or dominated concept c whose identifier is the variable x. If m is a type, the condition $x \sqsubseteq m$ belongs to the context of c. If that context is even, the condition is true and so is $x \sqsubseteq m'$ where m' is a supertype of m. If the context is odd, the condition is false and restricting m to some subtype doesn't make it true. Table 1 summarizes all these conditions.

There is one more situation where markers can be restricted. Consider a concept c_1, with marker m_1 and identifier x_1, dominating a concept c_2 with referent m_2 and associated variable x_2. Then, the condition $x_2 \doteq x_1$ enables one to iterate any condition on x_1 from c_1's context to c_2's context. This corresponds to the replacement of m_2 by m_{12}, the greatest lower bound of m_1 and m_2: if $x_1 \sqsubseteq m_1$ and $x_2 \sqsubseteq m_2$ then $x_2 \sqsubseteq m_{12}$[7] (assuming $x_2 \doteq x_1$). Notice however that the restriction on m_2 can only be done if the result isn't the absurd marker, because a false graph might be obtained if c_2 is evenly enclosed. If c_2 were oddly enclosed the last line of Table 1 would apply and therefore this new rule, which finds an upper limit for the value x_2 that satisfies the formula, wouldn't be needed.

As for logical axioms, Sowa only uses the empty set of graphs. As seen in the previous section, some predicate $true(x)$ is needed in order to be able to represent all closed formulas of first-order logic. That predicate turned out to be the equality $\doteq$. Therefore, the new axioms are graphs whose translation is some tautology based on $x \doteq x$. Looking at Definition 1 the possibilities listed in Table 2 are obtained, where m and m' are any markers different from ⊼ and t, t' are any concept types (although the given translations assume that they are non-relational).

It is obvious that the graphs involving $\top_r$ and $\bot_r$ may have any arity. However, there is a subtle difference. The mere presence of the absurd type $\bot_r$ automatically makes the graph false, and therefore the axiom true. The concepts used as relation arguments are thus irrelevant. But the same does not happen with the universal type $\top_r$. The concepts to which it is linked must be true too for the whole graph to be true. It is also worth noticing that ϕ translates the empty context in the same way as $\boxed{\top_c}$. The rules of inference will of course

---

[7] See Example 9.

$$\phi(\boxed{}) \quad = \quad \exists x \; x \doteq x$$

$$\phi(\boxed{T_c : m}) \quad = \quad \exists x \; x \doteq x \wedge x \doteq m$$

$$\phi(\neg\boxed{\boxed{\bot_c : m}}) \quad = \quad \neg\exists x \; \neg x \doteq x \wedge x \doteq m$$

$$\phi(\neg\boxed{\boxed{t : \overline{\ast}}}) \quad = \quad \neg\exists x \; Holds_1(t, x) \wedge \neg x \doteq x$$

$$\phi(\neg\boxed{\boxed{\bot_r}\!\!-\!\!\boxed{t : m}}) \quad = \quad \neg\exists x \; \neg x \doteq x \wedge Holds_1(t, x) \wedge x \doteq m$$

$$\phi(\neg\boxed{\boxed{t : m}\!\!-\!\!\boxed{\bot_r}\!\!-\!\!\boxed{t' : m'}}) \quad = \quad \neg\exists x \exists y \; \neg x \doteq x \wedge Holds_1(t, x) \wedge x \doteq m \wedge$$
$$Holds_1(t', y) \wedge y \doteq m'$$

$$\phi(\boxed{T_r}\!\!\rightarrow\!\!\boxed{T_c : m}) \quad = \quad \exists x \; x \doteq x \wedge x \doteq x \wedge x \doteq m$$

$$\phi(\boxed{T_c : m}\!\!-\!\!\boxed{T_r}\!\!-\!\!\boxed{T_c : m'}) \quad = \quad \exists x \exists y \; x \doteq x \wedge x \doteq x \wedge x \doteq m \wedge y \doteq y \wedge y \doteq m'$$

Table 2. Logical axioms

allow one to insert and erase logical axioms from any context. The empty context becomes therefore obsolete because it can be derived from any other axiom by erasure. However, it will be kept for convenience. There are other redundant graphs in the above table. For example, any $\neg\boxed{\boxed{\bot_c : m}}$ can be obtained from $\neg\boxed{\boxed{\bot_c : \ast}}$ by restricting the referent (see the fourth line of Table 1). In the same way, $\neg\boxed{\boxed{\bot_r}\!\!-\!\!\boxed{t : m}}$ can be derived from $\neg\boxed{\boxed{\bot_r}\!\!-\!\!\boxed{T_c : \ast}}$. The final set of logical axioms can be found in Definition 2.

Finally, the rules for handling coreference links are basically the same as in [Sowa, 1984] when the dominated concept is first-order. Otherwise a coreference link can't be inserted or removed in the general case. Let us see why. Consider again concepts c_1 and c_2 mentioned before. When a coreference link is drawn the following happens:

1. The condition $x_2 \doteq x_1$ is added to the context of c_2.
2. The condition $x_2 = m_2$ becomes $x_2 \sqsubseteq m_2$ if $m_2 \notin T_0^{nc} \cup \{\ast, \overline{\ast}\}$.
3. The condition $x_1 = m_1$ becomes $x_1 \sqsubseteq m_1$ if $m_1 \notin T_0^{nc} \cup \{\ast, \overline{\ast}\}$.

The erasure of a coreference link consists in doing the opposite actions. Due to step 1, coreference links can't be inserted when c_2 is evenly enclosed, and they can't be removed if c_2 is in an odd context. Additionally, if c_2 is a higher-order concept, step 2 applies. In this case inserting a coreference link relaxes the condition (i.e., it could become true), and therefore c_2 can't be oddly enclosed. Inversely, erasing a coreference link makes the condition stronger ($x_2 \sqsubseteq m_2$

Context of c_1	Context of c_2	$m_1 \in T_0^{nc} \cup \{*, \bar{*}\}$	$m_2 \in T_0^{nc} \cup \{*, \bar{*}\}$	Action
any	even	yes	yes	erasure
any	odd	yes	yes	insertion
odd	even	no	yes	erasure
even	odd	no	yes	insertion
odd	even	no	no	erasure if $m_1 = m_2$
even	odd	no	no	insertion if $m_1 = m_2$

Table 3. Conditions for changing coreference links

becomes $x_2 \doteq m_2$) which prevents c_2 from being evenly enclosed[8]. Steps 1 and 2 thus impose contradictory restrictions on the context of c_2.

Fortunately, there is an exception, namely if $m_1 = m_2$. Consider the case where c_2 is evenly enclosed and the coreference link has therefore been removed. The new conditions are $x_2 \doteq m_2$ and $x_1 \doteq m_1$. Due to our assumption, the conditions are equivalent and therefore the erasure corresponds to the iteration of a condition from the dominating to the dominated context. The other possibility is to insert a coreference link if c_2 is oddly enclosed. The new conditions are $x_1 \sqsubseteq m_1$ in c_1's context and $x_2 \sqsubseteq m_2 \wedge x_2 \doteq x_1$ for c_2. Again, due to their equivalence, insertion of a coreference link corresponds to an iteration.

Table 3 summarizes the preceding observations. No action is possible for the unlisted cases. It should be obvious that it is not necessary to check any restrictions if the coreference link to be inserted or removed is an exact copy of another existing one. Also, since coreference links represent equalities, they may be inserted or removed according to the transitivity rule.

The inference rules can at last be presented.

Definition 2. Let S be a set of conceptual graphs in the outer context. Any graph derived from S by the following *first-order rules of inference* is said to be *provable* from S.

Equivalence In any context, a logical axiom may be inserted or removed, a double negation may be drawn or erased around any set of graphs.

Generalization In an evenly enclosed context, any type or marker may be unrestricted, and any graph or coreference link may be deleted, as long as the conditions in Tables 1 and 3 are obeyed.

Specialization In an oddly enclosed context, any type or marker may be restricted, and any graph or coreference link may be inserted, as long as the conditions in Tables 1 and 3 are obeyed.

Iteration A graph may be copied from context c to any context dominated by c, and coreference links may be drawn between the original concepts and their copies.

Deiteration The result of some possible iteration may be deleted.

---

[8] The same reasoning applies to c_1 if it isn't a first-order concept.

Transitivity If concept c_1 dominates concept c_2 which in turn dominates $c_3 \neq c_1$, then a coreference link between c_1 and c_3 may be drawn or erased. If it is inserted, then the coreference link between c_2 and c_3 may be erased.

Individuation If concept c_1 dominates concept c_2 then $referent(c_2)$ may be replaced by the greatest lower bound of $referent(c_1)$ and $referent(c_2)$ if the result is different than *.

A graph provable from the following *logical axioms* is called a *theorem*.

- The empty set of graphs $\{\}$;
- $\boxed{\top_c: m}$ for any $m \in \mathcal{M} - \{\text{*}\}$;
- $\neg\ \boxed{\boxed{\bot_c: \text{*}}}$;
- $\neg\ \boxed{\boxed{\top_c: \text{*}}}$;
- $\neg\ \boxed{\boxed{\bot_r}\!\!\rightarrow\!\!\boxed{\top_c: \text{*}}}$ and for any other arity;
- $\boxed{\top_r}\!\!\rightarrow\!\!\boxed{\top_c: m}$ for any $m \in \mathcal{M} - \{\text{*}\}$ and any arity.

Example 9. Applying the individuation rule twice, and considering Example 2, the graph $\boxed{\text{SHAPE: \#rectangle}}\!\cdots\!\boxed{\text{SHAPE: \#rhombus}}$ is first transformed to $\boxed{\text{SHAPE: \#rectangle}}\!\cdots\!\boxed{\text{SHAPE: \#square}}$ and then $\boxed{\text{SHAPE: \#square}}\!\cdots\!\boxed{\text{SHAPE: \#square}}$ is derived. Notice that this graph doesn't necessarily imply $\boxed{\text{SHAPE: \#square}}$ because the former states that there exists a *subtype* of SQUARE which is a shape while the latter states that SQUARE *itself* is a shape.

Example 10. The subtype relationships CAT < FELINE < ANIMAL can be stated by the graph

Erasing the double negation (an equivalence rule), it can be simplied to

$$\neg\boxed{\boxed{\text{CAT}}\!\cdots\!\boxed{\text{FELINE}}\!\cdot\!\neg\!\cdots\!\boxed{\text{ANIMAL}}}$$

and applying the transitivity rule one gets

$$\neg\boxed{\boxed{\text{FELINE}}\!\cdots\!\boxed{\text{CAT}}\!\cdot\!\neg\!\cdots\!\boxed{\text{ANIMAL}}}$$

The first graph corresponds indeed to the given type hierarchy fragment, as can be easily seen by the translation of it:

$$\neg\exists x\ Holds_1(cat, x) \wedge \neg\neg\exists y\ Holds_1(feline, y) \wedge y \doteq x \wedge$$
$$\neg\exists z\ Holds_1(animal, z) \wedge z \doteq y$$

or more simply

$$\forall x \ Holds_1(cat, x) \rightarrow \forall y Holds_1(feline, y) \wedge x \doteq y \rightarrow \exists z \ Holds_1(animal, z) \wedge z \doteq y$$

This formula is equivalent to

$$\forall x \forall y \ Holds_1(cat, x) \wedge Holds_1(feline, y) \wedge x \doteq y \rightarrow \exists z \ Holds_1(animal, z) \wedge z \doteq y$$

which is the translation of the second graph. Obviously, it can be rewritten as

$$\forall x \forall y \ Holds_1(cat, x) \wedge Holds_1(feline, y) \wedge x \doteq y \rightarrow \exists z \ Holds_1(animal, z) \wedge z \doteq x$$

corresponding to the last graph.

5 Conclusions

This paper has provided a closer look at the logical foundations of Conceptual Structures Theory. It was shown that the original formal definitions of [Sowa, 1984] are incomplete: on one hand, some closed first-order formulas can't be represented with conceptual graphs, on the other hand the universal instantiation rule is missing. Therefore, the definitions of the ϕ operator and of the first-order inference rules have been corrected. Furthermore, they have been extended to handle higher-order types.

It is hoped that this paper provides a first step towards a meta-level reasoning engine and a deeper investigation of the model-theoretic and proof-theoretic properties of Conceptual Structures.

References

[Genesereth and Fikes, 1992] Michael R. Genesereth and Richard E. Fikes. Knowledge interchange format version 3.0 reference manual. Technical Report Logic-92-1, Computer Science Department, Stanford University, June 1992. "Living document" of the Interlingua Working Group of the DARPA Knowledge Sharing Effort.

[Hamilton, 1988] A. G. Hamilton. *Logic for Mathematicians*. Cambridge University Press, 1988. Revised edition.

[Sowa, 1984] John F. Sowa. *Conceptual Structures: Information Processing in Mind and Machine*. The System Programming Series. Addison-Wesley Publishing Company, 1984.

[Sowa, 1992] John F. Sowa. Conceptual graph summary. In Timothy E. Nagle, Janice A. Nagle, Laurie L. Gerholz, and Peter W. Eklund, editors, *Conceptual Structures: Current Research and Practice*, Ellis Horwood Series in Workshops, pages 3–51. Ellis Horwood, 1992.

[Wermelinger and Lopes, 1994] Michel Wermelinger and José Gabriel Lopes. Basic conceptual structures theory. In William M. Tepfenhart, Judith P. Dick, and John F. Sowa, editors, *Conceptual Structures: Current Practices — Proceedings of the Second International Conference on Conceptual Structures*, number 835 in Lecture Notes in Artificial Intelligence, pages 144–159, College Park MD, USA, 16–19 August 1994. University of Maryland, Springer-Verlag.

[Wermelinger, 1995] Michel Wermelinger. Teoria Básica das Estruturas Conceptuais. Master's thesis, Universidade Nova de Lisboa, 1995.

Existential Graphs and Dynamic Predicate Logic

Harmen van den Berg

Department of Applied Mathematics
University of Twente, P.O. Box 217
7500 AE Enschede, The Netherlands
e-mail: H.vandenBerg@math.utwente.nl

Abstract. Using first-order logic to represent the meaning of natural language sentences forces a non-compositional analysis of cross-sentential anaphora and donkey-anaphora. Groenendijk and Stokhof developed Dynamic Predicate Logic, in syntax equal to first-order logic, but with a different semantics. Using Dynamic Predicate Logic, anaphora can be analyzed in a compositional way.

The same results can be achieved using Existential Graphs, developed by C.S. Peirce. The advantages of using Existential Graphs are that Existential Graphs are less complicated and easier to use than Dynamic Predicate Logic, and that Existential Graphs do not require a change of semantics. We discuss two types of anaphora, describe Dynamic Predicate Logic and Existential Graphs, give logical properties of both formalisms, and compare the two approaches.

1 Introduction

Many theories of discourse semantics within the paradigm of model-theoretic semantics rely on the *principle of compositionality*. Unfortunately, several phenomena in natural language are of a non-compositional nature. Semantic theories often have to sacrifice the compositionality principle in order to analyze these phenomena. Recently, Groenendijk and Stokhof [5] have developed an alternative compositional semantics of discourse, which is nevertheless equivalent to some non-compositional theories. In this paper we will show that their approach is in some ways contained in Existential Graphs, as developed by C.S. Peirce [7]. However, Existential Graphs are easier and less complicated than the alternative semantic theory of Groenendijk and Stokhof.

2 Language and Compositionality

The term *compositionality* is used by many people in many different ways. Compositionality (in this broad sense) can be decomposed into several subprinciples (see [1, 2, 6]). Among these subprinciples is the *principle of invariance*, stating that

> if an expression E is formed from certain simpler expressions $e_1, e_2, \ldots, e_n$, these are constituent parts (subexpressions) of E, that is, they may not be changed when E is built up out of them. [2]

In other words: if a formula is constructed from a number of subformulas, these subformulas must be (unchanged) parts of the formula.

It is this subprinciple of compositionality which is important in this paper, and which is used by Groenendijk and Stokhof [5]. We will use their terminology, and thus use the term 'compositionality' to denote the principle of invariance.

In the rest of this section we will give examples of natural language phenomena which often force a non-compositional analysis. We will use first-order predicate logic to represent the meaning of these sentences, since logic is the basis of many theories of discourse semantics.

Two well-known phenomena in language are the occurrences of so-called donkey anaphora and cross-sentential anaphora. Both types of anaphora force a non-compositional analysis. When using first-order predicate logic (henceforth, PL) to represent the meaning of natural language sentences, anaphoric pronouns will show up as bound variables. To arrive at logical expressions that correspond in meaning (or truth conditions) to the natural language sentences, one often is forced to execute this translation in a non-compositional way. The following three sentences illustrate this.

(1) A man walks in the park. He whistles.
(2) If a farmer owns a donkey, he beats it.
(3) Every farmer who owns a donkey, beats it.

In (1), we come across a *cross-sentential anaphor*: *he* in the second sentence is anaphorically linked to *a man* in the first sentence. To express this anaphoric linking in PL, the scope of the existential quantifier has to extend over both sentences:

(4) $\exists x[\text{man}(x) \wedge \text{walk\_in\_the\_park}(x) \wedge \text{whistle}(x)]$

But the translation of *A man walks in the park*, which would be $\exists x[\text{man}(x) \wedge \text{walk\_in\_the\_park}(x)]$, is not a subformula of (4) anymore. Apparently, the translation from (1) to (4) is non-compositional, otherwise the translation would have been

(5) $\exists x[\text{man}(x) \wedge \text{walk\_in\_the\_park}(x)] \wedge \text{whistle}(x)$

which is not a proper translation of (1) in PL.

In (2), the sentence *A farmer owns a donkey* is part of the sentence; it translates in PL to $\exists x[\text{farmer}(x) \wedge \exists y[\text{donkey}(y) \wedge \text{owns}(x, y)]]$. Sentence (2) as a whole, however, translates into

(6) $\forall x \forall y[(\text{farmer}(x) \wedge \text{donkey}(y) \wedge \text{owns}(x, y)) \rightarrow \text{beat}(x, y)]$,

which does not contain an expression representing *A farmer owns a donkey*. Furthermore, in (2) only existentially quantified phrases occur, whereas in (6) universal quantification is needed to account for their meaning. Moreover, the universal quantifiers have wide scopes, whereas the indefinite terms in (2) appear in the antecedent of an implication. Again, by using predicate logic, one is forced to translate natural language sentences in a non-compositional way.

In (3), the sentence *(a) farmer who owns a donkey* is part of the sentence. Sentence (3) as a whole translates to (6), which does not contain an expression representing *farmer who owns a donkey*. And again, compositionality does not hold in this case.

3 Dynamic Predicate Logic

To rescue compositionality in the logical translations of these sentences, Groenendijk and Stokhof [5] propose a variant of predicate logic, *dynamic predicate logic* (DPL), in which the semantics of PL is changed in such a way that sentences (1)–(3) can be analyzed in a compositional way. Sentence (2), for instance, can be translated in DPL as

(7) $\exists x[\text{farmer}(x) \wedge \exists y[\text{donkey}(y) \wedge \text{owns}(x, y)]] \rightarrow \text{beat}(x, y)$,

with the same interpretation (semantics) as (6). In this section we will briefly describe DPL.

In DPL, the same meaning is assigned to (5) as to (4). This means that, in DPL, there is no difference in meaning between $\exists x P(x) \wedge Q(x)$ and $\exists x(P(x) \wedge Q(x))$. In other words, binding of variables can be passed on from the left conjunct to the right one. This motivates the following definitions [5].

Definition 3.1

- Conjunction ($\wedge$) is *internally dynamic* if it has the power to pass on variable bindings from its left conjunct to the right one. Conjunction is *externally dynamic* if it has the power to pass on variable bindings to conjuncts yet to come.
- Disjunction ($\vee$) is *internally dynamic* if it has the power to pass on variable bindings from its left disjunct to the right one. Disjunction is *externally dynamic* if it has the power to pass on variable bindings to disjuncts yet to come.
- Implication ($\rightarrow$) is *internally dynamic* if it has the power to pass on variable bindings from its antecedent to its consequent. Implication is *externally dynamic* if it has the power to pass on variable bindings outside the implication.
- Negation ($\neg$) is *internally dynamic* if it has the power to pass on variable bindings inside its scope. Negation is *externally dynamic* if it has the power to pass on variable bindings outside its scope.
- A quantifier ($\forall, \exists$) is *internally dynamic* if it can bind variables to the right inside its scope. A quantifier is *externally dynamic* if it can bind variables to the right outside its scope.
- A connective or quantifier that is not internally/externally dynamic is called *internally/externally static*.

◁

Groenendijk and Stokhof [5] show by a number of examples from natural language that the existential quantifier and conjunction are both internally and externally dynamic, and that the universal quantifier, implication, and negation are only internally dynamic. Disjunction is neither internally, nor externally dynamic. For instance, the combination of the sentences

(8) If a farmer owns a donkey, he beats it. *He hates it.

is ungrammatical, since the pronouns *he* and *it* in the second sentence cannot be anaphorically linked to the indefinite terms in the preceding implication. From examples such as these one can conclude that implication is not externally dynamic, otherwise this linking should be possible.

Groenendijk and Stokhof change the rules of interpretation (the semantics) for the logical connectives and quantifiers in such a way that the dynamics of connectives and quantifiers are according to what has been described above. The syntax of DPL is that of PL. The models they use for the interpretation of DPL consist (as usual) of a domain of individuals D and an interpretation function F. The assignments are (as usual) total functions from the set of variables to the domain, denoted by g, h, and so on, where $g[x]h$ denotes that g differs from h at most with respect to the value it assigns to x. In the standard semantics of PL, the interpretation of a formula is the set of assignments that verify the formula. In the semantics of DPL, the interpretation of a formula is a set of ordered pairs of assignments. Such pairs can be regarded as possible 'input-output' pairs: a pair $\langle g, h \rangle$ is in the interpretation of a formula φ if and only if when φ is evaluated with respect to g, h is a possible outcome of the evaluation procedure. The formal definition is as follows [5].

Definition 3.2 (Semantics of DPL) A model M is a pair $\langle D, F \rangle$, where D is a non-empty set of individuals, F an interpretation function, having as its domain the individual constants and predicates. If α is an individual constant, then $F(\alpha) \in D$; if α is an n-place predicate, then $F(\alpha) \subseteq D^n$. An assignment g is a function assigning an individual to each variable: $g(x) \in D$. G is the set of all assignment functions. We define $[t]_g = g(t)$ if t is a variable, and $[t]_g = F(t)$ if t is an individual constant. The interpretation function $[\ ]_M^{DPL} \subseteq G \times G$ is defined as follows.

- $[R(t_1,\ldots,t_n)] = \{\langle g,h \rangle \mid h = g \,\&\, \langle [t_1]_h \ldots [t_n]_h \rangle \in F(R)\}$.
- $[t_1 = t_2] = \{\langle g,h \rangle \mid h = g \,\&\, [t_1]_h = [t_2]_h\}$.
- $[\neg\varphi] = \{\langle g,h \rangle \mid h = g \,\&\, \neg\exists k : \langle h,k \rangle \in [\varphi]\}$.
- $[\varphi \wedge \psi] = \{\langle g,h \rangle \mid \exists k : \langle g,k \rangle \in [\varphi] \,\&\, \langle k,h \rangle \in [\psi]\}$.
- $[\varphi \vee \psi] = \{\langle g,h \rangle \mid h = g \,\&\, \exists k : \langle h,k \rangle \in [\varphi] \vee \langle h,k \rangle \in [\psi]\}$.
- $[\varphi \rightarrow \psi] = \{\langle g,h \rangle \mid h = g \,\&\, \forall k : \langle h,k \rangle \in [\varphi] \Rightarrow \exists j : \langle k,j \rangle \in [\psi]\}$.
- $[\exists x\varphi] = \{\langle g,h \rangle \mid \exists k : k[x]g \,\&\, \langle k,h \rangle \in [\varphi]\}$.
- $[\forall x\varphi] = \{\langle g,h \rangle \mid h = g \,\&\, \forall k : k[x]h \Rightarrow \exists j : \langle k,j \rangle \in [\varphi]\}$.

◁

As a result of this definition, the formulas $\exists x P(x) \wedge Q(x)$ and $\exists x(P(x) \wedge Q(x))$ have the same interpretation, and similarly for $\exists x P(x) \rightarrow Q(x)$ and $\forall x(P(x) \rightarrow Q(x))$. For more examples, see [5].

The following theorem holds [5].

Theorem 3.3 In DPL, with semantics as defined in Definition 3.2, the existential quantifier and conjunction are both internally and externally dynamic, the universal quantifier, implication, and negation are internally dynamic and externally static, and disjunction is internally and externally static. ◁

In this way, DPL is capable of giving a compositional account of cross-sentential and donkey anaphora.

4 Existential Graphs

In our view, however, such a (severe) change in the semantics of predicate logic is too high a price to pay for rescuing the principle of compositionality. Moreover, by using Existential Graphs [7, 8], such a change is not necessary either. The reason is that by using negative contexts (see Definition 4.2) to express negation and implication, most problems concerning anaphora disappear, and connectives and quantifiers have the required dynamics. To substantiate this claim, we will first briefly describe (our notational variant of) Existential Graphs. For a more detailed description, see [1, 2, 7, 8].

4.1 First-Order Logic in Existential Graphs

Propositional symbols are represented by graphs containing one node and no links. So if p_1 is a propositional symbol, the graph G with $V(G) = \{p_1\}$ and $E(G) = \varnothing$ represents p_1 ($V(G)$ is the set of vertices and $E(G)$ is the set of edges of G).

Drawing a graph G (on the 'sheet of assertion') amounts to asserting the truth of G, denoted by $\boxed{G}$. The negation of a formula is represented by drawing a *negative context* (Definition 4.2) around the corresponding graph G, denoted[1] by $\boxed{G}$. The conjunction of two formulas is represented by drawing the corresponding graphs within one context. So if φ_1 and φ_2 are propositional formulas with corresponding graphs g_1 and g_2, $\varphi_1 \wedge \varphi_2$ is represented by $\boxed{g_1\ g_2}$, the negation $\neg\varphi_1$ is represented by $\boxed{g_1}$, and $\neg(\varphi_1 \wedge \varphi_2)$ by $\boxed{g_1\ g_2}$.

Peirce chose conjunction and negation as 'primitive' connectives, that is, the other connectives are represented in terms of conjunction and negation. Disjunction of two propositional formulas is represented via the equivalence of $(\varphi_1 \vee \varphi_2)$ with $\neg(\neg\varphi_1 \wedge \neg\varphi_2)$, and the implication via the equivalence of $(\varphi \rightarrow \psi)$ with $\neg(\varphi \wedge \neg\psi)$. More complex formulas are treated according to the same principles.

---

[1] To be precise, one has to draw $\boxed{G}$ to denote this formula; however, if nothing occurs on the sheet of assertion outside the outer negative context, we usually omit the box with thin lines around it.

Variables are denoted by *tokens* (□), which are nodes in the graph with a special label '□'. Names of variable, like x, y, z, are not necessary in our notation. Those names merely indicate correspondences, which will be indicated by *coreference arcs* between tokens. Peirce used *undirected* lines of identity to denote quantifiers and variables. We will use *directed* coreference arcs between tokens instead, because in this way quantifiers and variables are explicitly represented. Furthermore, using directed arcs and tokens allows for a more detailed representation of quantifiers and anaphora than using undirected lines of identity.

A predicate[2] P with n variables is denoted by a frame named P with n tokens in it:

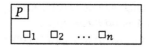

The numbers on the tokens indicate the order of the arguments in P. If a predicate has just one variable we will often omit the number.

A token without incoming coreference arc (a *quantified* token) is supposed to be existentially quantified. The scope of such a quantifier is restricted to the enclosing context (Definition 4.2). To give an example:

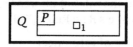

is equivalent to $\neg(Q \wedge \neg(\exists x P(x)))$.

Definition 4.1 The set QT(G) of *quantified tokens* of a graph G is the set of all tokens in G without incoming coreference arcs. ◁

The binding of variables by a quantifier is indicated by an coreference arc between the quantified token and the tokens denoting the variables. The universal quantifier can be represented using the equivalence of $\forall x P(x)$ with $\neg\exists x \neg P(x)$.

Substitution of a constant for a variable corresponds to drawing an coreference arc between the constant and the token, like in

$$\square \longleftarrow \boxed{p_1}$$

To give an example, the formula $\exists x \forall y P(x, y)$ will be represented by

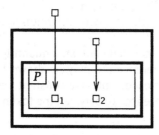

---

[2] In Existential Graphs, a n-ary predicate is just a symbol P with n lines of identity attached to it.

(because $\exists x \forall y P(x, y) \leftrightarrow \exists x \neg \exists y \neg P(x, y)$).

The definitions of (negative) context, enclosure, and dominance are necessary to describe the rules of inference, and are adapted from Sowa and Peirce [7, 9].

Definition 4.2 (Context)

- G appears in a *negative context* if $\boxed{\ G\ }$.
- If a graph does not occur in a negative context, it is *enclosed* at depth 0. The collection of all these graphs is called the *outermost context*.
- If C is a negative context that is enclosed at depth n, then any graph that occurs in the context of C is enclosed at depth $n + 1$.
- For any $n \geq 0$, a graph enclosed at depth $2n$ is *evenly enclosed*, and a graph enclosed at depth $2n + 1$ is *oddly enclosed*.
- If a negative context D occurs in a negative context C, then C *dominates* D.
- A $\neg$-*double context* (or *double negation*) is a negative context inside an negative context, with nothing enclosed in the outer area except for coreference arcs which pass from inside the inner area to outside the outer area.

◁

The axiomatization of first-order logic consists of one axiom, the *empty set of graphs*, denoted by {}, and the rules of inference (Definition 4.3). These rules are the (slightly modified) Peirce's Beta Rules [7–9].

Definition 4.3 (Predicate Rules of Inference)
Erasure. In an evenly enclosed context, any graph may be erased, and any incoming coreference arc (to an evenly enclosed token) may be erased. In any context, a token without incoming or outgoing coreference arcs may be erased.

Insertion. In an oddly enclosed context, any graph may be inserted, and two quantified tokens may be joined with a coreference arc. In any context, a token without incoming or outgoing coreference arcs may be inserted.

Iteration. A copy of any graph G may be inserted into the same context in which G occurs or in any context dominated by G. A coreference arc may be drawn from any quantified token in G to the corresponding token in the copy of G, provided that both tokens occur in the same context, or that the context in which the first token occurs dominates the context in which the second token occurs.

Deiteration. Any graph or coreference arc whose occurrence could be the result of Iteration may be erased. Duplicate arcs may be erased from any graph.

Double Negation. A double negation may be drawn around or removed from any graph in any context, without paying attention to coreference arcs crossing both contexts.

Substitution. In an oddly enclosed context, a quantified token may be replaced by a constant (i.e., a coreference arc may be drawn between the token and a constant). In an evenly enclosed context, a constant may be replaced by a quantified token (i.e., the coreference arc between the token and the constant may be removed).

◁

For an explanation of these rules and a proof of the equivalence of this axiomatization to standard first-order logic, see [1, 2, 4].

Is is straightforward to translate a formula from first-order logic into an equivalent existential graph. Of course, only formulas without free variables can be translated into graphs.

Definition 4.4 The mapping Ψ from first-order logic to Existential Graphs is defined by

- $\Psi(\varphi) = \varphi$ for φ atomic,

- $\Psi(P(x_1,\ldots,x_n)) = \boxed{\begin{array}{l} \boxed{P} \\ \square_1 \quad \square_2 \quad \ldots \quad \square_n \end{array}}$ where each token has an incoming coreference arc from the quantifier which binds it,

- $\Psi(\neg\varphi) = \boxed{\boxed{\Psi(\varphi)}}$,

- $\Psi(\varphi \wedge \psi) = \boxed{\Psi(\varphi)\ \Psi(\psi)}$,

- $\Psi(\varphi \rightarrow \psi) = \boxed{\Psi(\varphi)\ \boxed{\Psi(\psi)}}$,

- $\Psi(\varphi \vee \psi) = \boxed{\boxed{\Psi(\varphi)}\ \boxed{\Psi(\psi)}}$,

- $\Psi(\exists x\varphi) = \boxed{\square\ \Psi(\varphi)}$, where all tokens in $\Psi(\varphi)$ that correspond to variables in φ bounded by $\exists x$ are related to $\square$ using coreference arcs,

- $\Psi(\forall x\varphi) = \boxed{\square\ \boxed{\Psi(\varphi)}}$, where all tokens in $\Psi(\varphi)$ that correspond to variables in φ bounded by $\forall x$ are related to $\square$ using coreference arcs.

◁

4.2 Anaphora in Existential Graphs

In Existential Graphs, anaphoric linking can be expressed by coreference arcs between tokens, which is the equivalent of bound variables in PL.

Let us first consider cross-sentential anaphora, as exemplified by sentence (1). In an existential graph, the *conjunction* of two sentences is expressed by drawing the corresponding graphs within the same context. Anaphoric linking can be accomplished by drawing coreference arcs between those tokens that correspond to the pronouns in the sentences, so in (1) between the token representing *a man* and the token representing *he*. This yields the following (very) schematic picture of (1) (the graph on the right in Figure 1). It is clear that the graphs of the sentences *A man walks in the park* and *He whistles* (shown on the left in Figure 1) are subgraphs of this graph. Unlike predicate logic, the representation of sentence (1) using Existential Graphs is compositional. In the rest of this section we will argue that this is not a coincidence, but characteristic of Existential Graphs.

The aim of DPL was to give a semantics that distinguishes grammatical and ungrammatical anaphoric linking. Using Existential Graphs, it turns out to be the case that grammatical anaphoric linking can be syntactically characterized in terms of (dominating) contexts.

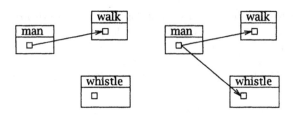

Fig. 1. A schematic picture of (1)

Definition 4.5 The linking of anaphora in an existential graph G is *first-order* if[3] every coreference arc in G is between two tokens in the same context or from a token in context C_1 to a token in context C_2, where C_1 dominates C_2.   ◁

To give an example, in Figure 2 the left graph is first-order, whereas the right graph is not first-order; in Figure 1, all graphs have first-order anaphoric linking.

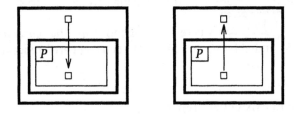

Fig. 2. (Non) first-order anaphoric linking

Allowed (i.e., grammatical) anaphoric linking corresponds to first-order anaphoric linking. So sentences with grammatical anaphors will yield first-order graphs, and first-order graphs express grammatical anaphors. Groenendijk and Stokhof have shown, as mentioned in Section 3, that grammaticality of anaphors can be characterized in terms of dynamics of logical connectives and quantifiers, and they argued that connectives and quantifiers should have the dynamics as mentioned in Theorem 3.3. We will argue that in (first-order) Existential Graphs connectives and quantifiers have the required dynamics.

A *conjunction* $\varphi \wedge \psi$ is represented in Existential Graphs by drawing the conjuncts within one context, as in $\boxed{\varphi\ \psi}$. (A box with thin lines denotes any

---

[3] It is called first-order because drawing coreference arcs using the *first-order* inference rules of Definition 4.3 yields only coreference arcs that obey the conditions of Definition 4.5.

kind of context; usually it denotes the sheet of assertion.) Anaphoric linking between conjuncts is expressed by drawing coreference arcs between tokens contained in the graphs corresponding to these conjuncts. Since these graphs appear in the same context, the coreference arcs will be between tokens in that context, of from a token in that context into a deeper nested context (in case ψ contains one or more contexts). In both cases, the newly drawn coreference arcs obey the conditions of Definition 4.5, thus showing that conjunction in Existential Graphs is internally dynamic. The addition of another conjunct χ (resulting in $\varphi \wedge \psi \wedge \chi$) corresponds to adding that graph within the same context, as in $\boxed{\varphi\ \psi\ \chi}$. Also in this case, anaphoric linking corresponds to first-order linking. This process of adding graphs can be repeated as often as necessary, thus showing that conjunction in Existential Graphs is also externally dynamic.

An *existential quantifier* $\exists x \varphi$ is represented in Existential Graphs by a single token in an even context (in particular in the outermost context), as in $\boxed{\square\ \varphi}$. Binding of variables by an existential quantifier is denoted by an coreference arc from this token to the tokens inside φ representing these variables. As these tokens occur in the same context or in a context dominated by that context (in case φ contains one or more contexts), the newly drawn coreference arcs obey the conditions of Definition 4.5. Therefore the existential quantifier is internally dynamic. The addition of another conjunct ψ (resulting in $\exists x \varphi \wedge \psi$) corresponds to adding that graph within the same context, as in $\boxed{\square\ \varphi\ \psi}$. Also in this case, anaphoric linking corresponds to first-order linking. This process of adding graphs can be repeated as often as necessary, thus showing that the existential quantifier in Existential Graphs is also externally dynamic.

A *universal quantifier* $\forall x \varphi$ is represented in Existential Graphs by a quantified token in an odd context (in particular a context of depth one) with φ contained in a negative context, as in

Binding of variables by an universal quantifier is denoted by an coreference arc from this token to the tokens inside φ representing these variables. As these tokens occur in a context dominated by that context, the newly drawn coreference arcs obey the conditions of Definition 4.5. Therefore the universal quantifier is internally dynamic. The addition of another conjunct ψ (resulting in $\forall x \varphi \wedge \psi$) corresponds to adding that graph in the outer context, as in

In this case, anaphoric linking between the universal quantifier and ψ does not correspond to first-order linking, because the conditions of Definition 4.5 are violated. This shows that the universal quantifier in Existential Graphs is externally static.

Since $\varphi \rightarrow \psi$ is equivalent to $\neg(\varphi \wedge \neg\psi)$, we represent the *implication $\varphi \rightarrow \psi$* by a nesting of negative contexts, as in

Anaphoric linking between the antecedent and the consequent of the implication is expressed by drawing coreference arcs between tokens contained in φ and tokens contained in ψ. Since ψ occurs in a context dominated by the context in which φ occurs, the newly drawn coreference arcs obey the conditions of Definition 4.5. Therefore the implication is internally dynamic. The addition of another conjunct χ (resulting in $\varphi \to \psi \wedge \chi$) corresponds to adding that graph in the outer context, as in

In this case, anaphoric linking between tokens in φ or ψ and χ does not correspond to first-order linking, because the conditions of Definition 4.5 are violated. This shows that implication in Existential Graphs is externally static.

A *negation* $\neg\varphi$ is represented in Existential Graphs by a negative context around φ, as in . Anaphoric linking inside the scope of the negation corresponds to drawing coreference arcs between tokens contained in φ, which obeys the conditions of Definition 4.5. Therefore the negation is internally dynamic. The addition of another conjunct ψ (resulting in $\neg\varphi \wedge \psi$) corresponds to adding that graph in the outer context, as in

In this case, anaphoric linking between tokens in φ and tokens in ψ does not correspond to first-order linking, because the conditions of Definition 4.5 are violated. This shows that negation in Existential Graphs is externally static.

Since $\varphi \vee \psi$ is equivalent to $\neg(\neg\varphi \wedge \neg\psi)$, we represent the *disjunction* $\varphi \vee \psi$ by a nesting of negative contexts, as in

Anaphoric linking between the two disjuncts is expressed by drawing coreference arcs between tokens contained in φ and tokens contained in ψ. Since ψ occurs in a context not dominated by the context in which φ occurs, these coreference arcs violate the conditions of Definition 4.5. Therefore disjunction is internally static. The addition of a conjunct χ (resulting in $\varphi \vee \psi \wedge \chi$) corresponds to adding that graph in the outer context, as in

In this case, anaphoric linking between tokens in φ or ψ and χ does not correspond to first-order linking, because the conditions of Definition 4.5 are violated. This shows that disjunction in Existential Graphs is externally static.

Above we have (informally) proven the following theorem, which tells us that by using Existential Graphs the dynamics of logical connectives and quantifiers are similar to those of DPL. Moreover, in Existential Graphs, these dynamics are obtained for free, without the need for changing the semantics of the formalism.

Theorem 4.6 Consider the class of existential graphs that have first-order anaphoric linking. Then conjunction and the existential quantifier are both internally dynamic and externally dynamic, implication, negation, and the universal quantifier are internally dynamic and externally static, disjunction is both internally static and externally static. ◁

Having shown this, we consider the donkey-sentences (2) and (3). Both sentences are translated in PL to

(9) $\forall x \forall y[(\text{farmer}(x) \wedge \text{donkey}(y) \wedge \text{owns}(x,y)) \rightarrow \text{beat}(x,y)]$

We represent these sentences in Existential Graphs as in Figure 3. The subsen-

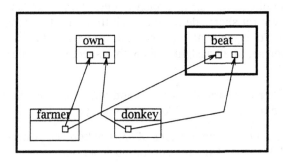

Fig. 3. Representation of (2) and (3)

tence *(a) farmer (who) owns a donkey* is represented in Figure 4. It is clear that

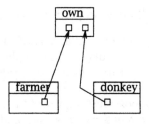

Fig. 4. Representation of *(a) farmer (who) owns a donkey*

the graph in Figure 4 is a subgraph of Figure 3, showing that also in the case of sentences (2) and (3) the use of Existential Graphs yields a compositional analysis.

In the linear form of Conceptual Graphs [9], sentence (2) can be represented as

```
IF:    [FARMER: x]->(OWN)->[DONKEY: y]
THEN:  [?x]->(BEAT)->[?y].
```

which has a subgraph `[FARMER: x]->(OWN)->[DONKEY: y]` which represents the sentence *a farmer owns a donkey.*

5 Logical Properties of DPL and Existential Graphs

In [5], Groenendijk and Stokhof prove several logical facts concerning DPL, which illustrate various properties of DPL. In DPL, it is possible to distinguish two kinds of logical equivalence, s-equivalence and full equivalence. Two formulas are s-equivalent, denoted by $\cong_s$, if they have the same truth conditions. This is the usual definition of equivalence in standard logic. Two formulas are full equivalent (or equivalent for short), denoted by $\cong$, if they have the same interpretation in every model. To formally define these notions we first have to define the satisfaction set, the set of all assignments with respect to which a formula is true in a model.

Definition 5.1 For any formula $\varphi \in$ DPL and any model M, the *satisfaction set* $\backslash\varphi\backslash_M$ is defined by $\backslash\varphi\backslash_M = \{g \mid \exists h : \langle g, h \rangle \in [\varphi]_M\}$. ◁

s-equivalence can be defined in terms of satisfaction sets, whereas equivalence can be defined in terms of the interpretation function.

Definition 5.2 (Equivalences in DPL) Two formulas φ and ψ are *s-equivalent*, notation $\varphi \cong_s \psi$, if $\forall M \backslash\varphi\backslash_M = \backslash\psi\backslash_M$. Two formulas φ and ψ are *equivalent*, notation $\varphi \cong \psi$, if $\forall M [\varphi]_M = [\psi]_M$. ◁

If two formulas are equivalent, they are also s-equivalent.

Several properties of DPL can be given in terms of these notions, as exemplified by the following theorem.

Theorem 5.3 In DPL, the following properties can be proven.

1. Negation, conjunction, and existential quantification can be used to define the other logical connectives (disjunction, implication, and universal quantification) in the usual way: $\varphi \rightarrow \psi \cong \neg(\varphi \wedge \neg\psi)$, $\varphi \vee \psi \cong \neg(\neg\varphi \wedge \neg\psi)$, and $\forall x \varphi \cong \neg\exists x \neg\varphi$.
   A different choice is not possible (because of the dynamics of the operators): $\varphi \wedge \psi \not\cong \neg(\varphi \rightarrow \neg\psi)$, $\varphi \wedge \psi \not\cong \neg(\neg\varphi \vee \neg\psi)$, and $\exists x \varphi \not\cong \neg\forall x \neg\varphi$.
2. $\varphi \not\cong \neg\neg\varphi$ and $\varphi \cong_s \neg\neg\varphi$.
3. $\varphi \wedge \psi \cong_s \neg(\varphi \rightarrow \neg\psi)$ and $\varphi \wedge \psi \not\cong_s \neg(\neg\varphi \vee \neg\psi)$.

4. $\varphi \vee \psi \cong \neg\varphi \to \psi$.

5. $\varphi \to \psi \not\cong \neg\varphi \vee \psi$ and $\varphi \to \psi \not\cong_s \neg\varphi \vee \psi$.

6. $\neg\varphi \to \psi \cong \neg\psi \to \varphi$, and $\varphi \to (\psi \to \chi) \cong (\varphi \wedge \psi) \to \chi$.

7. $\exists x\varphi \cong_s \neg\forall x\neg\varphi$, $\exists x\varphi \to \psi \cong \forall x(\varphi \to \psi)$, and $\neg\exists x\varphi \cong \forall x\neg\varphi$.

$\triangleleft$

Using a suitable definition of equivalences in Existential Graphs, one can prove that the same equivalences hold in Existential Graphs.

Definition 5.4 (Equivalences in Existential Graphs) Let g_1 and g_2 be two existential graphs. g_1 and g_2 are *s-equivalent*, denoted by $g_1 \cong_s g_2$, if there are two deductions $g_1 \dashv\vdash g_2$, and during these deductions no possible internal anaphoric linking is blocked. g_1 and g_2 are *equivalent*, denoted by $g_1 \cong g_2$, if there are two deductions $g_1 \dashv\vdash g_2$, and during these deductions no possible (internal or external) anaphoric linking is blocked. $\triangleleft$

In particular, two graphs are equivalent if they are the same.

Theorem 5.5 Given the definition of (s-)equivalence in Existential Graphs (Definition 5.4), the (non-)equivalences listed in Theorem 5.3 also hold in Existential Graphs. $\triangleleft$

Proof. By translating two formulas into Existential Graphs (using Ψ, see Definition 4.4), the (non-)equivalence will be clear from these two graphs. In most cases, the use of a double negation blocks anaphoric linking, and therefore prevents an equivalence. We only give the proof for 3, the proofs of the other (non-)equivalences can be found in [3].

- $\Psi(\varphi \wedge \psi) = \boxed{\varphi\ \psi} \cong_s$ $= \Psi(\neg(\varphi \to \neg\psi))$, because in the second existential graph only external anaphoric linking is blocked by the outer negation; internal anaphoric linking between φ and ψ is possible in both graphs.

- $\Psi(\varphi \wedge \psi) = \boxed{\varphi\ \psi} \not\cong_s$ $= \Psi(\neg(\neg\varphi \vee \neg\psi))$, because in the second existential graph also internal anaphoric linking is blocked by the negations.

$\blacktriangleleft$

Theorem 5.5 is an indication of the truth of the next claim.

Claim 5.6 Let φ, ψ be two (closed) formulas from DPL. Then $\varphi \cong \psi$ if and only if $\Psi(\varphi) \cong \Psi(\psi)$ and $\varphi \cong_s \psi$ if and only if $\Psi(\varphi) \cong_s \Psi(\psi)$. $\triangleleft$

In particular, if two formulas are translated into the same existential graph, they are equivalent.

This claim shows that Existential Graphs have the same logical properties as DPL, but without redefining its semantics. Both the logical properties and the dynamics of logical operators indicate that Existential Graphs are a more natural representation formalism to deal with anaphora. Since both Conceptual Graphs [9] and Knowledge Graphs [2, 4] use the logical framework of Existential Graphs, the above indicates that these graph-representations can deal with anaphora in a more natural way than (dynamic) logic.

6 Conclusions

The occurrence of cross-sentential anaphora and donkey-anaphora in natural language is problematic while giving a compositional analysis of natural language meaning in terms of predicate logic. Either one has to give up compositionality, or one has to change the semantics of predicate logic. Groenendijk and Stokhof opted for the second solution, resulting in Dynamic Predicate Logic (DPL). In this paper we have argued that there exists a third solution, the use of Existential Graphs. Existential Graphs, developed by C.S. Peirce almost 100 years ago, do have the characteristics and logical properties that are needed for a compositional treatment of anaphora. This is a consequence of the use of negative contexts instead of negations, the use of tokens instead of quantifiers, and the use of coreference arcs for variable bindings. Therefore, Existential Graphs are more suited for analyzing anaphora. Since Conceptual Graphs [9] and Knowledge Graphs [2, 4] use the logical framework of Existential Graphs, these formalisms are also more suited for analyzing anaphora.

References

[1] H. van den Berg. *Logic, Language, and Knowledge Graphs: An Interdisciplinary Approach to Natural Language Understanding.* Lecture Notes in Artificial Intelligence. Springer-Verlag, Berlin. To appear.

[2] H. van den Berg. *Knowledge Graphs and Logic: One of Two Kinds.* PhD thesis, University of Twente, Enschede, 1993.

[3] H. van den Berg. Existential graphs and dynamic predicate logic. Memorandum no. 1227, Department of Applied Mathematics, University of Twente, Enschede, 1994.

[4] H. van den Berg. First-order logic in knowledge graphs. In C. Martín-Vide, editor, *Current Issues in Mathematical Linguistics* (North-Holland Linguistic Series Volume 56), pages 319–328. North-Holland, Amsterdam, 1994.

[5] J. Groenendijk and M. Stokhof. Dynamic predicate logic. *Linguistics and Philosophy*, 14:39–100, 1991.

[6] J. Hintikka. *The Game of Language*, volume 22 of *Synthese Language Library*. D. Reidel, Dordrecht, 1983.

[7] D.D. Roberts. *The Existential Graphs of Charles S. Peirce.* Mouton, The Hague, 1973.

[8] D.D. Roberts. The existential graphs. In F. Lehmann, editor, *Semantic Networks in Artificial Intelligence*, pages 639–663. Pergamon Press, Oxford, 1992.

[9] J.F. Sowa. *Conceptual Structures: Information Processing in Mind and Machine.* Addison-Wesley, Reading, 1984.

Author Index

Springer-Verlag
and the Environment

We at Springer-Verlag firmly believe that an international science publisher has a special obligation to the environment, and our corporate policies consistently reflect this conviction.

We also expect our business partners – paper mills, printers, packaging manufacturers, etc. – to commit themselves to using environmentally friendly materials and production processes.

The paper in this book is made from low- or no-chlorine pulp and is acid free, in conformance with international standards for paper permanency.

Lecture Notes in Artificial Intelligence (LNAI)

Lecture Notes in Computer Science